Thematic Origins of Scientific Thought

Kepler to Einstein

THEMATIC ORIGINS OF SCIENTIFIC THOUGHT

Kepler to Einstein

GERALD HOLTON

Harvard University Press
Cambridge, Massachusetts
1973

Library of Congress Catalog Card Number 72-88126
Printed in the United States of America

ACKNOWLEDGMENTS

The author wishes to express appreciation to the following periodicals and publishers in whose publications most of these essays have previously appeared: *American Journal of Physics, American Scholar, American Scientist, Bulletin of the Institute of Physics and the Physical Society, Daedalus, Isis, Physics Education,* Columbia University Press, Hermann (Paris), and Houghton Mifflin Co. The provenance and subsequent publication history is noted at the bottom of the first page of each essay. The generous assistance of *The Graduate Journal* of the University of Texas, which undertook to set type for most of the book, is also gratefully acknowledged.

CONTENTS

Introduction

Introduction

THIS group of essays, selected from among those of my publications which lie outside research physics itself, is concerned with what is often called "the history of science and related studies." The clumsiness of that phrase itself indicates the need for the new approach to the study of the history of science that has been emerging, one that looks for fruitful ideas in fields ranging from the philosophy and sociology of science to psychology and aesthetics. Some new name may soon be required for this expanded field of scholarship; but more important, of course, are its new questions, conceptions, and models of approach.

With few exceptions, the essays span about a decade; some are detailed and formal, others much less so, and several were triggered by some accident of personal history discussed in the Acknowledgments. By and large, the later essays in this collection follow as development or consequence from the earlier ones. Throughout, a particular aim has been to show, by specific case studies of the growth of ideas of physicists from Kepler to Einstein and Bohr, in what respects the traditional views of the way the scientific mind works have to be changed and supplemented. For this purpose we have to introduce new concepts, such as that of the "thematic" content of science, a dimension that can be conceived as orthogonal to the empirical and analytical content. Themata are shown to play a dominant role in the initiation and acceptance of certain individual scientific insights. The case of the origins of relativity theory is singled out for study in depth, drawing in part on unpublished documents in the extensive *Nachlass* of Albert Einstein.

If these findings have consequences beyond the study of the history of science itself, it will be chiefly in three respects: they may provide the philosopher of science with the kind of raw material of actual cases on which some of the better work in that field is based; they may help the intellectual historian to redefine the place of modern science in our culture by identifying the contributions to the scientific imagination owing to the whole spectrum of influences, from the general literary and epistemological currents to the detailed sociology governing group work in the laboratory; and they may prompt the educator to reexamine the conventional concepts of education in science, both for scientists and for nonscientists.

I shall discuss each of these points more fully, but, first of all, let me take note of the historical problems treated here that seem to me of interest in their own right.

Problems in the History of Physics

A few examples of the specific questions that generated each of the essays must suffice at this point. In *Johannes Kepler's Universe: Its Physics and Metaphysics*, the nagging problem was why Kepler's astronomy worked so well although his pre-Newtonian physics was so ineffective. Though his instinct for physical problems was sound, his tools were not, and the success of his astronomy depended on his ability to shift to frankly metaphysical presuppositions when his physical ones gave out. Thus on February 10, 1605—a date that might be taken to be historic for physics—he revealed for the first time his devotion to the image of the universe as a physical machine in which universal terrestrial force laws would hold for the operation of the whole cosmos (see his letter to Herwart von Hohenburg). But his effort would have been doomed if he had not supplemented the mechanistic image with two other, very different ones: the universe as a mathematical harmony, and the universe as a central theological order. These three themes continued to echo in the work of the seventeenth-century scientists who followed Kepler, and indeed up to the delayed triumph of the purely mechanistic view in the completion of Newton's work by Laplace.

A little earlier, in the very first of the essays, the question raised is why

Newton chose to suppress his "Fifth Rule of Reasoning," one which not only would be accepted today, but, it would seem, was a logical consequence of Newton's own avowed dislike of feigning hypotheses. We are led there to introduce the recognition of the existence, and even the necessity at certain stages in the growth of science, of precisely such unverifiable, unfalsifiable, and yet not-quite-arbitrary hypotheses. This class of hypotheses, referred to as thematic hypotheses or thematic propositions, is developed at some length in the first, third, and sixth essays. The analysis has, I believe, significance also for the younger sciences that are now (erroneously, in my opinion) trying to emulate the older physical sciences by restricting their area of investigation, even if artificially, purely to the "contingent" plane of phenomenic (empirical) and analytic statements.

The third essay examines some specific themes that are active both in the sciences as narrowly constructed and in work outside the sciences (e.g., juxtaposition of the thema-antithema (or $\theta\bar{\theta}$) couple of atomism and the continuum; methodological thema-antithema of projection [externalization] and retrojection [internalization]). This prepares for the recognition that themata not merely belong to a pool of specifically scientific ideas but spring from the more general ground of the imagination.

In the fourth essay, *The Roots of Complementarity,* we first review the wave-particle paradox and the related separation between the observer and the observed. Historically this development took place in an era of thematic conflict between scientists with antithetical presuppositions. Some regarded Schrödinger's introduction of wave mechanics as "a fulfillment of a long baffled and insuppressible desire," as one physicist expressed it in 1927, while others abhorred this continuum-based approach and found satisfaction only in fundamental explanations rooted in the thema of discreteness. Both groups faced the same experimental data; but their allegiances were to conjugate notions, and in the passionate motivation behind their antithetical quests we recognize one of the chief properties of the thematic attachment.

At that point in history, it was Bohr's genial recommendation to accept both members of the $(\theta\bar{\theta})$ couple as equally valid but complementary pictures of nature, instead of following the traditional path of persistently trying to dissolve or conquer over one by means of the other. The historical question that arises here is how Bohr may have been prepared for this remarkable innovation in physics—and we follow the hints left by Bohr himself to the effect that this new thema may have had some of its roots in philosophical and literary works.

The fifth essay, *On the Origins of the Special Theory of Relativity,* poses a set of historical questions that are then successively attacked in the next five papers: what are the sources for the study of the origins of relativity theory, and what is their probable reliability? What was the state of science around 1905, what were the contributions which prepared the field, and what did Einstein know about them? By what steps may Einstein have reached the conclusions he published in his first, basic paper? To what extent was this work a member of a continuous chain with immediate predecessors, and to what extent may it be considered "revolutionary"? What were the roles of experiments and of speculative hypotheses in the genesis of relativity theory? What part did epistemological analysis play in Einstein's thought, and what was his influence in turn upon the epistemology of his time? What may we say about the style of his work, and how may it be connected with his personal orientation? What methodological principles for the study of the history of science itself emerge from such a case study?

On the way, a number of old questions are reexamined and new ones introduced in this section of the volume: the philosophical pilgrimage of Einstein himself, starting from an allegiance to Machist phenomenalism and ending primarily with a rationalistic realism; Einstein's frequent preoccupation with what he called "the nature of mental processes," and his conclusions concerning them; the fact that historical statements, like those in physics, have meaning only relative to a specifiable framework—and the unfortunate consequence of this fact when pedagogic statements are confused with quasi-historical statements; the sometimes quite flagrant neglect of "experimental evidence" when such evidence is contrary to a given thematic commitment; the "relativistic" context of a "crucial" experiment that may indeed be crucial in the setting of one theory but, in the setting of another, may be trivially true, and not even worth a specific mention; an attempt to wrestle more concretely with the old problem of the quasi-aesthetic choices which some scientists make, for example in rejecting as merely "ad hoc" an hypothesis which, to others, may appear to be a necessary doctrine; and finally, an attempt to trace the growth of a scientist's thematic commitment from childhood, and to see the work of a genial contributor in terms of a correspondence between his personal style and the structure of the laws of nature themselves.

Since a chief aim of these essays is to raise new questions for research rather than merely to answer old ones, there has been no attempt to come to solutions that pretend to finality; rather, we shall be satisfied if we can open the topic wider and find interesting questions spread over several, previously separate, specialty disciplines. Then, too, some problems are not by any means ready to be solved—such as that of the meaning of "genius"—and so it is best to leave them unashamedly open. This applies also, for example, to the role of verbal versus visual thinking in scientific theory building, and indeed to the question that Einstein himself, near the beginning of his *Autobiographical Notes,* struggled with: "What, precisely, is 'thinking'?"

The two following essays are specifically devoted to the construction of models for understanding the growth of science. Thus, the subject of the eleventh essay is the relation between the work of the individual scientist and that of the mass of scientific workers who are his contemporaries and his successors. It defines the crucial distinction between two different activities—related to each other and with a fuzzy border between them, but still quite different—that are nevertheless denoted by the same term, science. One is the private aspect, science-in-the making, the speculative, perhaps largely nonverbal activity, carried on without self-consciously examined methods, with its own motivations, its own vocabulary, and its own modes of progress. The other is the public aspect, science-as-an-institution, the inherited world of clarified, codified, refined concepts that have passed through a process of scrutiny and have become part of a discipline that can be taught, no longer showing more than some traces of the individual struggle by which it had been originally achieved. This, roughly, can be characterized as the difference between the "subjective" and the "objective" aspects of truth-seeking. The former aspect does not have to remain tacit, nor should it be discredited as inaccessible to rational study. On the contrary, it should be ferreted out, and, as Kepler and Einstein showed, it can sometimes be quite eloquently defended.

Attending in more detail to the public aspect of scientific development, the twelfth essay deals with the essentially qualitative aspects of scientific growth and change in recent decades. Here a key concept is not so much the escalation of knowledge as the escalation of ignorance—

the sometimes discontinuous process of breaking into new areas of work where very little is yet known. Other aspects discussed include the processes of diffusion of scientific information, branching of new fields, nonlinearity of potential in team work, and recruitment. Despite the faults of the system that are being nowadays quite amply discussed, the sociology of basic scientific research groups indicates that more than ever these groups may provide useful models for the conduct of work in certain other fields of scholarship also.

In the remaining essays I have attempted to draw some practical conclusions from these findings, particularly insofar as they pertain to two aspects of the public understanding of science. One is the actual design of curricula for colleges and schools; hence, the final essays of this collection are devoted to a detailed examination of the consequences for education, including setting criteria for curriculum design. The temptation today is still strong to set up academic courses as if the aim were to maximize the separateness of disciplines. The student sees the strands of the curriculum hanging down separately, one labelled physics, another psychology, a third economics, another political science. And he learns each from a different person, from a different book. On his own, he is supposed to make sense out of it all, to weave himself a tapestry from the strands handed to him. This is putting much too large a burden on the student. Moreover, the separateness of fields in the classroom does not correspond to the intellectual realities of work in those fields. We ought to show, particularly in the introductory or general-education courses, those coherences that in fact do exist.

The other aspect of the public understanding of science of deep concern to me is the manner in which science is understood or misunderstood among adult intellectuals, including those who deny that science can be considered an essential and positive part of the cultural achievement of our time. This concern is by no means new. Thus the paper, *Modern Science and the Intellectual Tradition,* now more than a decade old and written at the height of the postwar appreciation of science, warned that underneath the euphoria there was an old, underlying disease that spelled trouble ahead. And while some things have changed in the meantime—mostly the absolute values of some numbers, though hardly their ratios and percentages—the conclusions of that generally pessimistic article are perhaps even more appropriate now. New forces will have to be introduced into the cultural dynamics to change the continuing atrophy of the mechanism by which, in the more distant past, schisms of the present kind were averted. A clearer understanding of what science is and is not, which I see as one of the foremost purposes in these essays, will, one dares to hope, help make more explicit the bonds that could keep science in reciprocal contact with the rest of our culture.

The Nascent Moment

Apart from the search for interesting problems in the general area of the history of physics, many of these essays share in the attempt at least to begin the long-overdue work of understanding the "nascent moment" or nascent phase in scientific work. Einstein himself pointed frequently to both the interest and the difficulty in any such discussion. For example, he wrote, "Science as an existing, finished [corpus of knowledge] is the most objective, most unpersonal [thing] human beings know, [but] science as something coming into being, as aim, is just as subjective and psychologically conditioned as any other of man's efforts," and its study is what one should "permit oneself also." Elsewhere, Einstein used the suggestive phrase "the personal struggle" to describe what seemed to him to deserve central attention in the analysis of scientific development.

This advice is, of course, exactly counter to that of many other scientists, historians, and philosophers of science. Among the last, Hans Reichenbach's dictum is typical: "The philosopher of science is not much interested in the thought processes which lead to scientific discoveries . . . . that is, he is interested not in the context of discovery, but in the context of justification." Historians of science also have not paid much attention to the nascent phase because, as one of them put it, the reasons why scientists embrace their guiding ideas in individual cases "lie outside the apparent sphere of science entirely"; therefore they are all too easy to dismiss as not leading to certain knowledge. Those few philosophers of science who have looked at such problems have tended to label them as "metascience problems," hidden at the basis of science, and not really part of it.

Scientists themselves, by and large, have traditionally helped to derogate or avoid discussions of the personal context of discovery in favor of the context of justification. They would still agree with Robert Hooke's draft preamble to the statutes of the Royal Society: "The business of the Royal Society is: To improve the knowledge of natural things . . . (not meddling with Divinity, Metaphysics, Morals, Politics, Grammar, Rhetorick, or Logicks)." This is of course on the whole as it should be. Yet even the necessary few who are sympathetic to an analysis of the context of discovery use imagery that shows how skeptical they are of ever finding ways of understanding the "personal struggle." Thus Gunther Stent writes that the domain of the sciences "is the outer objective world of physical phenomena. Scientific statements therefore pertain mainly to relations between public events," and he goes on to warn that all else is of the nature of artistic statements which "pertain mostly to private events of affective significance." And Max Born has written, "I believe that there is no philosophical high-road in science, with epistemological signposts. No, we are in a jungle and find our way by trial and error, building our roads behind us as we proceed. We do not find signposts at cross-roads, but our own scouts erect them, to help the rest."

The intent of such rather typical responses may be benign, perhaps to indicate the scientist's disagreement either with popular conceptions of science as an impersonal, machinelike success story, or with those overbearing epistemological treatises that claim to put order into the turbulent work of the individual scientist and help him to decide which of his works may be good or bad. Nevertheless, the chief result is to discourage the study of this "jungle."

The detailed analysis of published scientific contributions generally only reinforces this feeling. Most of the publications are fairly straightforward reconstructions, implying a story of step-by-step progress along fairly logical chains, with simple interplays between experiment, theory, and inherited concepts. Significantly, however, this is not true precisely of some of the most profound and most seminal work. There we are more likely to see plainly the illogical, nonlinear, and therefore "irrational" elements that are juxtaposed to the logical nature of the concepts themselves. Cases abound that give evidence of the role of "unscientific" preconceptions, passionate motivations, varieties of temperament, intuitive leaps, serendipity or sheer bad luck, not to speak of the incredible tenacity with which certain ideas have been held despite the fact that they conflicted with the plain experimental evidence, or the neglect of theories that would have quickly solved an experimental puzzle. None of these elements fit in with the conventional model of the scientist; they seem unlikely to yield to rational study; and yet they play a part in scientific work.

This is much more obvious to one who actually lives in the middle of the doing of science itself. Scientists generally have been reluctant (perhaps even embarrassed) to discuss frankly this state of affairs in the published work on which scholars outside the sciences generally rely. It is only relatively recently that it has become more common for working scientists to allow themselves to be interrogated by well prepared and sensitive historians, or to make available their draft, notes, research apparatus, personal letters, and other documents.[1]

We have here a curious standoff. While many and perhaps most scientists are still skeptical about the possibility of understanding better the nascent phase, they are equally uncomfortable with having to leave it simply at references to "intuition" or other presently inexplicable mechanisms. I regard many of the confessions by scientists of how their work is really done not as attempts to discredit their own mode of progressing in the nascent phase, but rather as invitations to others to bring to bear some systematic thought in this area of the "personal struggle." If in the past historians and philosophers of science have, on the whole, been skeptical of accepting such problems into their area of scholarly work, it may have been that the silence of most scientists and the frank but unstructured testimony offered by the rest have frightened investigators away rather than attracted them, and caused them to dismiss as unworthy of examination a fledgling area of scholarship that, in my view, will sooner or later turn out to be a central part of their studies. Both from the point of view of drawing attention to areas of fruitful new questions and from that of a wider, more humanistic attitude toward the growth of science itself, it seems to me that the study of the personal context of discovery is about to come into its own.

Appropriate first steps in this direction are indicated in these pages; they are of two kinds. First, one must delineate more clearly science in the sense of the personal struggle, by distinguishing it from a different activity, also called "science," which is its public, institutional aspect. The two activities may be labeled S_1 and S_2 respectively. It is for science in the sense of S_2 that the Royal Society edict, quoted above, has in fact worked so successfully, even though for S_1 it has never applied. A scientist, whose external justification and approbation comes from S_2, the arena in which his published work is of prime importance, has generally little reason or incentive selfconsciously to examine S_1, the arena of his own imaginative processes within which he in fact lives from day to day. Unless specifically urged, he is likely to adopt in all his discussions of science the vocabulary and attitude of S_2, dry-cleaned of the personal elements. In this way he becomes an ally of those historians and philosophers of science who, for other reasons (including the fact that they themselves do not live in the world of S_1), have cause to neglect the nascent moment as a problem of research.

Once the distinctions between S_1 and S_2 are made, one is ready to work in the world of S_1 without offending judgments that more properly belong in the S_2 region; the apparent contradiction between the often "illogical" nature of discovery and the logical nature of physical concepts is resolved; and, without making light of the difficulties, one is ready to find that a very different set of rules holds in S_1 than in S_2. I do not doubt that solid knowledge about S_1 can be achieved. A science of S_1 must be possible. Though scientists themselves may for a time frown on such an enterprise, one may take comfort that the best of them —for example Einstein and Bohr, as demonstrated in these pages— would not.

Nor do we have to look only to them for an understanding of the difficulty and necessity of such an enterprise. It was the philosopher-psychologist William James, who as long ago as 1880 said in a context that can be stretched from philosophy to other branches of knowledge:

> Pretend what we may, the whole man within us is at work when we form our philosophical opinions. Intellect, will, taste, and passion co-operate just as they do in practical affairs; and lucky it is if the passion be not something as petty as a love of personal conquest over the philosopher across the way. The absurd abstraction of an intellect verbally formulating all its evidence, and carefully estimating the probability thereof by a vulgar fraction, by the size of whose denominator and numerator alone it is swayed, is ideally as inept as it is actually impossible. It is almost incredible that men who are themselves working philosophers should pretend that any philosophy can be, or ever has been, constructed without the help of personal preference, belief, or divination. How have they succeeded in so stultifying their sense for the living facts of human nature as not to perceive that every philosopher, or man of science either, whose initiative counts for anything in the evolution of thought, has taken his stand on a sort of dumb conviction that the truth must lie in one direction rather than another, and a sort of preliminary assurance that his notion can be made to work; and has borne his best fruit in trying to make it work?[8]

The Thematic Component

While the existence of presuppositions in S_1 cannot be denied—indeed, some of their aspects have been central preoccupations of philosophers for more than three centuries—they make it puzzling how scientific work can succeed at all. Would such preconceptions not hobble one in the search for the objective state of affairs? How can science change directions, and yet also preserve continuities? How has the scientific profession managed to construct a corpus that is largely so successful and so beautiful, despite this and other limitations on the individual scientific contributor?

Portions of the essays discussing a particular type of preconception which I have called thematic have been mentioned above. But since the concept of thematic analysis will perhaps be the least familiar and the one most easily confused with other current conceptions, we shall here go into a little more detail on the role themata play in scientific work. This may also help the reader by providing the outlines of the theoretical framework within which the material here being surveyed can be more readily accommodated.

All philosophies of science agree on the meaningfulness of two types of scientific statements, namely, propositions concerning empirical matters of fact (which ultimately boil down to meter readings) and propositions concerning logic and mathematics (which ultimately boil down to tautologies). To be sure, observation is now often carried on at the output end of a complex of devices. The observation that counts in, say, an experiment such as the first "observation" of the antiproton seems completely buried under both the total output of data obtainable from the mountain of special equipment and the mass of sophisticated technological knowledge and physical theory without which one neither could set up the conditions for the observation in the first place, nor would know what one is looking for, nor should be able to interpret the thin trace on an oscilloscope or in a photographic emulsion under the microscope by which one finally "sees" the action of the antiproton. But "propositions concerning empirical matters of fact" can be interpreted to be propositions concerning this final stage, protocol sentences in common language that command the general assent (i.e., assent in S_2) by specialists concerned with this type of "empirical matter of fact"; and this is what makes them "meaningful."

The propositions concerning logic and mathematics are analytical propositions. They are meaningful insofar as they are consistent within the system of accepted axioms, though they may or may not turn out to be more widely useful. Thus the algebra of ordinary commutative groups suffices for Newtonian mechanics, but not for quantum mechanics.

These two types of meaningful propositions may be called phenomenic and analytic and by way of analogy one may imagine them roughly as corresponding to a set of orthogonal x- and y-axes that represent the dimensions of the plane of usual scientific discourse.

One may name the x-y plane the *contingent plane*. The word *contingent* has been used[3] in a sense that is supposed to be more subtle than the term *empirical*: a contingent proposition is one "to whose truth or falsity experience is relevant"—as against "logically necessary."[4] But this is not the sense in which I intend "contingent" to be used; for, on the one hand, it unnecessarily introduces by the back door arguments concerning the nature and warrant of truth; and, on the other hand, a proposition can be contingent not only on empirical evidence but also on analytical evidence. The concept of the electron as part of the nucleus was discarded not on empirical grounds (on the contrary, electrons appear to "come out" of decaying nuclei all too conspicuously), but rather because it was thought wiser to retain the then new formal system of quantum-mechanical analysis, according to which an electron bound in the nucleus could be calculated to require utterly unreasonably large energies.

I therefore define the contingent plane as the plane in which a scientific concept or a scientific proposition has both empirical and analytical relevance. Contingency analysis is the study of the relevance of concepts and propositions in the x- and y-dimensions. It is a term equivalent to operational analysis in its widest sense.

All concepts and propositions can in general be subjected to contingency analysis. And we can reformulate the claim of the modern philosophies of science that are rooted in empiricism or positivism that those concepts or propositions are "meaningless" which have zero or nearly zero components in the x- or y-dimension (or in both x- and y-dimensions), that all meaningful science therefore happens in the x-y plane. This, in brief, was the content of Newton's public pronouncements against the postulation of innate properties and occult principles. It also lies behind Hume's exhortation, the persistent attacks of Comte and

Mach and their followers, and, outside science itself, the fury of Locke against the doctrine of innate principles, the suspiciousness of J. S. Mill against the intuitionism of the Scottish school, the reduced role that Ayer assigns to philosophy when he writes that the function of philosophy is "to clarify the propositions of science by exhibiting their logical relationships, and by defining the symbols which occur in them," and the fear of many modern scientists that going outside the contingent plane necessarily means opening the gates to a flood of obscurantism.

It is indeed one of the great advantages of the scheme that in the x-y plane many questions (e.g., concerning the reality of scientific knowledge) cannot be asked. The existence of such questions is not denied; but they do not have to be admitted into scientific discussions, since the possible answers are not verifiable or falsifiable, having no component that can be projected on the phenomenic dimension of empirical (observational) fact, and obeying no established logical calculus (beyond that of grammar) in which the analytic projection of the statement can be examined for consistency.

In fact, this attitude is one reason why science has grown so rapidly since the early part of the seventeenth century; keeping the discourse consciously in the contingent plane means keeping it in the arena of S_2, where statements can be shared and publicly verified or falsified. This habit has minimized prolonged disagreement or ambiguity or the mere authority of personal taste. It has helped expel certain metaphysical propositions which were masquerading as empirical or analytical ones. And in these ways, it has also helped forge a strong and wonderfully successful profession.

These successes do not, however, hide the puzzling fact that contingency analysis excludes an active and necessary component that is effective in scientific work, both on the personal and on the institutional level; that is, it neglects the existence of preconceptions that appear to be unavoidable for scientific thought, but are themselves not verifiable or falsifiable. Their existence has long been commented upon, and some of their properties have been examined from different points of view[5]; but much more can be said on this subject. Related to the first is a second puzzle, namely, why contingency analysis helps us to understand neither how the individual scientific mind arrives at the products that later can be fitted into the contingent plane, nor how science as an historical enterprise grows and changes. Thus in his influential book, SCIENTIFIC

EXPLANATION, the philosopher R. B. Braithwaite offers the rather typical confession that an explanation of such matters is beyond the realm not only of the scientist but also of the philosopher:

> The history of a science is the history of the development of scientific systems from those containing . . . few generalizations . . . into imposing structures with a hierarchy of hypotheses. . . . The problems raised by this development are of many different kinds. These are historical problems, both as to what causes the individual scientist to discover a new idea, and as to what causes the general acceptance of scientific ideas. The solution of these historical problems involves the individual psychology of thinking and the sociology of thought. *None of these questions are our business here.*[6]

If, however, we want to make them our business, it is at this point —to return to our analogy—that we define a third, or z-axis, perpendicular to the x- and y-axes of the contingent plane. It is the dimension of themata, of those fundamental preconceptions of a stable and widely diffused kind that are not resolvable into or derivable from observation and analytic ratiocination. They are often found in the initial or continuing motivation of the scientist's actual work, and also in the end product to which his work reaches out. Thus, while the two-dimensional x-y plane may suffice for most discourse within science in the sense of S_2, it is in the three-dimensional, x-y-z space within which a more complete analysis—whether historical, philosophical or psychological—of scientific statements and processes should proceed. For example, the study of the rise or fall of a thematic preoccupation is among the most interesting problems for the historian. Some themata grow slowly, as the result of a sequence of local successes—e.g., the thema of strict conservation, as embodied in the laws of conservation of mass and of energy, explained chemical reactions (such as the formation of HCl) better than the earlier use of material "principles" (such as the acidifying principles). The chemical ideas of material change became by and by so successful that whereas in Newton's time chemical reactions were understood in terms of organic digestive processes, a century and a half later the arrow of explanation had turned around and organic digestive processes were explained in terms of chemical reactions.[7] Some thematic concepts found their place more rapidly, perhaps as a result of stunning virtuoso demonstrations (e.g., the concept of a causal, mechanistic universe which was at least the external result of Newton's system of the world). Other themata have atrophied or are now discredited as explanatory devices—ideas such as macrocosmic-microcosmic correspondence, inherent principles, teleological drives, action at a distance, space filling media, organismic interpretations, hidden mechanisms, or absolutes of time, space, and simultaneity.[8]

Yet other themes are long lived and apparently stable. To find examples, we need only glance at some reports of current research in the physical sciences and related areas, selected almost at random and quoted or paraphrased from recent issues of research reports in the journals of the profession.

At the top of the pile of journals on my desk at the moment is a report of deep inelastic scattering of nuclear particles observed at the Stanford University linear accelerator, which has drawn comment from theorists. The problem is: what is the structure of the proton—which to an historian of science sounds like a contradiction in terms. The current candidates are pointlike hypothetical constituents, be they quarks, dions, partons, or stratons. One thing is clear: the antiquity of this quest for an elementary—although perhaps protean—constituent of all matter, a quest that has made sense to scientists all the way back to Thales. It is nothing less than an a priori commitment that deserves to be called thematic.

Almost invariably, for every thematically informed theory used in any science, there may also be found a theory using the opposite thema, or antithema. (Sometimes we find not merely opposing ($\theta\bar{\theta}$) dyads, but even triads.) Thus opposition exists to current theories that believe all hadrons to be dynamical constructs, satisfying self-consistency conditions. There are publications insisting, for example, that nature exists in an infinite number of strata with different qualities, each stratum being governed by its own laws of physics and each always in the midst of creation and annihilation. Still another view of the matter is hinted at in work such as that of G. F. Chew, who has speculated that the way to enlarge current ideas in elementary particle physics is to break out in entirely new directions: "Such a future step would be immensely more profound than anything comprising the hadron bootstrap [approach]; we would be obliged to confront the elusive concept of observation and, possibly, even that of consciousness. Our current struggle . . . may thus be only a foretaste of a completely new form of human intellectual endeavor, one that will not only lie outside physics but will not even be describable as 'scientific.' "

What seems certain is that regardless of temporary victories for one side or another, the dialectic process of this sort between a thema and its anti-thema, and hence between the adherents of two or more theories embodying them respectively, is almost inevitable, and is perhaps among the most powerful energizers of research. If the past is a guide, this process will last as long as there are scientists interested in putting questions to nature and to one another.

Next I find a report of a conference of physicists on fundamental interactions at high energy. P. A. M. Dirac opened the conference with the question, "Can equations of motion be used?" Although agreeing that Heisenberg's 1925 view can be a good guiding principle (e.g., that only observable quantities should be used in formulating a physical theory), he felt it nevertheless unlikely that the analytic S-matrix description would be the final answer in high-energy physics. Some day, according to Dirac, we would be discussing equations of motion of entities only remotely related to experimental quantities. He thought this prophecy would come to pass because of his "feeling for the unity of physics" and because of the important role played by equations of motion in all other branches of physics. This confidence, somewhat in the face of current fashion and experimental evidence, led Dirac to say, "a theory that has some mathematical beauty is more likely to be correct than an ugly one that gives a detailed fit to some experiments."

Such a quasi-aesthetic judgment is a form of thematic commitment with deep psychological roots. It is frequently the basis for choices made in actual scientific work (for example, when one ad hoc hypothesis is accepted and another is rejected, or when a whole approach to a scientific field is adopted or dismissed), though it is not common to see this confessed in public print. Thus, in 1926, Heisenberg wrote to Pauli, "The more I ponder the physical part of Schrödinger's theory, the more disgusting it appears to me." At about the same time, Schrödinger in his turn wrote about Heisenberg's approach: "I was frightened away [by it], if not repelled." And Fermi wrote to Enrico Persico in the same vein about what he called "the formal results in the zoology of spectroscopic terms achieved by Heisenberg. For my taste, they have begun to exaggerate their tendency to give up understanding things." At least since Copernicus defended his theory as "pleasing to the mind," it has been an everyday fact in the life of scientists that some of the terms and attributes they use have for them great motivating power, but cannot be subjected to contingency analysis.

Further on in a recent issue of PHYSICS TODAY, I find the report of a delicate analysis made on a meteorite that fell on Australia. A team assembled from several institutions reports finding 16 amino acids, at least five of which are common in living systems: these five would constitute an essential part of any chain in the chemical evolution toward living forms. In a sense, this is merely another report in a recent series that suddenly has shown the "dead interstellar spaces" to be populated with more and more complex materials, from hydroxyl radicals to formaldehyde. But what now excites the interest of the scientists most is the evidence of equal quantities of laevogyrate (left-handed) and dextrogyrate (right-handed) amino acids in the samples. The fact that the chirality in these samples is of both kinds increases by far the likelihood that these amino acids are not merely contaminations from handling; for reasons that are still entirely mysterious, the amino acids in living things on earth are almost all left-handed. Now it has become quite likely that the ideas of chemical basis of evolution are entering a new phase of elaboration: a search for clouds of amino acids in space becomes sensible. Chemistry, biology, geology, physics, and astronomy are being brought together in a remarkable interdisciplinary attempt to understand the historic, and perhaps continuing, evolution of life.

Triggered in part by this finding and this way of thinking is another review, *Chirality, Broken Symmetry, and the Origin of Life*. The title itself alerts us to the whole set of thematic elements that are basic in major areas of research today, as they were in other areas in the past: the efficacy of geometry as an explanatory tool; the conscious and unconscious preoccupation with symmetries; the use of the themata of evolution and devolution that might have been taken from the ordinary life cycle but that have become, in any case, fundamental tools of scientific thought (as much in psychological and sociological research as in genetics and astrophysics). It is the interdisciplinary spread or sharing of such fundamental themata that has produced something like a scientific imagination shared by all scientists, forming one of the bonds among them, and making possible the interdisciplinary approach that characterizes so many of the new developments.

Though certain themata are developed in detail in some of the essays that follow, it will be fruitful here to make distinctions between three different uses of the concept of themata:

(1) A *thematic concept* is analogous to a line element in space which has a significant projection on the z- or thematic dimension. Purely thematic concepts seem to be rare in established science. What is more significant is the thematic component of concepts such as force or inertia, which have strong x and y components also. Therefore, when we speak of force or inertia as a thematic concept we mean the thematic component of this concept.

(2) A *thematic position* or *methodological thema* is a guiding theme in the pursuit of scientific work, such as the preference for seeking to express the laws of physics whenever possible in terms of constancies, or extrema (maxima or minima), or impotency ("It is impossible that . . .").

(3) Between these two is the *thematic proposition* or *thematic hypothesis,* e.g., a statement or hypothesis with predominant thematic content, or the thematic component of a statement or hypothesis. A thematic proposition contains one or more thematic concepts, and may be a product of a methodological thema. Thus, the principle of constancy of the velocity of light in relativity theory is a thematic proposition, and it also expresses the constancy-seeking methodological thema.

Without preempting the discussion in the essays that follow, it will be useful here to touch briefly on a few other properties of themata. One is the question concerning their source. They are certainly not unapproachably synthetic a priori, in the eighteenth-century sense; nor is it necessary to associate them with Platonic, Keplerian, or Jungian archetypes, or with images, or with myths (in the nonderogatory sense, so rarely used in the English language), or with irreducibly intuitive apprehensions.[9]

It is likely that the origin of themata will be best approached through studies concerned with the nature of perception, and particularly of the psychological development of concepts in young children. Another direction that seems to have promise is the work building on Kurt Lewin's dynamic theory of personality. But, pending reliable results,[10] the most fruitful stance to take for the moment seems to me akin to that of a folklorist or anthropologist, namely, to look for and identify recurring general themes in the preoccupation of individual scientists and of the profession as a whole, and to identify their role in the development of science.

Another point concerns the antiquity and paucity of themes—the remarkable fact that the range and scale of recent theory, experience, and experimental means have multiplied vastly over the centuries while the number and kind of chief thematic elements have changed little. Since Parmenides and Heraclitus, the members of the thematic dyad of constancy and change have vied for loyalty, and so have, ever since Pythagoras and Thales, the efficacy of mathematics versus the efficacy of materialistic or mechanistic models. The (usually unacknowledged) presuppositions pervading the work of scientists have long included also the thematic couples of experience and symbolic formalism, complexity and simplicity, reductionism and holism, discontinuity and the continuum, hierarchical structure and unity, the use of mechanisms versus teleological or anthropomorphic modes of approach.

These, together with others further discussed in the essays and perhaps a few more—a total of fewer than 50 couples or triads—seem historically to have sufficed for negotiating the great variety of discoveries. Both nature and our pool of imaginative tools are characterized by a remarkable parsimony at the fundamental level, joined by fruitfulness and flexibility in actual practice. Only occasionally (as in the case of Niels Bohr) does it seem necessary to introduce a qualitatively new theme into science.

Thematic Preparations for Relativity

Perhaps the most pervasive characteristic of modern science is the generally accepted thema of the unlimited possibility of *doing* science, the belief that nature is inexhaustibly knowable. Kepler found support for this belief in equating the mind of God and the mind of man on those subjects which can be understood in the exact sciences. In our day, Heisenberg has said: "Exact science also goes forward in the belief that it will be possible in every new realm of experience to understand nature."

To recapture the exuberant enthusiasm of science, one should go not to a well-established contemporary physical science, but perhaps tc a field when it was still young. In the journals of the seventeenth and eighteenth centuries, we can find, side by side, what we would now consider very heterogeneous material—descriptions of a violent thunderstorm, statistics of and speculations on the causes of death in a certain village, notes on microscopic or telescopic wonders, or on the colors in chemical reactions, observations on the propagation of light, on the growth and types of reptiles, on the origin of the world. The heterogeneity speaks of a marvelous and colorful efflorescence of interests and an unselfconscious exuberance that verges sometimes on aimless play. The scientists of the time seem to us to have run from one astonishing and delightful discovery to the next like happy children surrounded by gifts.

As the terminology has grown more sophisticated, this wonderful feeling has become less evident to those not directly involved in science. And it also has become less wide-ranging in scientific circles as the work of the scientist has become more and more specialized. But from the platform of his specialized science, he can more than ever feel that nature bars no questions, that what can be imagined also can be—no, must be—investigated.

But behind this apparently atomistic fragmentation of attention has been a monistic aim. The seventeenth- and eighteenth-century researcher did not see the myriads of separate and disparate investigations around him as unrelated items in a randomly built catalogue of natural knowledge. Coupled with the theme of the universal accessibility of nature has been the old motivating methodological theme of an underlying *Einheit der Naturwissenschaften*—both a unity and a singularity of natural knowledge. The paths to an understanding of nature may be infinite (as the successes of even the most specialized interests indicate), and each of these paths is expected to have difficult but not insurmountable barriers. But all the paths have been vaguely thought to lead to *a* goal, *an* understanding of *one* nature, a delimited though no doubt complex rational corpus which some day a man's mind would be able to make his own (as the layman today says, somewhat frightened, "one great formula" that tells everything there is to know about nature).[11]

These two connected themata of unlimited outer accessibility and delimited inner meaning can be vaguely depicted by the device of a maze having in its outer walls innumerable entrances, through each of which one can hope to reach, sooner or later, the one mystery which lies at the center.

But another possibility has suggested itself more and more insistently: that at the innermost chamber of the maze one would find *nothing*. Writing in the fateful year of 1905, Joseph Larmor, one of Newton's successors as Lucasian Professor of Mathematics at Cambridge, saw it coming:

> There has been of late a growing trend of opinion, prompted in part by general philosophical views in the direction that the theoretical constructions of physical science are largely factitious, that instead of presenting a valid image of the relation of things on which further progress can be based, they are still little better than a mirage.[12]

The final encounter, he seems to cry out, cannot be with a mere shadow, or, worse still, with a narcissistic self-reflection of one's own thought processes.

Yet, on the face of it, it is not necessary to believe that knowledge of nature must turn out to be organizable in a philosophically satisfactory way. "We have no right to assume that any physical law exists," Max Planck once said. From a suitable distance, we cannot soundly claim that the historic development of science has proved nature to be understandable in a unique way—as distinct from documentable, manipulable, predictable within limits, or technically exploitable. What has happened is that the ground of the unknown has continually been shifted, the allegory has continually changed. David Hume expressed this in 1773:

> While Newton seemed to draw off the veil from some of the mysteries of nature, he showed at the same time the imperfections of this mechanical philosophy; and thereby restored her ultimate secrets to that obscurity in which they ever did and ever will remain.[13]

In the empirical sciences, we are far from being able to prove that we have been approaching an increasing understanding of the type that characterized the development of, say, some branches of mathematics. Our interests and tools change, but not in a linear, inevitable way. For example, the historic development from organismic science to a mechanistic and then to the mathematical style *could* have taken place in the opposite direction. And the ontological status of scientific knowledge itself has been turned completely upside down since the beginning of the twentieth century. The experimental detail is now not simply the token of a real world; on the contrary, it is all that we can be more or less sure about at the moment.

Karl Popper summarizes this view in these words:

I think that we shall have to get accustomed to the idea that we must not look upon science as a "body of knowledge," but rather as a system of hypotheses; that is to say, as a system of guesses or anticipations which in principle cannot be justifiied, but with which we work as long as they stand up to tests, and of which we are never justified in saying that we know that they are "true" or "more or less certain" or even "probable."[14]

Our justification for these hypotheses is that they have a hold on our imagination and that they help us to deal with our experience. On this basis, all the scientist needs to say, if anyone should ask him what he is doing, is: *hypotheses fingo*. This—a new methodological theme fundamental to the scientific revolution of the first ten years of our century—was precisely what Lodge, Larmor, Poincaré, and so many others could not accept. Poincaré, who was perhaps technically the best-prepared scientist in the world to understand Einstein's relativity theory of 1905, did not deign to refer to it once in his large published output up to his death in 1912. This silence—of which more is said in the sixth essay—was not mere negligence; Poincaré, despite his silence, had understood Einstein's message only too well.

Until Einstein, the postulate of relativity had been built into a physics resting on some principle of the unattainable but underlying Absolute. In Newton's work, relativity was made understandable and respectable by pointing to a curtain behind which, in the Sensorium of God, the finally unknowable absolutes of space and time were said to be hidden. God moved bodies, whose merely relative motion was all we could see, as a hidden puppeteer moves objects that act out his thoughts. In nineteenth-century physics, including that of Poincaré and Lorentz, the ultimate ground of explanation shifted to the undiscoverable ether, which thereby took, so to speak, the place of the essentially unknowable Deity of the previous two centuries as the repository of unaskable questions. What Einstein did in 1905—with his brash sentence declaring the ether to be "superfluous" as a result of elevating the principle of relativity from a heuristic conjecture to a fundamental proposition—was to pull up the curtain, and to announce that there was nothing at all behind it.

The demand that absolutes in science should be considered to be meaningless, that the ether, the last refuge of inscrutable reality, be abandoned, was too overwhelming for most scientists at the time. Their writings show what outrage was being committed on their own thematic orientation, and perhaps never more clearly than in the paper on *The Geometrisation of Physics* by Sir Oliver Lodge as late as 1921. In that critique of relativity theory he wrote:

> To summarise, then:
>
> In such a system there is no need for Reality; only Phenomena can be observed or verified: absolute fact is inaccessible. We have no criterion for truth; all appearances are equally valid; physical explanations are neither forthcoming nor required: there need be no electrical or any other theory of the constitution of matter. Matter is, indeed, a mentally constructed illusion generated by local peculiarities of Space. It is unnecessary to contemplate a continuous medium as a universal connector, nor need we try to think of it as suffering modification transmitted from point to point from the neighborhood of every particle of gravitational or electrified matter: a cold abstraction like a space-time manifold will do all that is wanted, or at least all that the equations compel. And, as a minor detail, which will bring us to the point, it is not necessary to invoke a real FitzGerald contraction in order to explain the result of the Michelson Experiment.
>
> . . . Undoubtedly general relativity, not as a philosophic theory but as a powerful and comprehensive method, is a remarkable achievement; and an ordinary physicist is full of admiration for the equations and the criteria, borrowed from hyper-Geometers, applied by the genius of Einstein, and expounded in this country with unexampled thoroughness and clearness by Eddington. But notwithstanding any temptation to idolatry, a physicist is bound in the long run to return to his right mind; he must cease to be influenced unduly by superficial appearances, impracticable measurements, geometrical devices, and weirdly ingenious modes of expression; and must remember that his real aim and object is absolute truth, however difficult of attainment that may be; that his function is to discover rather than to create; and that beneath and above and around all Appearances there exists a universe of full-bodied, concrete, absolute Reality.[15]

It was the last, dying cry against the replacement of the theme of the once-created, real universe by that of an ontologically agnostic universe, the one in which, for better or worse, we have found ourselves since 1905.

Scientific relativism was not the only road by which we have come to the empty center of the labyrinth. Through statistical mechanics the gross behavior of matter was more and more explainable, from mid-nineteenth century on, in terms of the concatenation of large numbers of molecules whose individual moment-to-moment behavior was technically too difficult to predict. To this, quantum physics then added the recognition that individual events on the sub-atomic level (for example, the path of individual photons going through an opening in a screen) are not predictable in principle, rather than merely for reasons of computational and experimental difficulty.

And there was yet another route to the establishment in science of the self-justifying existence of individual "events." Roger Cotes, in the defense of Newton's PRINCIPIA against the Cartesian and Leibnizian heresies, wrote:

> He who is presumptuous enough to think that he can find the true principles of physics and the laws of natural things by the force alone of his mind, and the internal light of his reason, must either suppose that the world exists by necessity, and by the same necessity follows the laws proposed; or if the order of Nature was established by the will of God, that himself, a miserable reptile, can tell what was fittest to be done. All sound and true philosophy is founded on the appearance of things; and if these phenomena inevitably draw us, against our wills, to such principles as most clearly manifest to us the most excellent counsel and supreme dominion of the All-wise and Almighty Being, they are not therefore to be laid aside because some men may perhaps dislike them. . . . Philosophy must not be corrupted in compliance with these men; for the order of things will not be changed.[16]

But it was; and very soon, too. As Koyré reminds us

> . . . the mighty, energetic God of Newton who actually "ran" the universe according to His free will and decision, became, in quick succession, a conservative power, an *intelligentia supra-mundana*, a "Dieu fainéant". . . . The infinite Universe of the New Cosmology, infinite in Duration as well as in Extension, in which eternal matter in accordance with eternal and necessary laws moves endlessly and aimlessly in eternal space, inherited all the ontological attributes of Divinity. Yet only those—all the others the departed God took away with Him.[17]

In this God-empty universe, what is it that one can ultimately encounter? Heisenberg has written that "changes in the foundations of modern science may perhaps be viewed as symptoms of shifts in the fundamentals of our existence which then express themselves simultaneously in many places, be it in changes in our way of life or in our usual thought forms. . . ." And the shift he singles out is "that for the first time in the course of history, man on earth faces only himself, that he finds no longer any other partner or foe." In science, too, "the object of research is no longer nature in itself but rather nature exposed to man's questioning, and to this extent man here also meets himself.[18]

This, then, is a main trend in the contemporary philosophy of science—resulting from a merging of the two strongest philosophical movements, existentialism and positivism—which appears to me to underlie the uneasiness expressed by many who see in a science a disjunctive and alienating component. The physicist and mathematician, Hermann Weyl, expressed a similar conclusion in a moving way:

> In existentialism is proclaimed a philosophical position which perhaps is better coordinated with the structure of modern scientific knowledge than Kantian idealism in which the epistemological positions of Democritus, Descartes, Galileo, and Newton appeared to have found their full philosophical expression.
>
> . . . When Bertrand Russell and others tried to resolve mathematics into pure logic, there was still a remnant of meaning in the form of simple logical concepts; but in the formalism of Hilbert, this remnant disappeared. On the other hand, we need *signs*, real signs, as written with chalk on the blackboard or with pen on paper. We must understand what it means to place one stroke after the other. It would be putting matters upside down to reduce this naively and grossly understood ordering of signs in space to some purified spatial conception and structure, such as that which is expressed in Euclidean geometry. Rather, we must support ourselves here on the natural understanding in handling things in our natural world around us. Not pure ideas in pure consciousness, but concrete signs lie at the base, signs which are for us recognizable and reproducible despite small variations in detailed execution, signs which by and large we know how to handle.
>
> As scientists, we might be tempted to argue thus: "As we know" the chalk mark on the blackboard consists of molecules, and these are made up of charged and uncharged elementary particles, electrons, neutrons, etc. But when we analyzed what theoretical physics means by such terms, we saw that these physical things dissolve into a symbolism that can be handled according to some rules. The symbols, however, are in the end again concrete signs, written with chalk on the blackboard. You notice the ridiculous circle.[19]

In summary: from the beginning to the present day, science has been shaped and made meaningful not only by its specific, detailed findings, but even more fundamentally by its thematic content. The reigning themata until about the mid-nineteenth century have been expressed most characteristically by the mandala of a static, homocentric, hierarchically ordered, harmoniously arranged cosmos, rendered in sharply delineated lines as in those of Copernicus's own hand-drawing. It was a finite universe in time and space; a divine temple, God-given, God-expressing, God-penetrated, knowable through a difficult process similar to that necessary for entering the state of Grace—by the works of the spirit and of the hand. Though not complete knowledge, it was as complete as the nature of things admits in this mortal life.

This representation was gradually supplanted by another, particularly in the last half of the nineteenth century. The universe became unbounded, "restless" (to use the happy description by Max Born), a weakly coupled ensemble of infinitely many separate, individually sovereign parts and events. Though evolving, it is continually interrupted by random discontinuities on the cosmological scale as well as on the submicroscopic scale. The clear lines of the earlier mandala have been replaced by undelineated, fuzzy smears, similar perhaps to the representation of distribution of electron clouds around atomic nuclei.

And now a significant number of our most thoughtful scholars, joining those critics who have never forgiven science its demythologizing role, seem to fear that a third mandala is rising to take precedence over both of these—the labyrinth with the empty center, where the investigator meets only his own shadow and his blackboard with his own chalk marks on it, his own solutions to his own puzzles. And this philosophical threat is thought to be matched by the physical threat considered as originating from a blind, aimless, self-motivating, ever-growing engine of technology.

It is therefore not surprising that those who think of our culture and our persons as caught up by these two tendencies find little comfort in the beauty of scientific advances, in the recital of coherence-making forces, or in promises that the lamp of science will light the way to a better society. Not until their doubts are allayed will there be any hope and valid place for science in our culture.

The Dimensions of Modern Historical Scholarship

Neither the analysis of a case in the history of science into its S_1 and S_2 components nor thematic analysis, separately or together, can suffice for a full understanding of the case. These two techniques are themselves only part of a larger set of work tools that should at least be enumerated here. (The distinction often made between externalistic and internalistic approaches is also far too blunt for laying bare the fine structure.)

Although many individual historians of science are largely, and properly, still occupied predominantly along the chief traditional directions, a general awareness and tolerance have arisen in recent years that recognize the existence of nine somewhat overlapping but still sufficiently separable directions that most historical work can now take in principle. Far from reducing science to an epiphenomenon of the social and economic development of society, such a multidimensional attack ensures that science is not reduced to one or another of the limited but popular caricatures. Ideally, the full potential of a major case is not likely to be exhausted until attention has been paid to each of these components. For any event E in the history of science at a time t (for example, Einstein's announcement of relativity, the case on which I have attempted to make some beginning in these essays), the systematic list of chief elements that may fruitfully engage the attention of historians of science would run somewhat as follows:

1. The awareness within the area of public scientific knowledge at time t of the scientific "facts," data, techniques, theories, and technical lore concerning event E—both in the published work of a particular scientist who is being studied and in the work of others in his field whom he may or may not have known about. (In terms used earlier, this is knowledge of S_2 at t).

2. The establishment of the time trajectory of the state of public scientific knowledge, both leading up to and going beyond the time chosen above. This is, as it were, the tracing of the World Line of an idea, a line on which the previously cited element E is merely a point. Under this heading we are dealing with antecedents, parallel developments, continuities and discontinuities, and the tracing of the public acceptance or rejection of an idea. This is our stock-in-trade: telling the story of a conceptual development (though perhaps too often at the cost of other concerns, making it therefore almost appear as if scientific ideas had an independent life of their own).

3. The reconstruction of the less institutional, more ephemeral, personal aspect of the scientific activity E at t. We are now looking at the same event in S_1, reflected in letters, drafts, laboratory notebooks, abandoned equipment, interviews, and reminiscences. Here we deal with an activity that may be poorly documented, and not necessarily appreciated or well understood by the agent himself.

4. The establishment of the time trajectory of this largely private

scientific activity under study—the personal continuities and discontinuities in development. What I am pointing to here is the kind of development we discuss in the tenth essay which brought Einstein from his first encounter with magnetism as a young boy to his early elaboration of general relativity theory. Now event E at time t begins to be understood in terms of the intersection of two trajectories, two World Lines, one for public science and one for private science.

5. Coupled to the trajectory of S_1 is the tracing of another line, the psychobiographical development of the scientist whose work is being studied. Far from being a passing fashion, it is an aspect of historical studies that in some form has been urged since at least the days of Wilhelm Dilthey. Whether one wishes to cast one's lot with the point of view of Freud, Erikson, Piaget, or others, or with several of them in part, seeking correlations between the S_1 trajectory and the historico-psychological development will at least give rise to those most precious of commodities, new and interesting questions. For example, for at least one case of high achievement, we shall pursue the hypothesis that a person's public scientific work is an expression of this intimate style of thought and life.

6. A similar line may be traced that relates the trajectory of the ideological or political as well as the literary events of the time to the trajectories of S_1 and S_2. We need merely mention the fact that Einstein (to his own astonishment) became a charismatic figure, with influence far beyond physics, as a result of the experimental confirmation in 1919 of his general relativity prediction concerning the deflection of starlight passing near the sun's disk. It is not unlikely that this mass response to what was perceived as a scientific "revolution" of the old order in physics, seemingly certified by nothing less than the stars themselves, was to a degree prepared by the existence of political revolutionary situations in many parts of the world at the time. Similarly it has been argued that the widespread interest among German scientists in abandoning the principle of causality in the early 1920's (despite Einstein's opposition) is closely related to developments in the German intellectual environment, including the huge vogue then enjoyed by Oswald Spengler's pessimistic and intuitionistic book, THE DECLINE OF THE WEST. On the whole I agree with the implication that in retrospect some of the seeds of destruction of the Weimar Republic may be discerned in studying this episode in the history of quantum physics. (Conversely, the influ-

ence of the progress of physics upon political history can of course also be documented, for example in the role of the nuclear reactor.)

7. A further component essential to full understanding of an event refers unavoidably to the sociological setting, conditions, influences—arising from colleagueship or the dynamics of team work, the state of professionalization at the time, the link between science and public policy or between science and industry, or institutional channels for the funding, evaluation, and acceptance of scientific work. One partial expression of this component may be found in the current attempts to construct empirical and quantitative measures for the progress of the sciences. But the qualitative aspects are at least as important. One could not, for example, study the great differences in the rates of acceptance of relativity theory or quantum theory in different countries, without studying the differences in the structure of their educational systems. Similarly, it would not be possible to understand fully the philosophical progression of Einstein away from his early devotion to empiricism—unless one understood and made provision for the influence of such colleagues as Max Planck.

8. The previous point is a reminder, if one were required, of another aspect, namely, the need for an analysis of the epistemological and logical structure of the work under study. Here has been our chief point of contact with a variety of philosophers of science whose contributions have, on occasion, indeed been illuminating for the historian of science.

9. Finally, there is the analysis of the individual scientist's thematic presuppositions that motivate and guide his research, as amply discussed above.

To be sure, the separations between the nine components I have cited are not hard and fast. Any categorical list is, to some degree, artificial—a fact that goes back to the well-known failure of language fully to comprehend experience. But the categories chosen do have merit at least operationally: they coincide in most cases with the specific professional interests of academic groups to whom the modern historian of science can look for collaboration and instruction.

And on the whole, these nine major strands, composing a net with which to catch the spirit and meaning of a particular scientific work, do fit the shape of historic cases themselves. One might envisage such a case as being, so to speak, caught in the center of a net made of these strands. It is unlikely and unnecessary for any historian of science to perform by himself an analysis along all of these nine dimensions. But if we are to rise above a partial view, the overlap of individual contributions should eventually make it possible for a scientific work to be so studied in full. In this way, the history of science may reach its highest and most ambitious level—which I take to be the historically based analysis of culture, society, and personality, seen through the focus of a lens which the case study of a scientific work provides.

Acknowledgments

In a volume stressing both the origins and the personal component of scientific and scholarly work, I shall perhaps be forgiven for making some personal comments on those origins of my own interests of which I am most aware, and for acknowledging thereby at least a few of my debts. One's thoughts are of course fashioned to some degree by interactions with one's teachers and friends, by ideas delighted in and argued over in the company of others who may not by any means have shared one's own opinion—and mention of whom should not be construed as implicating them in any way.

Through a series of lucky accidents I benefited from an unusual variety of contradictory influences. At the beginning of my experimental work in the high-pressure-physics laboratory of my thesis professor, the crusty and constantly questioning P. W. Bridgman, the central reality was certainly science done at first hand, although the problems of philosophy of science were never left out of sight entirely. On the other hand, the interaction between science and epistemology was at the center of attention in discussions over many years with Philipp Frank, that most humanistic and conciliatory of logical empiricists, and biographer of Albert Einstein; under Frank, I acted as instructor, assisting in one of his courses at Harvard. It was he who provided me later with the first introduction to the Einstein archival materials at the Princeton Institute for Advanced Study. And it was while reading through that voluminous material as part of the effort to catalog it with the help of Einstein's knowledgable secretary, Miss Helen Dukas (to whom many acknowledgments will be found in the following pages), that I became aware of historical problems within the framework of S_1 which rarely get asked outside it.

A complementary influence was exerted through the accident of being intrigued into teaching in the General Education program by Harvard's

President, James B. Conant, and the physicist Edwin C. Kemble. There I not only had to struggle with the competing historiographic theories in trying to understand how science evolved in the large and what role it had in culture and society, but also came up against some of the raw pedagogic problems that are discussed in the essays toward the end of this volume. Frequently, I felt caught in the crossfire between colleagues outside the sciences who were rather skeptical of the field—such as Douglas Bush and Raphael Demos—and intelligent and eloquent students in the classroom who in those days were, if anything, all too ready to concede any claim made for science.

As much as any of these, however, I must acknowledge the influence of a number of "shop clubs" and other series of informal meetings which had formed spontaneously in and around the Cambridge area. In one of these, under the general leadership of Philipp Frank and P. W. Bridgman, fairly regular meetings of a "Unity of Science" group took place for many years, where the irrepressible individuality and passionate defense of well formed views were always in evidence. There were memorable encounters involving Henry Aiken, Karl Deutsch, Roman Jakobson, Gyorgy Kepes, Philippe LeCorbeiller, W. V. O. Quine, Giorgio de Santillana, Harlow Shapley, B. F. Skinner, S. S. Stevens, Lazlo Tisza, Richard von Mises, Norbert Wiener, as well as occasional visitors from far-away New York or New Haven, such as Ernest Nagel and Henry Margenau. The logical empiricists among all these, contrary to the doctrinaire days of the 1920's and 1930's, laid as much stress as did all the others on fundamental, interdisciplinary discussions. For example, far from wishing to denigrate other areas and extending the hegemony of science, Frank held that a central question was "How can we bring about the closest possible *rapprochement* between philosophy and science?"

But beyond that, the members of that group made no efforts to exact any agreements, and relished the most wide-ranging debates. As a result, I happily never had the opportunity or felt the need for a single, doctrinaire master. Rather, the educational model was that of allowing a wide variety of views from different subject-matter fields to be brought to bear on living questions, and measuring their value by the usefulness of their contributions.

A second shop club, started by J. B. Conant, and meeting a few times each year, involved those on the college faculty—scientists, social scientists and humanists—who participated in the General Education teaching program. This allowed the civilizing experience of discussion with Jerome Bruner, Harry Levin, Ernst Mayr, Harry Murray, Talcott Parsons, David Riesman, Paul Tillich, George Wald, Morton White, and others. At about that time E. C. Kemble also arranged frequent working lunch meetings of a group of junior staff interested in the history of science, including I B. Cohen, T. S. Kuhn, and L. K. Nash; unquestionably we all profited from these meetings. Later on, during the period when I brought together a group of scientists and other scholars in Cambridge to help design the Project Physics Course, I benefited from discussions with members of the staff and particularly with Alfred Bork, Stephen Brush, Banesh Hoffmann, Loyd Swenson, and Stephen Toulmin.

In the early days after World War II, there was, of course, still the largely unseen but intensely felt presence of George Sarton, and the electrifying effect of Cambridge visits by Alexandre Koyré. More recently, through one accident or another, I have had opportunities to benefit from the views or counter-arguments of such colleagues and friends as Rudolf Arnheim, Erwin Hiebert, Everett Mendelsohn, Robert S. Cohen, Marx Wartofsky, Joseph Agassi, Yehuda Elkana, and Charles Weiner. Some stimulating contacts were made through DAEDALUS conferences, for example, with Erik Erikson, who subsequently became a colleague from whom I have learned greatly.

In listing influences, I must not omit the *Eranos Tagungen*, organized by Arnold Portmann, at which I was allowed to make extensive presentations and then learn from other members of the "faculty," such as Portmann, Mircea Eliade, and Gershom Scholem.

Of course, the students in one's own courses are always the most consistent and long-suffering critics, and I have indicated in some essays those members of my seminars who should be singled out for their thoughtful responses.

Finally, it is a pleasant duty to acknowledge the funds which made some of these studies possible; this is done in several of the essays themselves, but I wish to acknowledge in particular the helpfulness of grants from the section for the History and Philosophy of Science in the National Science Foundation.

I owe special debts to W. Gordon Whaley and Mark Carroll. Dean Whaley of The University of Texas at Austin, as Editor of THE GRADUATE JOURNAL, and Mark Carroll, as Director of Harvard University

Press, both suggested publication of this set of selected essays, and they converted my initial skepticism into ready compliance. To Audrey Nelson Slate, Clare Y. Whaley, Joseph Elder, Helen Stewart, and Beverly Reaume, I express my gratitude for help with the editorial and production problems of this volume.

Most of the essays in this volume are substantially as given in their original form, except for minor changes in a few passages and some expansions or abbreviations to remove obvious infelicities or redundancies. Some overlap that had been necessary when publishing separately appearing essays in different journals has here been deleted; but I hope I have left enough, and I have provided cross references, so that skipping back and forth is possible for a reader who does not wish to go through the essays as here arranged in sequential order.

Jefferson Physical Laboratory *Gerald Holton*
Department of Physics,
Harvard University

NOTES

1. Great credit for the recent push in this direction must go to such projects and institutions as the Center for the History of Physics at the American Institute of Physics, the Joint American Physical Society-American Philosophical Society Project for the History of Quantum Physics, and the American Academy of Arts and Sciences-AIP Joint Study on the History of Nuclear Physics, to the opening up of such archives as that of Albert Einstein in Princeton, and to the financial support provided for many of these activities by such granting agencies as the National Science Foundation and, to a certain extent more recently, the National Endowment for the Humanities. These remarks refer to the history of contemporary physics alone; similar activities are beginning to be pursued on the recent history of the other sciences.
2. William James, The Will to Believe and Other Essays in Popular Philosophy (New York: Longmans, Green & Co., 1897), pp. 92–93.
3. E.g., by R. B. Braithwaite, Scientific Explanation (Cambridge: Cambridge University Press, 1953), p. 22.
4. *Ibid.*, p. 107.
5. E.g., Ludwig Wittgenstein, Tractatus Logico-Philosophicus (London: Routledge & Kegan Paul, 1961), 6.3211, 6.33, 6.34, 6.35; or Stephen Toulmin, Foresight and Understanding (Bloomington, Indiana: Indiana University Press, 1961), p. 100.
6. Braithwaite, *op. cit.*, pp. 20–21; italics added.
7. See Toulmin, *op. cit.*, p. 69.

8. Of course, not all themata are meritorious. As Bacon warned in discussing the four Idols that can trap the scientific mind, some have turned out to divert or slow the growth of science.

Nor do all sciences have to benefit equally; the holistic viewpoint introduced at the start of the nineteenth century may have had benefits in physics but was on the whole a handicap for biology. Nor is there any necessity that the allegiance by a given scientist to a given thema must be unshakable. Scientists can and occasionally do change loyalties to a thema, as is demonstrated in the discussion of the philosophical development of Einstein in the eighth essay.

9. Similarly, it should not be necessary to stress that thematic analysis is not an ideology, a school of metaphysics, a plea for irrationality, an attack on the undoubted effectiveness of empirical data and experimentation, or, a desire to teach scientists how to do their job better. Nor is it a theoretical framework for accommodating such notions as paradigms or research programs.

10. These may first be merely the establishment of typology. Just as Ostwald had discerned a dichotomy between classical and romantic styles within science (which Albert Szent-Gyorgyi recently renamed systematic and intuitive, or Apollonian and Dionysian), Lewin distinguished between the Aristotelian and Galilean modes of thought, and showed their persistence in contemporary scientific work.

11. In this way we may possibly have been hoping to return to the primitive or childlike state of *Einheitswirklichkeit,* a primitive *Ganzheitserfahrung der grossen Welt* as against the later-learned *Ich-Welt,* the fragmented *Partialwelt,* or *Objektwelt,* to use the terms of Erich Neumann in Der Schöpferische Mensch (Zurich: Rhein-Verlag, 1959), pp. 105–109.

12. Joseph Larmor, in his Preface to Henri Poincaré, Science and Hypothesis (London: Walter Scott Publishing Co., 1905), p. 12.

13. David Hume. The History of England, Volume 8 (1773), p. 332.

14. Karl R. Popper, The Logic of Scientific Discovery (London: Hutchinson, 1959), p. 317.

15. Oliver Lodge, *The Geometrisation of Physics, and Its Supposed Basis on the Michelson-Morley Experiment,* Nature 106, No. 2677, 1921.

16. Roger Cotes, Preface to Newton's Mathematical Principles of Natural Philosophy (Andrew Motte translation of 1729, revised by Florian Cajori, Berkeley: University of California Press, 1946), pp. xxxi sq.

17. Alexandre Koyré, From the Closed World to the Infinite Universe (New York: Harper & Brothers, 1958), p. 276.

18. Werner Heisenberg, *The Representation of Nature in Contemporary Physics,* Daedalus, Summer 1958, pp. 103–105.

19. Hermann Weyl, *Wissenschaft als symbolische Konstruction des Menschens,* Eranos Jahrbuch (Zurich: Rhein-Verlag, 1949), pp. 382, 427–428.

I *On the Thematic Analysis of Science*

1 THE THEMATIC IMAGINATION IN SCIENCE

CURRENT OPINION on the way scientific theories are constructed is by no means unanimous. We may, nevertheless, take the account given not too long ago by the physicist Friedrich Dessauer as a quite typical contemporary presentation of the so-called hypothetico-deductive, or inductive, method of science. His scheme[1] reflects both general professional and popular understanding.

There are, he reports, five steps. (1) Tentatively, propose as a hypothesis a provisional statement obtained by induction from experience and previously established knowledge of the field. An example, drawn from experimental work in physics, might be this: the observed large loss of sound energy when ultrasonic waves pass through a liquid such as water is possibly due to a structural rearrangement of the molecules as the sound wave passes by them. (2) Now, refine and structure the hypothesis—for example, by making a mathematical or physical analogon showing the way sound energy may be absorbed by clumps of molecules. (3) Next, draw logical conclusions or predictions from the structured hypothesis which have promise of experimental check—for example, if

Delivered at the Johns Hopkins University as a Shell Companies Foundation Lecture under the title *Presupposition in the Construction of Theories*, this essay was first published in SCIENCE AS A CULTURAL FORCE, ed. Harry Woolf (Baltimore: The Johns Hopkins Press, 1964) and reprinted in THE GRADUATE JOURNAL, Volume VII, No. 1, pp. 87–109, 1965–66.

more and more pressure is applied to the sample of water, it should be more and more difficult for the associated molecular groups to absorb sound so strikingly. (4) Then check the predicted consequences (deduced from the analogon) against experience, by free observation or experimental arrangement. (5) If the deduced consequences are found to correspond to the "observed facts" within expected limits—and not only these consequences, but all different ones that can be drawn (behavior at constant pressure but changing temperature, or similar effects in other liquids)—then a warrant is available for the decision that "the result obtained is postulated as universally valid" (p. 298). Thus, the hypothesis, or initial statement, is found to be scientifically "established."

But, popular opinion continues, until the facts support such a position any hypothetical statement is to be held scrupulously with open-minded skepticism. The scientist, Dessauer reports, "does not take a dogmatic view of his assumption, he makes no claim for it, he does not herald it abroad, but keeps the question open and submits his opinion to the decision of nature itself, prepared to accept this decision without reserve" (p. 296). This, he concludes, is "the inductive method, the fundamental method of the entire modern era, the source of all our knowledge of nature and power over nature" (p. 301).

We note that this account fits in well with a widespread characterization of a supposed main difference between scientists and humanists: the former, it is often said, do not preempt fundamental decisions on aesthetic or intuitive grounds; they do not make a priori commitments, and only let themselves be guided by the facts and the careful process of induction. It is, therefore, not surprising that in this, as in most such discussions, nothing was said about the *source* of the original induction, or about the criteria of *preselection* which are inevitably at work in scientific decisions. Attention to these would seem to be as unimportant or fruitless as a discussion, say, of the "reality" of the final result.

This account of scientific procedure is not wrong; it has its use, for example, in broadly characterizing certain features of science as a public institution. But if we try to understand the actions and decisions of an actual contributor to science, the categories and steps listed above are deficient because they leave out an essential point: to a smaller or larger degree, the process of building up an actual scientific theory requires explicit or implicit decisions, such as the adoption of certain hypotheses and criteria of preselection that are not at all scientifically "valid" in the sense previously given and usually accepted.[2] One result of this recognition will be that the dichotomy between scientific and humanistic scholarship, which is undoubted and real at many levels, becomes far less impressive if one looks carefully at the construction of scientific theories. This will become evident first at the place where explicit and implicit decisions are most telling—namely in the formation, testing, and acceptance or rejection of hypotheses.

I

To illustrate this point as concretely as possible, let us look at a case for which it has long been thought the last word had been said. As is well known, Book III of Newton's PRINCIPIA, which was supposed to use the principles and mathematical apparatus developed in Books I and II to "demonstrate the frame of the System of the World," opens with a section that is as short as it is initially surprising: the four rules of reasoning in philosophy, the Regulae Philosophandi. At any rate, they appear so in the third edition, of 1726, known to us usually through Motte's translation of 1729. These are, of course, well-known rules, and I need remind you of them only briefly. They can be paraphrased as follows:

I. Nature is essentially simple; therefore, we should not introduce more hypotheses than are sufficient and necessary for the explanation of observed facts. This is a hypothesis, or rule, of simplicity and *verae causae*.

II. Hence, as far as possible, similar effects must be assigned to the same cause. This is a principle of uniformity of nature.

III. Properties common to all those bodies within reach of our experiments are to be assumed (even if only tentatively) as pertaining to all bodies in general. This is a reformulation of the first two hypotheses and is needed for forming universals.

IV. Propositions in science obtained by wide induction are to be regarded as exactly or approximately true until phenomena or experiments show that they may be corrected or are liable to exceptions. This principle states that propositions induced on the basis of experiment should not be confuted merely by proposing contrary hypotheses.

It has been justly said that these epistemological rules are by no means a "model of logical coherence."[3] They grew in a complex way, starting from only two rules (I and II) in the first edition of the PRINCIPIA (1687) where they were still called Hypotheses I and II. As Newton, with growing dislike for controversy, came to make the corrections for the third edition, he added the polemical rule IV which is a counter-attack on the hypotheses-laden missiles from the Cartesians and Leibnizians.

But it turns out that Newton at one time was on the verge of going further. It was discovered only recently in a study of Newton's manuscripts by Alexander Koyré[4] that Newton had written a lengthy *Fifth Rule,* and then had suppressed it. The significant parts of it for our purpose are the first and last sentences of this rule, and the likely reasons why it had to be suppressed.

"Rule V. Whatever is not derived from things themselves, whether by the external senses or by internal cogitation, is to be taken for hypotheses. . . . And what neither can be demonstrated from the phenomena nor follows from them by argument based on induction, I hold as hypotheses."

To us, even as to Newton's contemporaries, disciples, and defenders, the sense in which Newton uses here the word "hypothesis" in the suppressed rule is clearly pejorative. It was after all Newton himself who, in 1704, had written as the first sentence of the OPTICKS, "My design in this Book is not to explain the Properties of Light by Hypotheses, but to propose them, and prove them by Reason and Experiment." And in this and other ways, he had begun to sound the declaration *hypotheses non fingo* in the second, 1713, edition of the PRINCIPIA. We are apt to remember this slogan rather than the fact that in Newton's work from beginning to end, and even in the last edition of the PRINCIPIA itself, one can readily find explicit hypotheses as well as disguised ones. And we are apt to overlook that rules against hypotheses are themselves methodological hypotheses of considerable complexity.

But, then, is it not strange that Newton after all *did* suppress this Fifth Rule which the Newtonians after him, his modern, empiricist disciples, from Cotes to Dessauer, would accept readily? To understand why Newton may have done this is of importance if we want to understand the cost of having so long been the philosophical heirs of the victorious side in that seventeenth-century quarrel concerning what science should be like.

The answer has, I think, several elements, but one is surely an ancient one: that disciples are usually eager to improve on the master, and that the leader of a movement sometimes discovers he cannot or does not wish to go quite as fast to the Promised Land as those around him. (Thus, it was not Cortés but the men he had left in charge of Mexico who, as soon as his back was turned, tried to press the victory too fast to a conclusion and began to slaughter the Aztecs, with disastrous consequences.)

Here it is significant that Newton had only said, in one draft of his *General Scholium*: "I avoid hypotheses"; and in the final version, "I do not feign hypotheses," i.e., I make no false hypotheses. But his spokesman and friend, Samuel Clarke, translated him to read: "And hypotheses I *make* not"; and Andrew Motte rendered it as the famous "I frame no hypotheses." In this, as in several other places, Newton's protagonists went much further than he did and seemed to ask for a Baconian sense of certainty in science which Newton knew did not exist.

Newton had indeed exposed and rejected certain hypotheses as detrimental; he knew how to tolerate others as being at least harmless; and he, like everyone else, knew how to put to use those that are verifiable or falsifiable. But the fact is that Newton also found one class of hypotheses to be impossible to avoid in his pursuit of natural philosophy—a class that shared with Cartesian hypotheses the characteristic of neither being demonstrable from the phenomena nor following from them by an argument based on induction, to use the language in Newton's suppressed Fifth Rule itself. The existence, nay, the necessity, at certain stages, of entertaining such unverifiable and unfalsifiable, and yet not quite arbitrary, hypotheses—that is an embarrassing conception which did not and does not fit into a purely positivistically oriented philosophy of science. For the decision whether to entertain such hypotheses is coupled neither to observable facts nor to logical argument.

In Newton's case, two obvious examples of his use of this class of hypotheses—to which I refer as "thematic" propositions or thematic hypotheses, for reasons to be discussed later—appear in his theory of matter and his theory of gravitation. On the latter, A. Rupert Hall and Marie Boas Hall, in their book UNPUBLISHED SCIENTIFIC PAPERS OF SIR ISAAC NEWTON, have printed the first manuscript draft of the *General Scholium* (written in January, 1712–13) in which Newton very plainly confesses his inability to couple the hypothesis of gravitational forces with observed phenomena: "I have not yet disclosed the cause of gravity, nor have I undertaken to explain it, since I could not understand it from the phenomena. For it does not arise from the centrifugal force of any vortex, since it does not tend to the axis of a vortex but to the center of a planet."[5] And speaking of Newton's inability to arrive at the cause of gravity from phenomena, the Halls add: "In one obvious sense, this is true, and in that sense it knocks the bottom out of the aethereal hypothesis. In another sense it is false: Newton knew that God was the cause of gravity, as he was the cause of all natural forces. . . ."[6]

Exactly so—for this indeed was Newton's central presupposition in the theory of gravitation. The Halls continue, "That this statement could be both true and false was Newton's dilemma: In spite of his confident expectations, physics and metaphysics (or rather theology) did not smoothly combine. In the end, mechanism and Newton's conception of God could not be reconciled . . . . Forced to choose, Newton preferred God to Leibniz."

That Newton could not bring himself to announce this hypothesis in the PRINCIPIA is not strange since the grounds of the hypothesis are so foreign to the avowed purpose of this book on the MATHEMATICAL PRINCIPLES OF NATURAL PHILOSOPHY. And also, a thematic hypothesis becomes more persuasive the longer the period of unsuccessful attempts to use other hypotheses, namely, those that *are* coupled to phenomena. The thematic hypothesis is often an impotency proposition, in the sense that the search for alternatives has proved to be vain. The point when one is forced to rely on thematic hypotheses is exactly when one has to say, with Newton: "I could not understand it from the phenomena."

So when we approach the physics of a man like Newton, and even when we try to interpret his epistemological position, we must look beyond the explicit and obvious component of it, the basically operationist and relativistic physics of observable events. What made his work meaningful to Newton was surely that in his physics he was concerned with a God-penetrated, real world: God himself is standing behind the scenes, like a marionette player, moving the unseen strings of the puppets that merely act out the thoughts in His great sensorium. And this is a proposition which Newton tried to avoid having to state openly, where his friends and enemies would see it, though this reluctance accounts for some of the strange tension which pervades the PRINCIPIA and his other writing. Reading Newton, one is struck by the fact that below the surface the major problems which haunted him were very closely related. They are: (a) the cause of gravity, whose existence only he had "established from phenomena"; (b) the existence of other forces, e.g., short-range forces to explain cohesion, chemical phenomena, etc.; (c) the nature of space and time, what he called the "sensory" of God; (d) and last, but not least, the proofs for the existence of the Deity (namely, by showing that there can be no other final causes for demonstrated forces and motions than the Deity—that, therefore, the Deity not only has properties, but also "dominion").

In Newton's physics, the hypothesized "sensory" of God is the cut-off point beyond which it was unnecessary and inappropriate to ask further questions. And this is an important function of a thematic hypothesis, which by its very nature is not subject to verification or falsification. For unlike the usual class of hypothesis—which, to use Aristotle's formula, is a statement that may be "believed by the learner" but ultimately is "a matter of proof"—the thematic hypothesis is precisely built as a bridge over the gap of ignorance. Thus, as scientists, we cannot and need not ask *why it is* that we believe, with Descartes, in an "inescapably believable" proposition; or why it is that we can perceive correspondences between certain observations and the predictions that follow from a model; or nowadays, for that matter, with Niels Bohr, why we can "build up an understanding of the regularities of nature upon the consideration of pure number."

II

We have indeed left the recipe for a step-by-step construction of scientific theory far behind. Let us now turn from the specific example and attempt to discern in a schematic way what the analysis of scientific theories in terms of themata adds to the more conventional kind of analysis.

Regardless of what scientific statements they believe to be "meaningless," all philosophies of science agree that two types of proposition are *not* meaningless, namely, statements concerning empirical matters of "fact" (which ultimately boil down to meter readings), and statements concerning the calculus of logic and mathematics (which ultimately boil down to tautologies).

There are clearly difficulties here that we might well discuss. For example, the empirical matters of fact of modern science are not simply "observed," but are nowadays more and more obtainable only by way of a detour of technology (to use a term of Heisenberg's) and a detour of theory. But in the main we can distinguish between these two types of "meaningful" statements quite well. Let us call them respectively *empirical* (or *phenomenic*) and *analytical* statements, and think of them as if they were arrayed respectively on orthogonal x- and y-axes, thereby we can represent these two "dimensions" of usual scientific discourse by a frank analogy, and generate terminology which will be useful as long as we do not forget that all analogy has its limits.

Now we may use the x-y plane to analyze the concepts of science (such as force), and the propositions of science, e.g., a hypothesis (such as "X-rays are made of high energy photons") or a general scientific law (such as the law of universal gravitation). The *concepts* are analogous to points in the x-y plane, having x- and y-coordinates. The *propositions* are analogous to line elements in the same plane, having projected components along x- and y-axes.

To illustrate, consider a concept such as force. It has empirical, x-dimension meaning because forces can be qualitatively discovered and, indeed, quantitatively measured, by, say, the observable deflection of solid bodies. And it has analytical, y-dimension meaning because forces obey the mathematics of vector calculus (e.g., the parallelogram law of composition of forces), rather than, for example, the mathematics of scalar quantities.

Now consider a proposition (a hypothesis, or a law): the law of universal gravitation has an empirical dimension or x-component—for example, the observation in the Cavendish experiment where massive objects are "seen" to "attract" and where this mutual effect is measured. And the law of universal gravitation has an analytical or y-component, the vector-calculus rules for the manipulation of forces in Euclidean space.

An interpolation is here in order, to avoid the impression that there is some absolute meaning intended for the x- or y-components. Indeed, it is preferable to use the term "heuristic-analytic" for the y-dimension, on grounds which I can at least indicate by noting that there exist in principle infinitely many possible logical and mathematical systems, including mutually contradictory ones, from which we choose those that suit our purposes. On the x-axis we do not appear to have this degree of freedom to make "arbitrary" decisions on heuristic grounds. At least at first glance, we seem constrained to deal with the phenomena of our natural world as they present themselves to us, rather than with many mutually contradictory worlds of phenomena from which we might be free to select those to which we wish to pay attention. However, one can at least imagine worlds that are quite differently constructed, where on the one hand an infinitely large pool of phenomena contains "contradictory" sets (i.e., stones that sometimes fall and sometimes rise, in some random sequence), but where on the other hand our logical and mathematical tools are severely restricted—say, only to Aristotelian syllogisms and elementary arithmetic. Then we would be forced to select from all possible observables those which can be represented and discussed in terms of scalar quantities, and we would have to exclude forces, acceleration, momenta, etc. In that case, the x-dimension could be named the dimension of heuristic-empirical statements.

Now, to some extent we *are* in this situation even now in our "real" world. We get a hint of it when we think of the great number of phenomena that are thought to be important today, but that were unknown yesterday;[7] or if we think of the continual change in the allegory (for example, the allegory of motion itself), from the Aristotelian conception which equated motion and change of any kind, to the modern, much attenuated idea of motion as the rate of change of distance or displacement with respect to time, or quantifiable local motion.

We realize the same point also when we think of all the "phenomena" which at any time are simply not admitted into science—for example, heat and sound in Galileo's physics, or most types of single-event occurrences that do not promise experimental control or repetition in modern physical science. In short, we are always surrounded on all sides by far more "phenomena" than we can use, and which we decide —and must decide—to discard at any particular stage of science.

The choice of allowable *analytical* systems is in principle also very large. Thus, any point, on any object, could for purposes of kinematical description be regarded as the center of the world. But the choice, in practice, is quite restricted. Indeed, the reason that science, until the late nineteenth century, was so sure of the uniqueness of the given world is to be sought in the fact that the analytical systems then available were so simple and had so long remained without fundamental qualitative changes and alternatives. Thus Newton could say in the preface of the PRINCIPIA that geometry itself is "founded in mechanical practice and is nothing but that part of universal mechanics which accurately proposes and demonstrates the art of measuring." This impression helped to reinforce the feeling that the world, found and analyzed by science in terms of then current x- and y-components, existed in a unique, a priori way. In mathematics one calls such a situation, where the potential plurality of solutions shrinks to one or a very few, a "degenerate" case. It is only after the discovery of non-Euclidean mathematics that one begins to see the essential arbitrariness of the y-dimension elements in which our scientific statements are couched, and that one becomes open to the suggestion that there is also an arbitrariness in the decisions about what x-dimension elements to select. This recognition is perhaps at the heart of the current agnosticism concerning the old question as to the "reality" of the world described in the x-y plane.

But whether they are arbitrary or not, the x-y axes have, since the seventeenth and eighteenth centuries, more and more defined the total allowable content of science and even of sound scholarship generally. Hume, in a famous passage, expressed eloquently that only what can be resolved along x- and y-axes is worthy of discussion:

> If we take in our hands any volume; of divinity, or school metaphysics, for instance: Let us ask, Does it contain any abstract reasoning concerning quantity or number? No. Does it contain any experimental reasoning concerning matter of fact or criteria? No. Commit it then to the flames. For it can contain nothing but sophistry and illusion.

If we now leave the x-y, or *contingent,*[8] plane, we are going off in an undeniably dangerous direction. For it must be confessed at once that the tough-minded thinkers who attempt to live entirely in the x-y plane are more often than not quite justified in their doubts about the claims of the more tender-minded people (to use a characterization made by William James). The region below or above this plane, if it exists at all, might well be a muddy or maudlin realm, even if the names of those who have sometimes gone in this direction are distinguished. As Eduard Dijksterhuis has said:

> Intuitive apprehension of the inner workings of nature, though fascinating indeed, tend to be unfruitful. Whether they actually contain a germ of truth can only be found out by empirical verification; imagination, which constitutes an indispensable element of science, can never even so be viewed without suspicion.[9]

And yet, the need for going beyond the x-y plane in understanding science and, indeed, in doing science, had been consistently voiced long before Copernicus, who said that the ultimate restriction on the choice of scientific hypotheses is not only that they must agree with observation, but also "that they must be consistent with certain preconceptions called 'axioms of physics,' such as that every celestial motion is circular, every celestial motion is uniform, and so forth."[10] And if we look carefully, we can find even among the most hard-headed modern philosophers and scientists a tendency to admit the necessity and existence of a non-contingent dimension in scientific work. Thus Bertrand Russell[11] speaks of cases "where the premises of sciences turn out to be a set of pre-suppositions neither empirical nor logically necessary"; and in a remarkable passage, Karl R. Popper confesses very plainly to the impossibility of making a science out of only strictly verifiable and justifiable elements:

> Science is not a system of certain, or well-established, statements; nor is it a system which steadily advances towards a state of finality . . . . *We do not know: we can only guess.* And our guesses are guided by the unscientific . . . faith in laws, in regularities which we can uncover—discover. Like Bacon, we might describe our own contemporary science—"the method of reasoning which men now ordinarily apply to nature"—as consisting of "anticipations, rash and premature" and as "prejudices."[12]

One could cite and analyze similar opinions by a number of other scientists and philosophers. In general, however, there has been no systematic development of the point. But it is exactly here that we should discern the existence of a door at the end of the corridor through which the philosophy of science has recently been traveling. To supplement contingency analysis, I suggest a discipline that may be called thematic analysis of science, by analogy with thematic analyses that have for so long been used to great advantage in scholarship outside science. In addition to the empirical or phenomenic (x) dimension and the heuristic-analytic (y) dimension, we can define a third, or z-axis. This third dimension is the dimension of fundamental presuppositions, notions, terms, methodological judgments and decisions—in short, of themata or themes—which are themselves neither directly evolved from, nor resolvable into, objective observation on the one hand, or logical, mathematical, and other formal analytical ratiocination on the other hand. With the addition of the thematic dimension, we generalize the plane in which concepts and statements were previously analyzed. It is now a three-dimensional "space"—using the terms always in full awareness of the limits of analogy—which may be called *proposition space.* A concept (such as force), or a proposition such as the law of universal gravitation, is to be considered, respectively, as a point or as a configuration (line) in this threefold space. Its resolution and projection is in principle possible on each of the three axes.

To illustrate: the phenomenic and analytic-heuristic components of the physical concept force (its projections in the x-y plane) have been mentioned. We now look at the thematic component and see that throughout history there has existed in science a "principle of potency." It is not difficult to trace this from Aristotle's *ἐνέργεια,* through the neo-Platonic *anima motrix,* and the active *vis* that still is to be found in Newton's PRINCIPIA, to the mid-nineteenth century when "Kraft" is still used in the sense of energy (Mayer, Helmholtz). In view of the obstinate preoccupation of the human mind with the theme of the potent, active—some might have said masculine—principle, before and quite apart from any science of dynamics (and also with its opposite, the passive persisting principle on which it acts), it is difficult to imagine any science in which there would not exist a conception of force (and of its opposite, inertia).

It would also be difficult to understand certain conflicts. Scholastic physics defined "force" by a projection in the phenomenic dimension that concentrated on the observation of continuing terrestrial motions against a constantly acting obstacle; Galilean-Newtonian physics defined "force" quite differently, namely, by a projection in the phenomenic dimension that concentrated on a thought experiment such as that of an object being accelerated on a friction-free horizontal plane. The projections above the analytic dimension differed also in the two forms of physics (i.e., attention to magnitudes versus vector properties of forces). On these two axes, the concepts of force are entirely different. But the reason why natural philosophers in the two camps in the early seventeenth century thought they were speaking about the same thing, nevertheless, is that they shared the need or desire to incorporate into their physics the same thematic conception of *anima,* or *vis,* or *Kraft*—in short, of force.

A second example of thematic analysis might be the way one would consider not a concept but a general scientific proposition. Consider the clear thematic element in the powerful laws of conservation of physics, for example the law of conservation of momentum, as formulated for the first time in useful form by Descartes. In Descartes's physics, as Dijksterhuis wrote:

> All changes taking place in nature consist in motions of . . . three kinds of particles. The primary cause of these motions resides in God's *concursus ordinarius*, the continuous act of conservation. He so directs the motion that the total *quantitas motus* (momentum), i.e., the sum of all the products of mass and velocity, remain constant.[13]

This relation, $\Sigma\, mv = \text{const.}$, "constitutes the supreme natural law. . . ."[14] This law, Descartes shows, springs from the invariability of God, in virtue of which, now that He has wished the world to be in motion, the variation must be as invariable as possible.

Since then, we have learned to change the *analytic* content of the conservation law—again, from a scalar to a more complex calculus—and we have extended the phenomenic applicability of this law from impact between palpable bodies to other events (e.g., scattering of photons). But we have always been trying to cling to this and to other conservation laws, even at a time when the observations seem to make it very difficult to do so.[15] The *thema* of conservation has remained a guide, even when the language has had to change. We now do not say the conservation law springs from the "invariability of God"; but with that curious mixture of arrogance and humility which scientists have learned to put in place of theological terminology, we say instead that the law of conservation is the physical expression of the elements of constancy by which nature makes herself understood by us.

The strong hold that certain themes have on the mind of the scientist helps to explain his commitment to some point of view that may in fact run exactly counter to all accepted doctrine and to the clear evidence of the senses. Of this no one has spoken more eloquently and memorably than Galileo when he commented on the fact that to accept the idea of a moving earth one must overcome the strong impression that one can "see" that the sun is really moving:

> Nor can I sufficiently admire the eminence of those men's intelligence [Galileo's Salviati says in the Third Day of the *Dialogue Concerning the Two Principal Systems*], who have received and held it [the Copernican system] to be true, and with the sprightliness of their judgments have done such violence to their own senses that they have been able to prefer that which their reason dictated to them to that which sensible experience represented most

manifestly to the contrary . . . . I cannot find any bounds for my admiration how reason was able, in Aristarchus and Copernicus, to commit such rape upon their senses as, in spite of them, to make itself master of their belief.

Among the themata which permeate Galileo's work and which helped reason to "commit such rape upon their senses," we can readily discern the then widely current thema of the once-given real world which God supervises from the center of His temple; the thema of mathematical nature; and the thema that the behavior of things is the consequence of their geometrical shapes (for which reason Copernicus said the earth rotates "because" it is spherical, and Gilbert, following the lead, is said to have gone so far as to prove experimentally, at least to his own satisfaction, that a carefully mounted magnetized sphere keeps up a constant rotation). Thus too, Sigmund Freud in MOSES AND MONOTHEISM, after surveying the overwhelmingly unfavorable evidence standing against the central thesis in his book, would say in effect, "But one must not be misled by the evidence."

III

While developing the position that themata have as legitimate and necessary a place in the pursuit and understanding of science as have observational experience and logical construction, I should make clear that we need not decide now also on the *source* of themata. Our first aim is simply to see their role in science and to describe some of them, as a folklorist might when he catalogues the traditions and practices of a people. It is not necessary to go further and to make an association of themata with any of the following conceptions: Platonic, Keplerian, or Jungian archetypes or images; myths (in the nonderogatory sense, so rarely used in the English language); synthetic a priori knowledge; intuitive apprehension or Galileo's "reason"; a realistic or absolutistic or, for that matter, any other philosophy of science. To show whether any such associations do or do not exist is a task for another time.

I also do not want to imply that the occurrence of themata is characteristic only of science in the last centuries. On the contrary, we see the thematic component at work from the very beginning, in the sources of cosmogonic ideas later found in Hesiod's THEOGONY and in Genesis. Indeed, nowhere can one see the persistence of great questions and the obstinacy of certain preselected patterns for defining and solving problems better than in cosmologic speculations. The ancient Milesian cosmologic assumptions presented a three-step scheme: at the beginning, in F. M. Cornford's words, there was:

a primal Unity, a state of indistinction or fusion in which factors that will later become distinct are merged together. (2) Out of this Unity there emerge, by separation, parts of opposite things . . . . This separating out finally leads to the disposition of the great elemental masses constituting the world-order, and the formation of the heavenly bodies. (3) The Opposites interact or reunite, in meteoric phenomena, or in the production of individual living things . . . .[16]

Now the significant thing to notice is that when we move these conceptions from the animistic to the physical level, this formula of cosmogony recurs point for point, in our day, in the evolutionist camps of modern cosmology. That recent theory of the way the world started proposes a progression of the universe from a mixture of radiation and neutrons at time $t = 0$; through the subsequent stages of differentiation by expansion and neutron decay; and finally to the building up of heavier elements by thermonuclear fusion processes, preparing the ground for the later formation of molecules. And even the ancient main *opposition* to the evolutionary cosmology itself, namely, the tradition of Parmenides, has its equivalent today in the "steady-state" theory of cosmology.

So the questions persist (e.g., concerning the possibility of some "fundamental stuff," of evolution, of structure, of spatial and temporal infinities). And the choices between alternative problem solutions also persist. These thematic continuities indicate the obverse side of the iconoclastic role of science; for science, since its dawn, has also had its more general themata-creating and themata-using function. James Clerk Maxwell expressed this well a century ago in an address on the subject of molecular mechanics:

The mind of man has perplexed itself with many hard questions. Is space infinite, and in what sense? Is the material world infinite in extent, and are all places within that extent equally full of matter? Do atoms exist, or is matter infinitely divisible?

The discussion of questions of this kind has been going on ever since men began to reason, and to each of us, as soon as we obtain the use of our faculties, the same old questions arise as fresh as ever. They form as essential a part of the science of the nineteenth century of our era, as of that of the fifth century before it.[17]

We may add that thematic questions do not get solved and disposed of. Nineteenth-century atomism triumphs over the ether vortices of Kelvin—but then field theories rise which deal with matter particles again as singularities, now in a twentieth-century-type continuum. The modern version of the cosmological theory based on the thema of a life cycle (Beginning, Evolution, and End) may seem to triumph on experimental grounds over the rival theory based on a thema of Continuous Existence, and throw it out the window—but we can be sure that this thema will come in again through the back door. For contrary to the physical theories in which they find embodiment in x-y terms, themata are not proved or disproved. Rather, they rise and fall and rise again with the tides of contemporaneous usefulness or intellectual fashion. And occasionally a great theme disappears from view, or a new theme develops and struggles to establish itself—at least for a time.

Maxwell's is an unusual concession, but it is not difficult to understand why scientists speak only rarely in such terms. One must not lose sight of the obvious fact that science itself has grown strong because its practitioners have seen how to project their discourse into the x-y plane. This is the plane of public science,[18] of fairly clear conscious formulations. Here a measure of public agreement is in principle easy to obtain, so that scientists can fruitfully cooperate or disagree with one another, can build on the work of their predecessors, and can teach more or less unambiguously the current content and problems of the field. All fields which claim or pretend to be scientific try similarly to project their concepts, statements, and problems into the x-y plane, to emphasize the phenomenic and analytic-heuristic aspects.

But it is clear that while there can be automatic factories run by means of programmed computers and the feedback from sensing elements, there can be no automatic laboratory. The essence of the automaton is its success in the x-y plane at the expense of the z-direction; (hence automata do not make qualitatively new findings). And the essence of the genial contributor to science is often exactly the opposite—sensitivity in the z-direction even at the expense of success in the x-y plane. For while the z-dimension is never absent even in the most exact of the sciences as pursued by actual persons, it is a direction in which most of us must move *without* explicit or conscious formulation and without training; it is the direction in which the subject matter and the media for communication are entirely different from those invented specifically for discussion of matters in the x-y plane with which the scientist after long training can feel at home.

Therefore it is difficult to find people who are bilingual in this sense. I am not surprised that for most contemporary scientists any discussion which tries to move self-consciously away from the x-y plane is out of bounds. However, it is significant that even in our time the men of genius—such as Einstein, Bohr, Pauli, Born, Schrödinger, Heisenberg—have felt it to be necessary and important to try just that. For the others, for the major body of scientists, the plane of discourse has been progressively tilted or projected from x-y-z space into the x-y plane. (Perhaps prompted by this example, the same thing is happening more and more in other fields of scholarship.) The themata actually used in science are now largely left implicit rather than explicit. But they are no less important. To understand fully the role a hypothesis or a law has in the development of science we need to see it also as an exemplification of persistent motifs, for example, the thema of "constancy" or of "conservation"; of quantification; of atomistic discreteness; or inherently probabilistic behavior; or—to return to our example from Newton—of the interpenetration of the worlds of theology and of physics. Indeed, in this way we can make a useful differentiation that to my knowledge has not been noted before, namely, that Newton's *public, experimental,* and *mathematical* philosophy is science carried on in the x-y plane, whereas Newton's more covert and more general *natural* philosophy is science in the x-y-z proposition space.[19]

IV

I have spoken mostly of the physical sciences. I might, with equal or greater advantage, have dealt with the newer sciences, which do not have a highly developed corpus either of phenomena or of logical calculi and rational structure. In those cases, the z-elements are not only still relatively more prominent but also are discussed with much greater freedom—possibly because at its early stage a field of study still bears the overwhelming imprint of one or a few men of genius. It is they who, I believe, are particularly "themata-prone," and who have the necessary courage (or folly?) to make decisions on thematic grounds.

This was the case in early mechanics, chemistry and biology, and again, with relativity and the new quantum mechanics. I suspect that an analogous situation has held in early modern psychology and sociology. Moreover, in those fields, as in the natural sciences during a stage of transformation, the significance and impact of themata is indicated by the fact that they force upon people notions that are usually regarded as paradoxical, ridiculous, or outrageous. I am thinking here of the "absurdities" of Copernicus's moving earth, Bruno's infinite worlds, Galileo's inertial motion of bodies on a horizontal plane, Newton's gravitational action without a palpable medium of communication, Darwin's descent of man from lower creatures, Einstein's twin paradox and maximum speed for signals, Freud's conception of sexuality of children, or Heisenberg's indeterminacy conception. The wide interest and intensity of such debates, among both scientists and enraged or intrigued laymen, is an indication of the strength with which themata—and frequently conflicting ones—are always active in our consciousness.

And the thematic component is most obvious when a science *is* young, and therefore has not yet elaborated the complex hierarchical structure of hypotheses which Richard Braithwaite has pointed out to be the mark of an advanced science. As a result, the chain leading from observational "facts" to the most general hypotheses—those with a large thematic component—is not long, as in, say, modern physics or chemistry, but is fairly short. A physical scientist is used to having his most general and most thematic hypotheses safely out of sight, behind the clouds of a majestic Olympus; and so he is apt to smile when he sees that the altar of other gods stands on such short legs. When, for example, a chemist interprets a half-dozen clicks on a Geiger counter as the existence of a new chemical element at the end of the Periodic Table, he implicitly (and, if challenged, explicitly) runs up on a ladder of hierarchically connected hypotheses, each of which has *some* demonstrable phenomenic and heuristic-analytic component, until at the top he comes up to the general thematic hypotheses—which he is, by agreement of this fraternity, exempt from going into—namely, the thematic hypotheses of atomicity, of constancy, of the transformability of qualities, of the ordering role of integers. In contrast, the early psychoanalysts, for example, tried to go by a relatively untortuous route from the detail of observed behavior to the generality of powerful principles. Freud himself once warned of the "bad habit of psychoanalysis . . . to take trivia as evidence when they also admit of another, less deep explanatory scheme."[20]

I do not, of course, say this to condemn a science, but on the contrary to point out a difference between it and the physical sciences which, I hope, may help to explain the attitude of "hard" scientists to fields outside their own (or even of psychologists of one school to those of another). At the same time, it may help to elucidate why disciplines such as psychology (and certainly history) are so constructed that they are wrong to imitate the habit in the modern physical sciences to depress or project the discussion forcibly to the x-y plane. When the thematic component is as strong and as explicitly needed as it is in these fields, the criteria of acceptability should be able to remain explicitly in three-dimensional proposition space. To cite an instance, I am by no means impressed with the *Conclusion* at the end of R. G. Collingwood's influential book, ESSAY ON PHILOSOPHICAL METHOD:

> The natural scientist, beginning with the assumption that nature is rational, has not allowed himself to be turned from that assumption by any of the difficulties into which it has led him; and it is because he has regarded that assumption as not only legitimate but obligatory that he has won the respect of the whole world. If the scientist is obliged to assume that nature is rational, and that any failure to make sense of it is a failure to understand it, the corresponding assumption is obligatory for the historian, and this not least when he is the historian of thought.[21]

This is a statement of the most dangerous kind, not because it is so easy to show it is wrong, but because it is so difficult to show this.

V

Much could, and should, be said about other problems in the thematic analysis of science, such as the mechanisms by which themata change; or the way in which the choice of a thematic hypothesis governs what we are to look for in the x-y plane and what we do with the findings; or the remarkably small number of different themata that, over time, seem to have played the important roles in the development of science; or the fact, implied in the examples given, that most and perhaps all of these themata are not restricted merely to uses in scientific context, but seem to come from the less specialized ground of our general imaginative capacity.

But in closing I might best simply restate how these conceptions can help us to a view that goes beyond the usual antithetical juxtaposition between science and the humanities. For the much lamented separation between science and the other components of our culture depends on the oversimplification that science is done only in the contingent plane, whereas scholarly or artistic work involves basic decisions of a different kind, with predominantly aesthetic, qualitative, mythic elements. In my view this dichotomy is much attenuated, if not eliminated, if we see that in science, too, the contingent plane is not enough, and never has been.

It is surely unnecessary to warn that despite the appearance and reappearance of the same thematic elements in science and outside, we shall not make the mistake of thinking that science and nonscience are at bottom somehow the same activity. There are differences which we should treasure. As Whitehead once said about the necessity to tolerate, no, to *welcome* national differences: "Men require of their neighbors something sufficiently akin to be understood, something sufficiently *different* to provoke attention, and something great enough to command admiration." It is in the same sense that we should be prepared to understand the separateness that gives identity to the study of each field, as well as the kinship that does exist between them.

To return, therefore, to Newton's Fifth Rule of Reasoning: he surely must have known that he could not publish it and remain true to his own work and that of most major innovators. As Newton's suppressed rule stands, it ends, you will recall, with the words: "Those things which neither can be demonstrated from the phenomena nor follow from them by an argument of induction, I hold as hypotheses." To be justified in publishing this rule Newton would have had to add something—perhaps this sentence: "And such hypotheses, namely thematic hypotheses, do also have place in natural philosophy."

NOTES

1. Friedrich Dessauer, *Galilei, Newton und die Wendung des abendländischen Denkens,* ERANOS JAHRBUCH 14 (Zurich: Rhein-Verlag, 1946), pp. 282–331.

2. Some of the arguments presented here were first considered in my George Sarton Memorial Lecture, presented on December 28, 1962, at the meeting of the American Association for the Advancement of Science; others in a lecture of November 14, 1963, to the American Philosophical Society. See also my article, *Über die Hypothesen, welche der Naturwissenschaft zu Grunde liegen,* ERANOS JAHRBUCH 31 (Zurich: Rhein-Verlag, 1963), pp. 351–425.

3. Cf. Alexandre Koyré, *Etudes Newtoniennes I- Les Regulae philosophandi,* ARCHIVES INTERNATIONALES D'HISTOIRE DES SCIENCES, 13:6, 1960. See also Koyré's *L'Hypothèse et l'expérience chez Newton,* BULLETIN DE LA SOCIÉTÉ FRANÇAISE DE PHILOSOPHIE 50:60–97, 1956.

4. Koyré, *Etudes Newtoniennes I, loc. cit.*

5. A. Rupert Hall and Marie Boas Hall, UNPUBLISHED SCIENTIFIC PAPERS OF SIR ISAAC NEWTON (Cambridge: Cambridge University Press, 1962), p. 352.

6. *Ibid.*, p. 213. Or perhaps more precisely, in Newton's thought, as Koyré has said, the cause of gravity is the action of the " 'Spirit' of God." Alexandre Koyré, FROM THE CLOSED WORLD TO THE INFINITE UNIVERSE (New York: Harper & Brothers, 1958), p. 234.

7. Or, conversely, observables that were important may become unimportant, as in chemistry, where the fundamental attention to the appearance and color of the flame in violent chemical reactions was given up with the phlogiston theory.

8. One may call the x-y plane the *contingent* plane because the meaning of concepts and statements in it are contingent on their having both empirical and analytical relevance. Contingency analysis is thus the study of the relevance of concepts and propositions in x- and y-dimensions. It is a term equivalent to operational analysis in its wider sense.

9. Eduard Jan Dijksterhuis, THE MECHANIZATION OF THE WORLD PICTURE, trans. C. Dijksterhuis (Oxford: Clarendon Press, 1961), p. 304.

10. Quoted from Edward Rosen, THREE COPERNICAN TREATISES (New York: Dover Publications, 1959), p. 29.

11. Bertrand Russell, HUMAN KNOWLEDGE, ITS SCOPE AND LIMITS (London: Allen & Unwin, 1948), Part 6, Chapter 2.

12. Karl R. Popper, THE LOGIC OF SCIENTIFIC DISCOVERY (New York: Basic Books, 1959), p. 278.

13. Dijksterhuis, *op. cit.*, p. 410.

14. René Descartes, PRINCIPIA PHILOSOPHIAE II, c. 36; OEUVRES 8:62–65.

15. [See, for example, the frank admission by Henri Poincaré, SCIENCE AND HYPOTHESIS, quoted in Essay 6, note 24.]

16. F. M. Cornford, PRINCIPIUM SAPIENTIAE (London: Cambridge University Press, 1952), Chapter 11.

17. Quoted in C. C. Gillispie, THE EDGE OF OBJECTIVITY: AN ESSAY IN THE HISTORY OF SCIENTIFIC IDEAS (Princeton, New Jersey: Princeton University Press, 1960), p. 477.

18. For the distinction between public and private science, see Gerald Holton, *On the Duality and Growth of Science*, AMERICAN SCIENTIST, 41:89–99, 1953, and reprinted here as Essay 11.

19. As Newton warns in the *General Scholium* (PRINCIPIA, 3rd edition), "hypotheses, whether metaphysical or physical, whether of occult qualities or mechanical, have no place in *experimental* philosophy." But in the previous paragraph, at the end of a long passage on the properties of the Deity and His evidences through observable nature, Newton writes: "And thus much concerning God; to discourse of whom from the appearances of things, does certainly belong to *Natural* Philosophy." (Emphasis added.)

20. Sigmund Freud, *Ein religiöses Erlebnis*, IMAGO, 14:7–10, 1928.

21. R. G. Collingwood, AN ESSAY ON PHILOSOPHICAL METHOD (Oxford: Clarendon Press, 1933), pp. 225–226.

2 JOHANNES KEPLER'S UNIVERSE: ITS PHYSICS AND METAPHYSICS

THE important publications of Johannes Kepler (1571–1630) preceded those of Galileo, Descartes, and Newton in time, and in some respects they are even more revealing. And yet, Kepler has been strangely neglected and misunderstood. Very few of his voluminous writings have been translated into English.[1] In this language there has been neither a full biography[2] nor even a major essay on his work in over twenty years. Part of the reason lies in the apparent confusion of incongruous elements—physics and metaphysics, astronomy and astrology, geometry and theology—which characterizes Kepler's work. Even in comparison with Galileo and Newton, Kepler's writings are strikingly different in the *quality* of preoccupation. He is more evidently rooted in a time when animism, alchemy, astrology, numerology, and witchcraft presented problems to be seriously argued. His mode of presentation is equally uninviting to modern readers, so often does he seem to wander from the path leading to the important questions of physical science. Nor is this impression merely the result of the inevitable astigmatism of our historical hindsight. We are trained on the ascetic standards of presentation originating in Euclid, as reestablished, for example, in Books I and II of Newton's PRINCIPIA,[3] and are taught to hide behind a rigorous structure the actual steps of discovery—those guesses, errors, and oc-

This essay was originally published in AMERICAN JOURNAL OF PHYSICS, VOLUME XXIV, No. 5, pp. 340–351, May, 1956.

casional strokes of good luck without which creative scientific work does not usually occur. But Kepler's embarrassing candor and intense emotional involvement force him to give us a detailed account of his tortuous progress. He still allows himself to be so overwhelmed by the beauty and variety of the world as a whole that he cannot yet persistently limit his attention to the main problems which can in fact be solved. He gives us lengthy accounts of his failures, though sometimes they are tinged with ill-concealed pride in the difficulty of his task. With rich imagination he frequently finds analogies from every phase of life, exalted or commonplace. He is apt to interrupt his scientific thoughts, either with exhortations to the reader to follow a little longer through the almost unreadable account, or with trivial side issues and textual quibbling, or with personal anecdotes or delighted exclamations about some new geometrical relation, a numerological or musical analogy. And sometimes he breaks into poetry or a prayer—indulging, as he puts it, in his "sacred ecstasy." We see him on his pioneering trek, probing for the firm ground on which our science could later build, and often led into regions which we now know to be unsuitable marshland.

These characteristics of Kepler's style are not merely idiosyncrasies. They mirror the many-sided struggle attending the rise of modern science in the early seventeenth century. Conceptions which we might now regard as mutually exclusive are found to operate side-by-side in his intellectual make-up. A primary aim of this essay is to identify those disparate elements and to show that in fact much of Kepler's strength stems from their juxtaposition. We shall see that when his physics fails, his metaphysics comes to the rescue; when a mechanical model breaks down as a tool of explanation, a mathematical model takes over; and at its boundary in turn there stands a theological axiom. Kepler set out to unify the classical picture of the world, one which was split into celestial and terrestrial regions, through the concept of a universal physical *force*; but when this problem did not yield to physical analysis, he readily returned to the devices of a unifying *image*, namely, the central sun ruling the world, and of a unifying *principle*, that of all-pervading mathematical harmonies. In the end he failed in his initial project of providing the mechanical explanation for the observed motions of the planets, but he succeeded at least in throwing a bridge from the old view of the world as unchangeable *cosmos* to the new view of the world as the playground of dynamic and mathematical laws. And in the process he turned up, as if it were by accident, those clues which Newton needed for the eventual establishment of the new view.

Toward a Celestial Machine

A sound instinct for physics and a commitment to neo-Platonic metaphysics—these are Kepler's two main guides which are now to be examined separately and at their point of merger. As to the first, Kepler's genius in physics has often been overlooked by critics who were taken aback by his frequent excursions beyond the bounds of science as they came to be understood later, although his DIOPTRICE (1611) and his mathematical work on infinitesimals (in NOVA STEREOMETRIA, 1615) and on logarithms (CHILIAS LOGARITHMORUM, 1624) have direct appeal for the modern mind. But even Kepler's casually delivered opinions often prove his insight beyond the general state of knowledge of his day. One example is his creditable treatment of the motion of projectiles on the rotating earth, equivalent to the formulation of the superposition principle of velocities.[4] Another is his opinion of the *perpetuum mobile:*

> As to this matter, I believe one can prove with very good reasons that neither any never-ending motion nor the quadrature of the circle—two problems which have tortured great minds for ages—will ever be encountered or offered by nature.[5]

But, of course, on a large scale, Kepler's genius lies in his early search for a physics of the solar system. He is the first to look for *a universal physical law based on terrestrial mechanics* to comprehend the whole universe in its quantitative details. In the Aristotelian and Ptolemaic world schemes, and indeed in Copernicus's own, the planets moved in their respective orbits by laws which were either purely mathematical or mechanical in a nonterrestrial sense. As Goldbeck reminds us, Copernicus himself still warned to keep a clear distinction between celestial and merely terrestrial phenomena, so as not to "attribute to the celestial bodies what belongs to the earth."[6] This crucial distinction disappears in Kepler from the beginning. In his youthful work of 1596, the MYSTERIUM COSMOGRAPHICUM, a single geometrical device is used to show the necessity of the observed orbital arrangement of all planets. In this respect, the earth is treated as being an equal of the other planets.[7] In the words of Otto Bryk,

> The central and permanent contribution lies in this, that for the first time

the whole world structure was subjected to a single law of construction—though not a force law such as revealed by Newton, and only a non-causative relationship between spaces, but nevertheless one single law.[8]

Four years later Kepler meets Tycho Brahe and from him learns to respect the power of precise observation. The merely approximate agreement between the observed astronomical facts and the scheme laid out in the MYSTERIUM COSMOGRAPHICUM is no longer satisfying. To be sure, Kepler always remained fond of this work, and in the DISSERTATIO CUM NUNCIO SIDEREO (1610) even hoped that Galileo's newly-found moons of Jupiter would help to fill in one of the gaps left in his geometrical model. But with another part of his being Kepler knows that an entirely different approach is wanted. And here Kepler turns to the new conception of the universe. While working on the ASTRONOMIA NOVA in 1605, Kepler lays out his program:

I am much occupied with the investigation of the physical causes. My aim in this is to show that the celestial machine is to be likened not to a divine organism but rather to a clockwork . . . , insofar as nearly all the manifold movements are carried out by means of a single, quite simple magnetic force, as in the case of a clockwork all motions [are caused] by a simple weight. Moreover I show how this physical conception is to be presented through calculation and geometry.[9]

The celestial machine, driven by a single terrestrial force, in the image of a clockwork! This is indeed a prophetic goal. Published in 1609, the ASTRONOMIA NOVA significantly bears the subtitle PHYSICA COELESTIS. The book is best known for containing Kepler's First and Second Laws of planetary motion, but it represents primarily a search for one universal force law to explain the motions of planets—Mars in particular—as well as gravity and the tides. This breathtaking conception of unity is perhaps even more striking than Newton's, for the simple reason that Kepler had no predecessor.

The Physics of the Celestial Machine

Kepler's first recognition is that forces between bodies are caused not by their relative positions or their geometrical arrangements, as was accepted by Aristotle, Ptolemy, and Copernicus, but by mechanical interactions between the material objects. Already in the MYSTERIUM COSMOGRAPHICUM (Chapter 17) he announced "*Nullum punctum, nullum centrum grave est,*" and he gave the example of the attraction between a magnet and a piece of iron. In William Gilbert's DE MAGNETE (1600), published four years later, Kepler finds a careful explanation that the action of magnets seems to come from pole points, but must be attributed to the parts of the body, not the points.

In the spirited *Objections* which Kepler appended to his own translation of Aristotle's Περὶ οὐρανοῦ, he states epigrammatically "*Das Mittele is nur ein Düpfflin,*" and he elaborates as follows:

> How can the earth, or its nature, notice, recognize and seek after the center of the world which is only a little point [*Düpfflin*]—and then go toward it? The earth is not a hawk, and the center of the world not a little bird; it [the center] is also not a magnet which could attract the earth, for it has no substance and therefore cannot exert a force.

In the Introduction to the ASTRONOMIA NOVA, which we shall now consider in some detail, Kepler is quite explicit:

> A mathematical point, whether it be the center of the world or not, cannot move and attract a heavy object . . . . Let the [Aristotelian] physicists prove that such a force is to be associated with a point, one which is neither corporeal nor recognisable as anything but a pure reference [mark].

Thus what is needed is a "true doctrine concerning gravity"; the axioms leading to it include the following:

> Gravitation consists in the mutual bodily striving among related bodies toward union or connection; (of this order is also the magnetic force).

This premonition of universal gravitation is by no means an isolated example of lucky intuition. Kepler's feeling for the physical situation is admirably sound, as shown in additional axioms:

> If the earth were not round, a heavy body would be driven not everywhere straight toward the middle of the earth, but toward different points from different places.
>
> If one were to transport two stones to any arbitrary place in the world, closely together but outside the field of force [*extra orbe virtutis*] of a third related body, then those stones would come together at some intermediate place similar to two magnetic bodies, the first approaching the second through a distance which is proportional to the mass [*moles*] of the second.

And after this precursor of the principle of conservation of momentum, there follows the first attempt at a good explanation for the tides in terms of a force of attraction exerted by the moon.

But the Achilles' heel of Kepler's celestial physics is found in the very first "axiom," in his Aristotelian conception of the law of inertia, where inertia is identified with a tendency to come to rest—*causa privativa motus:*

> Outside the field of force of another related body, every bodily substance, insofar as it is corporeal, by nature tends to remain at the same place at which it finds itself.[10]

This axiom deprives him of the concepts of mass and force in useful form—the crucial tools needed for shaping the celestial metaphysics of the ancients into the celestial physics of the moderns. Without these concepts, Kepler's world machine is doomed. He has to provide separate forces for the propulsion of planets tangentially along their paths and for the radial component of motion.

Moreover, he assumed that the force which reaches out from the sun to keep the planets in tangential motion falls inversely with the increasing distance. The origin and the consequences of this assumption are very interesting. In Chapter 20 of the MYSTERIUM COSMOGRAPHICUM, he speculated casually why the sidereal periods of revolution on the Copernican hypothesis should be larger for the more distant planets, and what force law might account for this:

> We must make one of two assumptions: either the forces of motion [*animae motrices*] [are inherent in the planets] and are feebler the more remote they are from the sun, or there is only one *anima motrix* at the center of the orbits, that is, in the sun. It drives the more vehemently the closer the [moved] body lies; its effect on the more distant bodies is reduced because of the distance [and the corresponding] decrease of the impulse. Just as the sun contains the source of light and the center of the orbits, even so can one trace back to this same sun life, motion and the soul of the world . . . . Now let us note how this decrease occurs. To this end we will assume, as is very probable, that the moving effect is weakened through spreading from the sun in the same manner as light.

This suggestive image—with its important overtones which we shall discuss below—does, however, not lead Kepler to the inverse-square law of force, for he is thinking of the spreading of light *in a plane*, corresponding to the plane of planetary orbits. The decrease of light intensity is therefore associated with the linear increase in circumference for more distant orbits! In his pre-Newtonian physics, where force is proportional not to acceleration but to velocity, Kepler finds a ready use for the inverse first-power law of gravitation. It is exactly what he needs to explain his observation that the speed of a planet in its elliptical orbit decreases linearly with the increase of the planet's distance from the sun. Thus Kepler's Second Law of Planetary Motion—which he actually discovered *before* the so-called First and Third laws—finds a partial physical explanation in joining several erroneous postulates.

In fact, it is clear from the context that these postulates originally suggested the Second Law to Kepler.[11] But not always is the final outcome so happy. Indeed, the hypothesis concerning the physical forces acting on the planet seriously delays Kepler's progress toward the law of elliptical orbits (First Law). Having shown that "the path of the planet [Mars] is not a circle but an oval figure," he attempts (Chapter 45, ASTRONOMIA NOVA) to find the details of a physical force law which would explain the "oval" path in a quantitative manner. But after ten chapters of tedious work he has to confess that "the physical causes in the forty-fifth chapter thus go up in smoke." Then in the remarkable fifty-seventh chapter, a final and rather desperate attempt is made to formulate a force law. Kepler even dares to entertain the notion of combined magnetic influences and animal forces [*vis animalia*] in the planetary system. Of course, the attempt fails. The accurate clockwork-like celestial machine cannot be constructed.

To be sure, Kepler does not give up his conviction that a universal force exists in the universe, akin to magnetism. For example, in Book 4 of the EPITOME OF COPERNICAN ASTRONOMY (1620), we encounter the picture of a sun as a spherical magnet with one pole at the center and the other distributed over its surface. Thus a planet, itself magnetized like a bar magnet with a fixed axis, is alternately attracted to and repelled from the sun in its elliptical orbit. This is to explain the radial component of planetary motion. The tangential motion has been previously explained (in Chapter 34, ASTRONOMIA NOVA) as resulting from the drag or torque which magnetic lines of force from the rotating sun are supposed to exert on the planet as they sweep over it. But the picture remains qualitative and incomplete, and Kepler does not return to his original plan to "show how this physical conception is to be presented through calculation and geometry."[9] Nor does his long labor bring him even a fair amount of recognition. Galileo introduces Kepler's work into his discussion on the world systems only to scoff at Kepler's notion that the moon affects the tides,[12] even though Tycho Brahe's data and Kepler's work based on them had shown that the Copernican scheme which Galileo was so ardently upholding did not correspond to the experimental facts of planetary motion. And Newton manages to remain strangely silent about Kepler throughout Books I and II of the PRINCIPIA, by introducing the Third Law anonymously as "the phenomenon of the 3/2th power" and the First and Second Laws as "the *Copernican* hypothesis."[13] Kepler's three laws have come to be treated as essentially empirical rules. How far removed this achievement was from his original ambition!

Kepler's First Criterion of Reality: The Physical Operations of Nature

Let us now set aside for a moment the fact that Kepler failed to build a mechanical model of the universe, and ask why he undertook the task at all. The answer is that Kepler (rather like Galileo) was trying to establish a new philosophical interpretation for "reality." Moreover, he was quite aware of the novelty and difficulty of the task.

In his own words, Kepler wanted to "provide a philosophy or physics of celestial phenomena in place of the theology or metaphysics of Aristotle."[14] Kepler's contemporaries generally regarded his intention of putting laws of physics into astronomy as a new and probably pointless idea. Even Michael Mästlin, Kepler's own beloved teacher, who had introduced Kepler to the Copernican theory, wrote him on October 1, 1616:

> Concerning the motion of the moon you write you have traced all the inequalities to physical causes; I do not quite understand this. I think rather that here one should leave physical causes out of account, and should explain astronomical matters only according to astronomical method with the aid of astronomical, not physical, causes and hypotheses. That is, the calculation demands astronomical bases in the field of geometry and arithmetic . . . .

The difference between Kepler's conception of the "physical" problems of astronomy and the methodology of his contemporaries reveals itself clearly in the juxtaposition of representative letters by the two greatest astronomers of the time—Tycho Brahe and Kepler himself.

Tycho, writing to Kepler on December 9, 1599, repeats the preoccupation of two millennia of astronomical speculations:

> I do not deny that the celestial motions achieve a certain symmetry [through the Copernican hypothesis], and that there are reasons why the planets carry through their revolutions around this or that center at different distances from the earth or the sun. However, the harmony or regularity of the scheme is to be discovered only a posteriori . . . . And even if it should appear to some puzzled and rash fellow that the superposed circular movements on the heavens yield sometimes angular or other figures, mostly elongated ones, then it happens accidentally, and reason recoils in horror from this assumption. For one must compose the revolutions of celestial objects definitely from circular motions; otherwise they could not come back on the same path eternally in equal manner, and an eternal duration would be impossible, not to mention that the orbits would be less simple, and irregular, and unsuitable for scientific treatment.

This manifesto of ancient astronomy might indeed have been subscribed to by Pythagoras, Plato, Aristotle, and Copernicus himself. Against it, Kepler maintains a new stand. Writing to D. Fabricius on August 1, 1607, he sounds the great new *leitmotif* of astronomy: *"The difference consists only in this, that you use circles, I use bodily forces."* And in the same letter, he defends his use of the ellipse in place of the superposition of circles to represent the orbit of Mars:

> When you say it is not to be doubted that all motions occur on a perfect circle, then this is false for the composite, i.e., the real motions. According to Copernicus, as explained, they occur on an orbit distended at the sides, whereas according to Ptolemy and Brahe on spirals. But if you speak of components of motion, then you speak of something existing in thought; i.e., something that is not there in reality. For nothing courses on the heavens except the planetary bodies themselves—no orbs, no epicycles . . . .

This straightforward and modern-sounding statement implies that behind the word "real" stands "mechanical," that for Kepler the real world is the world of objects and of their mechanical interactions in the sense which Newton used; e.g., in the preface to the PRINCIPIA:

> Then from these [gravitational] forces, by other propositions which are also mathematical, I deduce the motions of the planets, the comets, the moon, and the sea. I wish we could derive the rest of the phenomena of nature by the same kind of reasoning from mechanical principles . . . .[15]

Thus we are tempted to see Kepler as a natural philosopher of the mechanistic-type later identified with the Newtonian disciples. But this is deceptive. Particularly after the failure of the program of the ASTRONOMIA NOVA, another aspect of Kepler asserted itself. Though he does not appear to have been conscious of it, he never resolved finally whether the criteria of reality are to be sought on the *physical* or the *metaphysical* level. The words "real" or "physical" themselves, as used by Kepler, carry two interpenetrating complexes of meaning. Thus on receiving Mästlin's letter of October 1, 1616, Kepler jots down in the margin his own definition of "physical":

> I call my hypotheses physical for two reasons . . . . My aim is to assume only those things of which I do not doubt they are real and consequently physical, where one must refer to the nature of the heavens, not the elements. When I dismiss the perfect eccentric and the epicycle, I do so because they are purely geometrical assumptions, for which a corresponding body in the heavens does not exist. The second reason for my calling my hypotheses physical is this . . . I prove that the irregularity of the motion [of planets] corresponds to the nature of the planetary sphere; i.e., is physical.

This throws the burden on the *nature* of heavens, the *nature* of bodies. How, then, is one to recognize whether a postulate or conception is in accord with the nature of things?

This is the main question, and to it Kepler has at the same time two very different answers, emerging, as it were, from the two parts of his soul. We may phrase one of the two answers as follows: *the physically real world, which defines the nature of things, is the world of phenomena explainable by mechanical principles.* This can be called Kepler's first criterion of reality, and assumes the possibility of formulating a sweeping and consistent dynamics which Kepler only sensed but which was not to be given until Newton's PRINCIPIA. Kepler's other answer, to which he keeps returning again and again as he finds himself rebuffed by the deficiencies of his dynamics, and which we shall now examine in detail, is this: *the physically real world is the world of mathematically expressed harmonies which man can discover in the chaos of events.*

Kepler's Second Criterion of Reality: The Mathematical Harmonies of Nature

Kepler's failure to construct a *Physica Coelestis* did not damage his conception of the astronomical world. This would be strange indeed in a man of his stamp if he did not have a ready alternative to the mechanistic point of view. Only rarely does he seem to have been really uncomfortable about the poor success of the latter, as when he is forced to speculate how a soul or an inherent intelligence would help to keep a planet on its path. Or again, when the period of rotation of the sun which Kepler had postulated in his physical model proved to be very different from the actual rotation as first observed through the motion of sunspots, Kepler was characteristically not unduly disturbed. The truth is that despite his protestations, Kepler was not as committed to mechanical explanations of celestial phenomena as was, say, Newton. He had another route open to him.

His other criterion, his second answer to the problem of physical reality, stemmed from the same source as his original interest in astronomy and his fascination with a universe describable in mathematical terms, namely from a frequently acknowledged metaphysics rooted in Plato and neo-Platonists such as Proclus Diadochus. It is the criterion of *harmonious regularity in the descriptive laws of science.* One must be careful not to dismiss it either as just a reappearance of an old doctrine or as an aesthetic requirement which is still recognized in modern scientific work; Kepler's conception of what is "harmonious" was far more sweeping and important than either.

A concrete example is again afforded by the Second Law, the "Law of Equal Areas." To Tycho, Copernicus, and the great Greek astronomers, the harmonious regularity of planetary behavior was to be found in the uniform motion in component circles. But Kepler recognized the orbits—after a long struggle—as ellipsi on which planets move in a nonuniform manner. The figure is lopsided. The speed varies from point to point. And yet, nestled within this double complexity is hidden a harmonious regularity which transports its ecstatic discoverer—namely, the fact that a constant area is swept out in equal intervals by a line from the focus of the ellipse, where the sun is, to the planet on the ellipse. For Kepler, the law is harmonious in three separate senses.

First, *it is in accord with experience.* Whereas Kepler, despite long and hard labors, had been unable to fit Tycho's accurate observations on the motion of Mars into a classical scheme of superposed circles, the postulate of an elliptical path fitted the observations at once. Kepler's dictum was: "harmonies must accommodate experience."[16] How difficult it must have been for Kepler, a Pythagorean to the marrow of his bones, to forsake circles for ellipsi! For a mature scientist to find in his own work the need for abandoning his cherished and ingrained preconceptions, the very basis of his previous scientific work, in order to fulfill the dictates of quantitative experience—this was perhaps one of the great sacrificial acts of modern science, equivalent in recent scientific history to the agony of Max Planck. Kepler clearly drew the strength for this act from the belief that it would help him to gain an even deeper insight into the harmony of the world.

The second reason for regarding the law as harmonious is its reference to, or discovery of, a *constancy*, although no longer a constancy simply of angular velocity but of areal velocity. The typical law of ancient physical science had been Archimedes' law of the lever: a relation of direct observables in static configuration. Even the world systems of Copernicus and of Kepler's MYSTERIUM COSMOGRAPHICUM still had lent themselves to visualization in terms of a set of fixed concentric spheres. And we recall that Galileo never made use of Kepler's ellipsi, but remained to the end a true follower of Copernicus who had said "the mind shudders" at the supposition of noncircular nonuniform celestial motion, and "it would be unworthy to suppose such a thing in a Creation constituted in the best possible way."

With Kepler's First Law and the postulation of elliptical orbits, the old simplicity was destroyed. The Second and Third Laws established the physical law of constancy as an ordering principle in a changing situation. Like the concepts of momentum and caloric in later laws of constancy, areal velocity itself is a concept far removed from the immediate observables. It was therefore a bold step to search for harmonies beyond both perception and preconception.

Thirdly, the law is harmonious also in a grandiose sense: the fixed point of reference in the Law of Equal Areas, the "center" of planetary motion, is the center of the *sun itself*, whereas even in the Copernican scheme the sun was a little off the center of planetary orbits. With this discovery Kepler makes the planetary system at last truly heliocentric, and thereby satisfies his instinctive and sound demand for some material object as the "center" to which ultimately the physical effects that keep the system in orderly motion must be traced.

A Heliocentric and Theocentric Universe

For Kepler, the last of these three points is particularly exciting. The sun at its fixed and commanding position at the center of the planetary system matches the picture which always rises behind Kepler's tables of tedious data—the picture of a centripetal universe, directed toward and guided by the *sun* in its manifold roles: as the *mathematical* center in the description of celestial motions; as the central *physical* agency for assuring continued motion; and above all as the *metaphysical* center, the temple of the Deity. The three roles are in fact inseparable. For granting the special simplicity achieved in the description of planetary motions in the heliocentric system, as even Tycho was willing to grant, and assuming also that each planet must experience a force to drag it along its own constant and eternal orbit, as Kepler no less than the Scholastics thought to be the case, then it follows that the common need is supplied from what is common to all orbits; i.e., their common center, and this source of eternal constancy itself must be constant and eternal. Those, however, are precisely the unique attributes of the Deity.

Using his characteristic method of reasoning on the basis of archetypes, Kepler piles further consequences and analogies on this argument. The most famous is the comparison of the world-sphere with the Trinity: the sun, being at the center of the sphere and thereby antecedent to its two other attributes, surface and volume, is compared to God the Father. With variations the analogy occurs many times throughout Kepler's writings, including many of his letters. The image haunts him from the very beginning (e.g., Chapter 2, MYSTERIUM COSMOGRAPHICUM) and to the very end. Clearly, it is not sufficient to dismiss it with the usual phrase "sunworship."[17] At the very least, one would have to allow that the exuberant Kepler is a worshipper of the whole solar system in all its parts.

The power of the sun-image can be traced to the acknowledged influence on Kepler by neo-Platonists such as Proclus (fifth century) and Witelo (thirteenth century). At the time it was current neo-Platonic doctrine to identify light with "the source of all existence" and to hold that "space and light are one."[18] Indeed, one of the main preoccupations of the sixteenth-century neo-Platonists had been, to use a modern term, the transformation properties of space, light, and soul. Kepler's discovery of a truly heliocentric system is not only in perfect accord with the conception of the sun as a ruling entity, but allows him, for the first time, to focus attention on the sun's position through argument from physics.

In the medieval period the "place" for God, both in Aristotelian and in neo-Platonic astronomical metaphysics, had commonly been either beyond the last celestial sphere or else all of space; for only those alternatives provided for the Deity a "place" from which all celestial motions were equivalent. But Kepler can adopt a third possibility: in a truly heliocentric system God can be brought back into the solar system itself, so to speak, enthroned at the fixed and common reference object which coincides with the source of light and with the origin of the physical forces holding the system together. In the DE REVOLUTIONIBUS Copernicus had glimpsed part of this image when he wrote, after describing the planetary arrangement:

> In the midst of all, the sun reposes, unmoving. Who, indeed, in this most beautiful temple would place the light-giver in any other part than that whence it can illumine all other parts.

But Copernicus and Kepler were quite aware that the Copernican sun was not quite "in the midst of all"; hence Kepler's delight when, as one of his earliest discoveries, he found that orbital planes of all planets intersect at the sun.

The threefold implication of the heliocentric image as mathematical, physical, and metaphysical center helps to explain the spell it casts on Kepler. As Wolfgang Pauli has pointed out in a highly interesting discussion of Kepler's work as a case study in "the origin and development of scientific concepts and theories," here lies the motivating clue: "It is because he sees the sun and planets against the background of this fundamental image [*archetypische Bild*] that he believes in the heliocentric system with religious fervor"; it is this belief "which causes him to search for the true laws concerning the proportion in planetary motion. . . ."[19]

To make the point succinctly, we may say that in its final version *Kepler's physics of the heavens is heliocentric in its kinematics, but theocentric in its dynamics,* where harmonies based in part on the properties of the Deity serve to supplement physical laws based on the concept of specific quantitative forces. This brand of physics is most prominent in Kepler's last great work, the HARMONICE MUNDI (1619). There the so-called Third Law of planetary motion is announced without any attempt to deduce it from mechanical principles, whereas in the ASTRONOMIA NOVA magnetic forces had driven—no, obsessed—the planets. As in his earliest work, he shows that the phenomena of nature exhibit underlying mathematical harmonies. Having not quite found the mechanical gears of the world machine, he can at least give its equations of motion.

The Source of Kepler's Harmonies

Unable to identify Kepler's work in astronomy with physical science in the modern sense, many have been tempted to place him on the other side of the imaginary dividing line between classical and modern science. Is it, after all, such a large step from the harmonies which the ancients found in circular motion and rational numbers to the harmonies which Kepler found in elliptical motions and exponential proportions? Is it not merely a generalization of an established point of view? Both answers are in the negative. For the ancients and for most of Kepler's contemporaries, the hand of the Deity was revealed in nature through laws which, if not qualitative, were harmonious in an essentially self-evident way; the axiomatic simplicity of circles and spheres and integers itself proved their deistic connection. But Kepler's harmonies reside in the very fact that the relations *are quantitative*, not in some specific simple *form* of the quantitative relations.

It is exactly this shift which we can now recognize as one point of breakthrough toward the later, modern conception of mathematical law in science. Where in classical thought the quantitative actions of nature were limited by a few necessities, the new attitude, whatever its metaphysical motivation, opens the imagination to an infinity of possibilities. As a direct consequence, where in classical thought the quantitative results of experience were used largely to fill out a specific pattern by a priori necessity, the new attitude permits the results of experience to reveal in themselves whatever pattern nature has in fact chosen from the infinite set of possibilities. Thus the seed is planted for the general view of most modern scientists, who find the world harmonious in a vague aesthetic sense because the mind can find, inherent in the chaos of events, order framed in mathematical laws—of whatever form they may be. As has been aptly said about Kepler's work:

> Harmony resides no longer in numbers which can be gained from arithmetic without observation. Harmony is also no longer the property of the circle in higher measure than the ellipse. Harmony is present when a multitude of

phenomena is regulated by the unity of a mathematical law which expresses a cosmic idea.[20]

Perhaps it was inevitable in the progress of modern science that the harmony of mathematical law should now be sought in aesthetics rather than in metaphysics. But Kepler himself would have been the last to propose or accept such a generalization. The ground on which he postulated that harmonies reside in the quantitative properties of nature lies in the same metaphysics which helped him over the failure of his physical dynamics of the solar system. Indeed, the source is as old as natural philosophy itself: *the association of quantity per se with Deity.* Moreover, as we can now show, Kepler held that man's ability to discover harmonies, and therefore reality, in the chaos of events is due to a direct connection between ultimate reality; namely, God, and the mind of man.

In an early letter, Kepler opens to our view this mainspring of his life's work:

> May God make it come to pass that my delightful speculation [the *Mysterium Cosmographicum*] have everywhere among reasonable men fully the effect which I strove to obtain in the publication; namely, that the belief in the creation of the world be fortified through this external support, that thought of the creator be recognized in its nature, and that His inexhaustible wisdom shine forth daily more brightly. Then man will at last measure the power of his mind on the true scale, and will realize that *God, who founded everything in the world according to the norm of quantity, also has endowed man with a mind which can comprehend these norms.* For as the eye for color, the ear for musical sounds, so is the mind of man created for the perception not of any arbitrary entities, but rather of quantities; the mind comprehends a thing the more correctly the closer the thing approaches toward pure quantity as its origin.[21]

On a superficial level, one may read this as another repetition of the old Platonic principle ὁ θεὸς ἀεὶ γεωμετεῖ; and of course Kepler does believe in "the creator, the true first cause of geometry, who, as Plato says, always geometrizes."[22] Kepler is indeed a Platonist, and even one who is related at the same time to both neo-Platonic traditions—which one might perhaps better identify as the neo-Platonic and the neo-Pythagorean—that of the mathematical physicists like Galileo and that of the mathematical mysticism of the Florentine Academy. But Kepler's God has done more than build the world on a mathematical model; he also specifically created man with a mind which "carries in it concepts built on the category of quantity," *in order that man may directly communicate with the Deity:*

> Those laws [which govern the material world] lie within the power of understanding of the human mind; God wanted us to perceive them when he created us in His image in order that we may take part in His own thoughts . . . . Our knowledge [of numbers and quantities] is of the same kind as God's, at least insofar as we can understand something of it in this mortal life.[23]

The procedure by which one apprehends harmonies is described quite explicitly in Book 4, Chapter 1, of HARMONICE MUNDI. There are two kinds of harmonies; namely, those in sense phenomena, as in music, and in "pure" harmonies such as are "constructed of mathematical concepts." The feeling of harmony arises when there occurs a matching of the perceived order with the corresponding innate archetype [*archetypus, Urbild*]. The archetype itself is part of the mind of God and was impressed on the human soul by the Deity when He created man in His image. The kinship with Plato's doctrine of ideal forms is clear. But whereas the latter, in the usual interpretation, are to be sought outside the human soul, Kepler's archetypes are within the soul. As he summarizes at the end of the discussion, the soul carries "not an image of the true pattern [*paradigma*], but the true pattern itself . . . . Thus finally the harmony itself becomes entirely soul, nay even God."[24]

This, then, is the final justification of Kepler's search for mathematical harmonies. The investigation of nature becomes an investigation into the thought of God, Whom we can apprehend through the language of mathematics. *Mundus est imago Dei corporea,* just as, on the other hand, *animus est imago Dei incorporea.* In the end, Kepler's unifying principle for the world of phenomena is not merely the concept of mechanical forces, but God, expressing Himself in mathematical laws.

Kepler's Two Deities

A final brief word may be in order concerning the psychological orientation of Kepler. Science, it must be remembered, was not Kepler's original destination. He was first a student of philosophy and theology at the University of Tübingen; only a few months before reaching the goal of church position, he suddenly—and reluctantly—found himself transferred by the University authorities to a teaching position in mathematics and astronomy at Graz. A year later, while already working on the MYSTERIUM COSMOGRAPHICUM, Kepler wrote: "I wanted to become a theologian; for a long time I was restless: Now, however, observe how through my effort God is being celebrated in astronomy."[25] And more than a few times in his later writings he referred to astronomers as priests of the Deity in the book of nature.

From his earliest writing to his last, Kepler maintained the direction and intensity of his religio-philosophical interest. His whole life was one of uncompromising piety; he was incessantly struggling to uphold his strong and often nonconformist convictions in religion as in science. Caught in the turmoil of the Counter-Reformation and the beginning of the Thirty Years' War, in the face of bitter difficulties and hardships, he never compromised on issues of belief. Expelled from communion in the Lutheran Church for his unyielding individualism in religious matters, expelled from home and position at Graz for refusing to embrace Roman Catholicism, he could truly be believed when he wrote, "I take religion seriously, I do not play with it,"[26] or "In all science there is nothing which could prevent me from holding an opinion, nothing which could deter me from acknowledging openly an opinion of mine, except solely the authority of the Holy Bible, which is being twisted badly by many."[27]

But as his work shows us again and again, Kepler's soul bears a dual image on this subject too. For next to the Lutheran God, revealed to him directly in the words of the Bible, there stands the Pythagorean God, embodied in the immediacy of observable nature and in the mathematical harmonies of the solar system whose design Kepler himself had traced—a God "whom in the contemplation of the universe I can grasp, as it were, with my very hands."[28]

The expression is wonderfully apt: so intense was Kepler's vision that the abstract and concrete merged. Here we find the key to the enigma of Kepler, the explanation for the apparent complexity and disorder in his writings and commitments. In one brilliant image, Kepler saw the three basic themes or cosmological models superposed: *the universe as physical machine, the universe as mathematical harmony, and the universe as central theological order*. And this was the setting in which harmonies were interchangeable with forces, in which a theocentric conception of the universe led to specific results of crucial importance for the rise of modern physics.

NOTES

1. Books 4 and 5 of the EPITOME OF COPERNICAN ASTRONOMY, and Book 5 of the HARMONIES OF THE WORLD, in GREAT BOOKS OF THE WESTERN WORLD (Chicago: Encyclopedia Britannica, 1952), Volume 16.

2. The definitive biography is by the great Kepler scholar Max Caspar, JOHANNES KEPLER, Stuttgart: W. Kohlhammer, 1950; the English translation is KEPLER, trans. and ed. C. Doris Hellman, New York: Abelard-Schuman, 1959. Some useful short essays are in JOHANN KEPLER, 1571–1630 (A series of papers prepared under the auspices of the History of Science Society in collaboration with the American Association for the Advancement of Science), Baltimore: Williams & Wilkins Co., 1931. [Since this article was written, a number of useful publications on Kepler have appeared —Ed.]

3. But Newton's OPTICKS, particularly in the later portions, is rather reminiscent of Kepler's style. In Book II, Part IV, Observation 5, there is, for example, an attempt to associate the parts of the light spectrum with the "differences of the lengths of a monochord which sounds the tones in an eight."

4. Letter to David Fabricius, October 11, 1605.

5. Letter to Herwart von Hohenburg, March 26, 1598, i.e., seven years before Stevinus implied the absurdity of perpetual motion in the HYPOMNEMATA MATHEMATICA (Leyden, 1605). Some of Kepler's most important letters are collected in Max Caspar and Walther von Dyck, JOHANNES KEPLER IN SEINEN BRIEFEN, Munich and Berlin: R. Oldenbourg, 1930. A more complete collection in the original languages is to be found in Vols. 13–15 of the modern edition of Kepler's collected works, JOHANNES KEPLERS GESAMMELTE WERKE, ed. von Dyck and Caspar, Munich: C. H. Beck, 1937 and later. In the past, these letters appear to have received insufficient attention in the study of Kepler's work and position. (The present English translations of all quotations from them are the writer's.) Excerpts from some letters were also translated in Carola Baumgardt, JOHANNES KEPLER, New York: Philosophical Library, 1951.

6. Ernst Goldbeck, *Abhandlungen zur Philosophie und ihrer Geschichte*, KEPLERS LEHRE VON DER GRAVITATION (Halle: Max Niemeyer, 1896), Volume VI—a useful monograph demonstrating Kepler's role as a herald of mechanical astronomy. The reference is to DE REVOLUTIONIBUS, first edition, p. 3. [The

main point, which it would be foolhardy to challenge, is that in the description of phenomena Copernicus still on occasion treated the earth differently from other planets.]

7. In Kepler's Preface to his DIOPTRICE (1611) he calls his early MYSTERIUM COSMOGRAPHICUM "a sort of combination of astronomy and Euclid's Geometry," and describes the main features as follows: "I took the dimensions of the planetary orbits according to the astronomy of Copernicus, who makes the sun immobile in the center, and the earth movable both round the sun and upon its own axis; and I showed that the differences of their orbits corresponded to the five regular Pythagorean figures, which had been already distributed by their author among the elements of the world, though the attempt was admirable rather than happy or legitimate . . . ." The scheme of the five circumscribed regular bodies originally represented to Kepler the *cause* of the observed number (and orbits) of the planets: "*Habes rationem numeri planetarium.*"

8. Johannes Kepler, DIE ZUSAMMENKLÄNGE DER WELTEN, Otto J. Bryk, trans. and ed. (Jena: Diederichs, 1918), p. xxiii.

9. Letter to Herwart von Hohenburg, February 10, 1605. At about the same time he writes in a similar vein to Christian Severin Longomontanus concerning the relation of astronomy and physics: "I believe that both sciences are so closely interlinked that the one cannot attain completion without the other."

10. Previously, Kepler discussed the attraction of the moon in a letter to Herwart, January 2, 1607. The relative motion of two isolated objects and the concept of inertia are treated in a letter to D. Fabricius, October 11, 1605. On the last subject see Alexandre Koyré, *Galileo and the Scientific Revolution of the Seventeenth Century*, THE PHILOSOPHICAL REVIEW, 52, No. 4: 344–345, 1943.

11. Not only the postulates but also some of the details of their use in the argument were erroneous. For a short discussion of this concrete illustration of Kepler's use of physics in astronomy, see John L. E. Dreyer, HISTORY OF THE PLANETARY SYSTEM FROM THALES TO KEPLER (New York: Dover Publications, 1953), second edition, pp. 387–399. A longer discussion is in Max Caspar, JOHANNES KEPLER, NEUE ASTRONOMIE (Munich and Berlin: R Oldenbourg, 1929), pp. 3*–66*.

12. Giorgio de Santillana, ed., DIALOGUE ON THE GREAT WORLD SYSTEMS (Chicago: University of Chicago Press, 1953), p. 469. However, an oblique compliment to Kepler's Third Law may be intended in a passage on p. 286

13. Florian Cajori, ed., NEWTON'S PRINCIPIA: MOTTE'S TRANSLATION REVISED (Berkeley: University of California Press, 1946), pp. 394–395. In

Book III, Newton remarks concerning the fact that the Third Law applies to the moons of Jupiter: "This we know from astronomical observations." At last, on page 404, Kepler is credited with having "first observed" that the 3/2th power law applies to the "five primary planets" and the earth. Newton's real debt to Kepler was best summarized in his own letter to Halley, July 14, 1686: "But for the duplicate proportion [the inverse-square law of gravitation] I can affirm that I gathered it from Kepler's theorem about twenty years ago."

14. Letter to Johann Brengger, October 4, 1607. This picture of a man struggling to emerge from the largely Aristotelian tradition is perhaps as significant as the usual one of Kepler as Copernican in a Ptolemaic world. Nor was Kepler's opposition, strictly speaking, Ptolemaic any longer. For this we have Kepler's own opinion (HARMONICE MUNDI, Book 3): "First of all, readers should take it for granted that among astronomers it is nowadays agreed that all planets circulate around the sun . . . ," meaning of course the system not of Copernicus but of Tycho Brahe, in which the earth was fixed and the moving sun served as center of motion for the other planets.

15. Cajori, *op. cit.*, p. xviii.

16. Quoted in Kepler, WELTHARMONIK, ed. Max Caspar (Munich and Berlin: R. Oldenbourg, 1939), p. 55*.

17. E.g., Edwin Arthur Burtt, THE METAPHYSICAL FOUNDATIONS OF MODERN SCIENCE (London: Routledge & Kegan Paul, 1924 and 1932), p. 47 ff.

18. For a recent analysis of neo-Platonic doctrine, which regrettably omits a detailed study of Kepler, see Max Jammer, CONCEPTS OF SPACE (Cambridge: Harvard University Press, 1954), p. 37 ff. Neo-Platonism in relation to Kepler is discussed by Thomas S. Kuhn, THE COPERNICAN REVOLUTION, Cambridge: Harvard University Press, 1957.

19. Wolfgang Pauli, *Der Einfluss archetypischer Vorstellungen auf die Bildung naturwissenschaftlicher Theorien bei Kepler*, in NATURERKLÄRUNG UND PSYCHE (Zurich: Rascher Verlag, 1952), p. 129.

An English translation of Jung and Pauli is THE INTERPRETATION OF NATURE AND THE PSYCHE, trans. R. F. C. Hull and Priscilla Silz, New York: Pantheon Books, 1955.

20. Hedwig Zaiser, KEPLER ALS PHILOSOPH (Stuttgart: E. Suhrkamp, 1932), p. 47.

21. Letter to Mästlin, April 19, 1597. (Italics supplied.) The "numerological" component of modern physical theory is in fact a respectable offspring from this respectable antecedent. For example, see Niels Bohr, ATOMIC THEORY AND THE DESCRIPTION OF NATURE (New York: Macmillan Co.,

1934), pp. 103–104: "This interpretation of the atomic number [as the number of orbital electrons] may be said to signify an important step toward the solution of one of the boldest dreams of natural science, namely, to build up an understanding of the regularities of nature upon the consideration of pure number."

22. HARMONICE MUNDI, Book 3.

23. Letter to Herwart, April 9/10, 1599. Galileo later expressed the same principle: "That the Pythagoreans had the science of numbers in high esteem, and that Plato himself admired human understanding and thought that it partook of divinity, in that it understood the nature of numbers, I know very well, nor should I be far from being of the same opinion." de Santillana, *op. cit.*, p. 14. Descartes's remark, "You can substitute the mathematical order of nature for 'God' whenever I use the latter term" stems from the same source.

24. For a discussion of Kepler's mathematical epistemology and its relation to neo-Platonism, see Max Steck, *Über das Wesen des Mathematischen und die mathematische Erkenntnis bei Kepler*, DIE GESTALT (Halle: Max Niemeyer, 1941), Volume V. The useful material is partly buried under nationalistic oratory. Another interesting source is Andreas Speiser, MATHEMATISCHE DENKWEISE, Basel: Birkhäuser, 1945.

25. Letter to Mästlin, October 3, 1595.

26. Letter to Herwart, December 16, 1598.

27. Letter to Herwart, March 28, 1605. If one wonders how Kepler resolved the topical conflict concerning the authority of the scriptures *versus* the authority of scientific results, the same letter contains the answer: "I hold that we must look into the intentions of the men who were inspired by the Divine Spirit. Except in the first chapter of Genesis concerning the supernatural origin of all things, they never intended to inform men concerning natural things." This view, later associated with Galileo, is further developed in Kepler's eloquent introduction to the ASTRONOMIA NOVA. The relevant excerpts were first translated by Thomas Salusbury, MATHEMATICAL COLLECTIONS (London: 1661), Part I, pp. 461–467.

28. Letter to Baron Strahlendorf, October 23, 1613.

3 THEMATIC AND STYLISTIC INTERDEPENDENCE

I

It is commonly acknowledged that a proposal of Plato set the style for one of the main traditions of classic scientific thought. As Blake, Ducasse, and Madden phrase the account of their book, THEORIES OF SCIENTIFIC METHOD, Plato "set his pupils in the Academy the task of working out a system of geometrical hypotheses which, by substituting uniform and circular movements for the apparently irregular movements of the heavenly bodies [that is, the planets, particularly during retrograde motion], would make it possible to explain the latter in terms of the former—in his own famous phrase, to 'save the phenomena.' "[1] Simplicius writes in his Commentary on Aristotle's DE CAELO: "For Plato, Sosigenes says, set this problem for students of astronomy: 'By the assumption of what uniform and ordered motions can the apparent motions of the planets be accounted for?' " This famous problem kept natural philosophers agitated for 2,000 years and was immensely influential in shaping science as we know it.

To this day, it still strikes us as a sound scientific question, and we are not surprised to hear that one of Plato's disciples produced a very creditable solution by proposing a geocentric system of homocentric spheres. Plato starts from puzzling observations—particularly the apparent halt-

This article was originally presented at The University of Florida, Gainesville, for the Thirtieth Anniversary of the University College and was published under the title *Science and New Styles of Thought* in THE GRADUATE JOURNAL Volume VII, No. 2, pp. 399–421, 1967.

ing and brief backward excursion which the paths of those wanderers, the planets, show at regular intervals during their otherwise predominantly forward, night-by-night progress against the background of the fixed stars. We still ask a similar type of question when a comet or asteroid is discovered, or when an artificial satellite suddenly is launched into our skies: what are the elements of a mathematical analogue (or equation) representing its motion? We no longer have to solve such a kinematic problem by the very tedious geometric methods—it can be translated into an equivalent language by which an electronic computer can give us a quick answer when presented with data derived from observation. But, qualitatively, the computer adds nothing new. In fact, the superposition of circular motions now has its exact mathematical equivalent in the treatment of periodic motions through the sum of a series of terms of a trigonometric function.

Moreover, the Platonic problem appears to be concerned with three elements that modern science still deals with: First, the "facts" of observation (here, the observed motion of the planets) impress themselves on our senses. Next, we find here a puzzling mixture of complexity and order which triggers the curiosity (we see not one planet but many, not a simple forward motion but an apparent retrogression, with a regular but unique pattern for each planet). And finally, we resolve the apparent puzzle by the imaginative construction of an analogon. The analogon can be either mathematical or physical; it is successful if it correlates convincingly the puzzling element in the observation with the consequences, perhaps unexpectedly but logically sound consequences, of the postulated structure of the analogon. In Plato's case, the analogon which he specifically invites is that of uniform motion along circular paths in an interrelated kinematical system. This analogon is put forward—hypothesized.

But the most significant element in moving from problem to solution in Platonic science is one that has not yet been mentioned—*preselection*. There are many constraints which have been placed from the very beginning on the possible solution in an almost imperceptible manner; constraints on the "facts," on the hypothesis, and the correspondence between analogon and facts by which we are relieved of the feeling of puzzlement, by which we "understand."

For of all the possible facts of observation, we are invited to look just at celestial ones; and of all those, at the planets. Of the facts of observation concerning planets, only one kind is selected and not, for example, their different colors or their changing brightness, both of which strike the eye much more quickly than their motion with respect to the fixed stars, certainly more quickly than any element of order within the "disorderliness" of retrogression. Of all the possible analogons (for example, animistic, or physically mechanical), we are given here a geometrical one; of all geometrical-kinematic ones, only uniform motion and only on circular paths. Of all the correspondence by which we "understand" the behavior of the planets, we must choose only the correlation of their point-by-point location in the sky with the progress of imaginary points in the geometrical system, and not, for example, a mere catalogue of positions, such as the Babylonians used to "understand" and predict celestial phenomena.

The issue now raises itself forcibly: what are the criteria which guide us in these preselections of facts, hypotheses, and explanatory methods? On the answer to this question, far more than on the "facts of nature" themselves, depends what kind of science we shall have, whether such a science itself is possible, and what this science can teach us. It is, therefore, one of the basic questions of any science or philosophy of science.

This is the point where the style of thought of the time enters our discussion. For I need only remind you of the main outlines of Plato's position which explain his criteria of preselection and consequently explain the style of Platonic science itself. In the LAWS, Book XII, the Athenian speaks to Clinias and Megillus:

May we say, then, that we know of two motives—those we have already rehearsed—of credibility in divinity? . . . One of them is our theory of the soul, our doctrine that it is more ancient and more divine than anything that draws perennial being from a motion that once had a beginning; the other our doctrine of the orderliness in the motion in the movements of the planets and other bodies swayed by the mind that has set this whole frame of things in comely array. No man who has once turned a careful and practiced gaze on this spectacle has ever been so ungodly at heart that its effect has not been the very reverse of that currently expected. 'Tis the common belief that men who busy themselves with such themes are made infidels by their astronomy and its sister sciences, with their disclosure of a realm where events happen by stringent necessity, not by the purpose of a will bent on the achievement of good. . . . The situation has been precisely reversed since the days when observers of these bodies conceived them to be without souls. Even then, they awakened wonder, and aroused in the breast of close students the suspicion, which has now been converted into accepted doctrine, that were they without souls, and by consequence without intelligence, they would never have conformed to such precise computation.[2]

Therefore, anyone aspiring to be "a sufficient magistrate of the whole community," Plato continues, must "possess the requisite preliminary sciences . . . and apply his knowledge meetly to his moral and legal behavior." The proper study of man, of which the "preliminary sciences" are only a stage, is, of course, ultimately the soul. And astronomy is used as an illustration again and again in Plato's work to make this point.[3]

Astronomy, we see, is not pursued properly if one studies only the minute precession of the equinox, or corrections of the calendar or other measurable detail—problems that only later come to be highly regarded. Rather, a man must "pursue his studies aright with his mind's eye fixed on their single end." [Epinomis, 991e.] As the Athenian had said earlier [Laws 7: 821d]: ". . . the reason why I am now insisting that our citizens and their young people must learn enough of all the facts about the divinities of the sky is to prevent blasphemy of them [such as the charge that they wander around, when in fact they can be shown to have regular motions; 822a], and to ensure a reverent piety in the language of all our sacrifices and prayers."

Astronomy is thus an adjunct to moral philosophy. And now we have no difficulty in reconstructing and understanding the criteria of preselection in Plato's astronomical problems. They are those criteria which assure that the subject matter and persuasiveness of this science shall contribute to moral education. To let scientists look for their own subject of study and fashion their own criteria for selecting facts, hypotheses, and explanatory method would be as absurd as giving the name "true musicians" [Republic 3: 402] to those who would invent their own instruments, and play on them any disharmonious tune or rhythm that pleases them, and who in general have not studied "the forms of soberness, courage, liberality, and high-mindedness."

Indeed, it is so easy to show how the moralistic setting surrounding Platonic science determined the science of its time that we sometimes assign it to students as an exercise, and as an exemplary warning that metaphysical presuppositions have had a powerful hold upon the sciences of the ancients.

II

Is the situation today entirely different? I do not believe it is, or could be. Although the scene, both in science and outside, has changed greatly since the classic Greek period, we have never had—and could not have imagined—a science separated from external involvements, and existing truly "for its own sake." The criteria of preselection change, the basic concerns shift, but the existence of a stylistic relationship among the different works of a given period remains constant.

An example will be helpful here. It is commonplace that the predilection for seeing problems in terms of a harmoniously ordered world was still characteristic even of the very language of scientific imagination in the classic period of the seventeenth and eighteenth centuries. How far we have come in science more recently—and not only in science—from that position is perhaps expressed most directly and simply in a passage that seems as strange now as it was congenial in 1681 when Thomas Burnet published THE SACRED THEORY OF THE EARTH; speaking of the annoying disorderliness of the distribution of the stars, he said:

> They lie carelesly scatter'd, as if they had been sown in the Heaven, like Seed, by handfuls; and not by a skilful hand neither. What a beautiful Hemisphere they would have made, if they had been plac'd in rank and order, if they had been all dispos'd into regular figures, and the little ones set with due regard to the greater. Then all finisht and made up into one fair piece or great Composition, according to the rules of Art and Symmetry.

We have not, of course, lost the concepts of hierarchy, continuity, and order in contemporary work. They stay in science, but mainly as inherited elements. They are not the new themes that correspond to the characteristic style of our own age—of which one of the most powerful and significant is the antithetical thema of disintegration, violence, and derangement.

Thus in the language of physics alone we find the rise in the last six decades of terms such as radioactive decay, or decay of particles; displacement law; fission; spallation; nuclear disintegration; discontinuity (as in energy levels of atoms); dislocation (in crystals); indeterminacy, uncertainty; probabilistic (rather than classically deterministic) causality; time reversal; strangeness quantum number; negative states (of energy, of temperature); forbidden lines and transitions; particle annihilation. I once wrote that it is not too farfetched to imagine that some physicist will propose to name a new particle the "schizoid particle"—and shortly thereafter I discovered that the term "schizon" was being introduced into the technical literature of particle physics.

It is as if after a successful search for simplicities and harmonies in science over the last three centuries, the search has turned to a more direct confrontation of complexity and derangement, of sophisticated and astonishing relationships among strangely juxtaposed parts. And if one is interested in the parallels between style in science and style outside science, it is not surprising to discover that this theme in the physical sciences has its counterpart in modern themes outside science, for example in the analogous preoccupation with the theme of apparent derangement in contemporary art.

To select one example among many, consider the work by the French artist M. Arman, who has called some of his work *colères* or *coupes*. In the words of Peter Jones, a critic who made a study of his work,[4] the aim of the *colère* is to "hold fast on a surface *one instant*, the explosive instant in which objects are violently disintegrated into a mass of pieces —action sculpture in the highly recalcitrant (and thus challenging) medium of objects that break the way the artist wants only with much flair and practice on his part. . . . Arman is fascinated by . . . the coordinated mastery of all the factors involved, brought to bear at one decisive point of space and time." In describing the work *Allegro Furioso* (see Figure 1), Jones gives this description:

> The *colère* looks spontaneous, but its construction was deliberate all through. Here is what Arman did. Having laid the black panel that serves as a base flat on the floor (and having built up temporary planks on the sides) he began by smashing a cello. This came first because it was to be the determining factor of the composition. Arman broke it diagonally, to divide his surface in two. Then he took the viola, an old and dry hand-made instrument which he knew would spread itself broadly on impact (while a newer one would have broken more compactly) : he broke the viola to left of centre in order to leave a "V" in the middle of the panel. Next he broke the two violins in such a way as to have them going in the same direction as the viola. In order not to have them too widely dispersed he did not swing them through the air and smash them on the board as he had done with the other instruments, but held them by the neck and scroll, stamped on them with his foot, and dropped the necks nearby. Thus he achieved a compact mass on the left side. Finally, to counterbalance fully the mass of the cello, he threw down the bows on the left.[5]

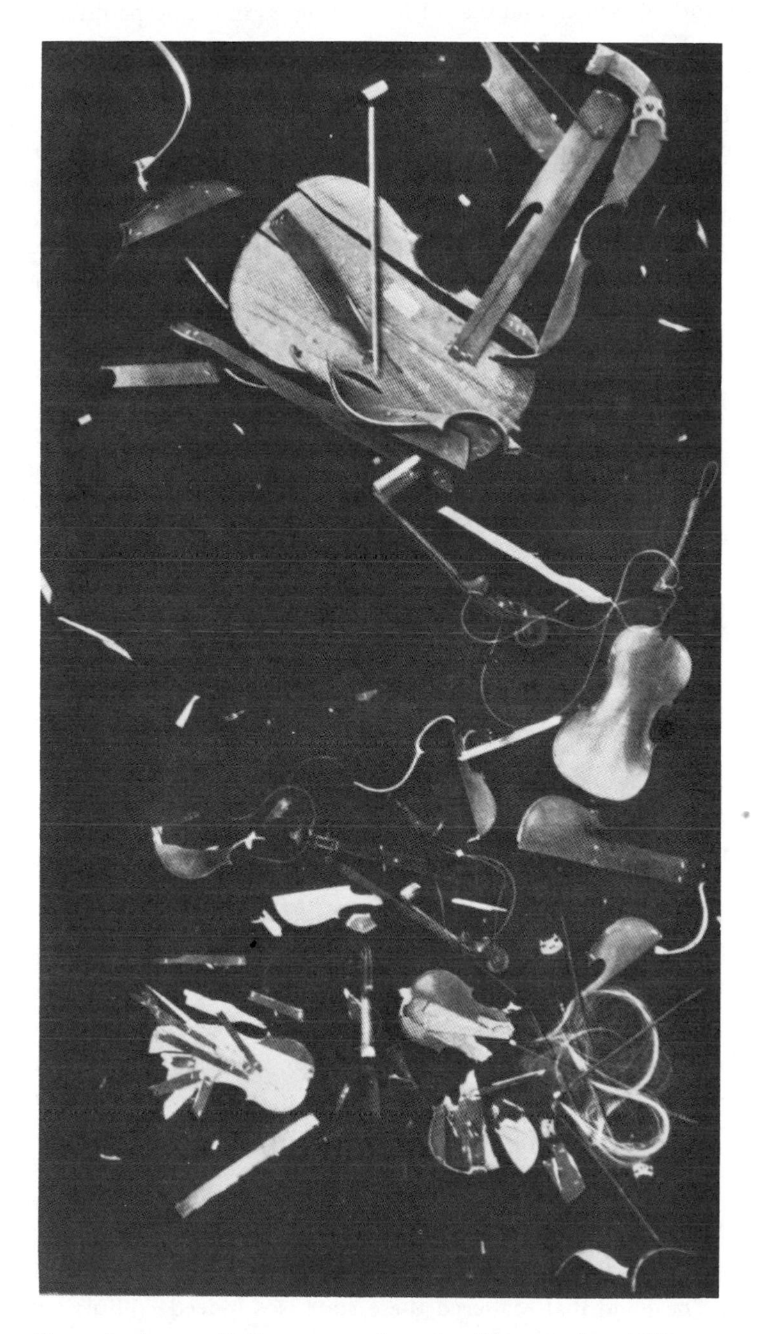

Figure 1. Arman, Colère quatuor à cordes *Allegro furioso,* 1962.

The work of Arman and that of his colleagues in action painting and action sculpture may or may not turn out to be good art. But despite our initial impulse to object, I believe we must take the intention seriously. It is through dismemberment of materials that one may hope—as one method among many others—to discover certain clues to the original, simple symmetry that is hidden in the wholeness of the object. Vesalius knew this, of course. And I might point out that a physicist interested in the orderly structure of nuclei or subnuclear particles often has to induce that structure and symmetry by means of an Arman-like process: he prepares the nucleus by first stripping away the atom's outer electrons that shield it, and then lets the nucleus, at the end of a violent journey through a particle accelerator, bombard a target. There, if the energy is high enough, the projectile nucleus and perhaps also the target will disintegrate, and the fragments will go off with momentum, energy, and spins that are full of fascination and meaning to those who can achieve a "coordinated mastery of all the factors involved." Meanwhile, to the uninitiated, it all looks merely like a ridiculous or dangerous destruction.

If this analogy is valid, it may be that the current attention to the thema of derangement in science as well as in art is at bottom an indication of the return of its antitheme of order in a new, more sophisticated guise. The simple harmonies, the simple symmetries, have been found out. How much more satisfying it will be if we can discern harmonies and simplicities directly, through a more highly trained vision, in complex, apparently broken, and deranged configurations! It may be that we are beginning to train new sensibilities which will set a new style.

In the meantime, the careful attention to disorderliness has yielded surprising new simplicities. For it is a very telling fact that the "carelesly scatter'd" appearance of the stars, which seemed so disorderly and irregular in the seventeenth century, has since then slowly provided data for an entirely different view of the earth, the sun, the solar system, and indeed our whole universe. First, it turned out that we are situated in one of the outer arms of a huge lens-shaped gathering of stars, many of which seem themselves to be the centers of their own solar-system-like planetary distributions; therefore, looking from our earth into different directions we see quite different densities—large numbers of stars when we look toward the center of the galaxy and beyond, few stars at right angles. The hand that scattered these stars was indeed skillful; but to recognize this we first had to get used to the idea that that hand did not also put *us* at the center of things, and that it did not place a higher value on simple lattice arrangements than on stochastic arrangements. And, secondly, looking beyond the stars to the nebulae, we again find them to be placed essentially at random as far as the telescopic eye can see. So, after all, we seem to be in a more or less isotropic and homogeneous, that is, in a most simple and symmetric, world. Or at least the style of thought of our time has made it easier to entertain the new model, one so greatly at variance with that of earlier centuries.

III

Let us turn briefly to another conception that has experienced a cycle of changing acceptability: atomism. This thema also has been important in scientific thought from the first—indeed, from the atomism of Pythagorean number physics—and usually it found itself co-existing with and arrayed against the equally ancient thema of the continuum. (As with order and disorder themata and other examples, it is pairing of opposing or complementary theme and antitheme.) In the early part of the nineteenth century, with Dalton, atomism ceased to be regarded as a "mere" philosophical position and began to go to the forefront of the stage. But there were continuing attacks, for example by Humphry Davy. And even when in the early part of the twentieth century the atomistic hypothesis was discussed as if it had become merely a phenomenic one (namely, borne out of experimental evidence), it was still not acceptable to all opponents, just as in Newton's day it was still possible to oppose Copernicanism.

Eventually, the basic hypotheses, such as heliocentricity and atomism, were accepted into science because they were regarded as phenomenic ones. But is this correct? The answer is no. They remain thematic propositions, and so not directly coupled to the phenomena.[6] A simple example will suffice to prove it. The atom which Wilhelm Ostwald accepted in 1909—one of the last to do so—was an atom *we* would now reject as incapable of explaining radioactivity, x-rays, spectra, valence, and so forth, not to mention the new discoveries made since that time, such as isotopism and space quantization. What Ostwald and others thought they were accepting was the "experimental fact" of atoms. What they really were accepting, however, was the thematic hypothesis of atomism. It is this hypothesis, of course, which has survived the recent advances, the new data, whereas the *experimental* atom of 1909 has long been proved to be "wrong."

Similarly, what Ernst Mach was attacking when he objected to the notion of atoms, saying they were not congruous with sensations, was not the phenomic hypothesis of the atom as an explanatory device to deal with, say, observed scintillations. Rather, he was attacking the conception of fundamental submicroscopic discreteness as against continuity.

The twentieth-century victory of discreteness was really the climax of a whole century of preparation for this new style of thinking in all branches of science. We see here rather beautifully a family of related developments—the theme of discreteness expressing itself in physics, biology, and chemistry. For between 1808 and 1905, physics, biology, and chemistry saw the introduction of remarkably similar conceptions. In each of these fields it was found fruitful to assume the existence of fundamental, discrete entities. Thus Dalton (1808) proposed that matter consists of atoms which maintain their integrity in all chemical reactions. In biology, Schleiden for plants (1838) and Schwann for animals (1839) proposed the theory of cells, by the various combinations of which living tissues were assumed to be built. Mendel's work (1865) led to the idea that the material governing heredity contains a structure of definite entities, or genes, which may be transmitted from one generation to the next without change.

Meanwhile, heat, electricity, and light, which were the parts of physics that the eighteenth century had visualized largely in terms of the actions of imponderable fluids, were being rephrased in a similar manner. In Joule's kinetic theory (1847), sensible heat was identified with the motions of discrete atoms and molecules. The electron, a particle carrying the smallest unit of negative charge, was discovered (1897). Finally, the energy of the sources of radiation and then of radiation itself was found to be quantized (1900 and 1905). It was as if these new views in the sciences stemmed from a similarly directed change in the mental models used to comprehend phenomena—a change in style where the guiding idea is no longer a continuum, but a particle, a discrete quantum.[7]

IV

The discussion of thematic analysis and antinomies comes to the point, then, where one can consider a pair of themes which are central to the question of the relationship of personal style and scientific achievement. I am referring to the methodological themes of projection (or externalization) on the one hand, and retrojection (or internalization) on the other.

We are led into them by noting a step basic to all scientific work, but rarely discussed: it is the process of removing the discourse from the personal level—the level on which the problem originally becomes of interest to the particular person who works on it, the level on which aesthetic and "private," sometimes not even easily communicable considerations may be important—to a second level, that of public science, where the discourse is more unambiguously understandable, being predominantly about phenomena and analytical schemes. [The differences between "private" and "public" science, or S_1 vs. S_2, are further dealt with in Essay 11.] This is a process which every scientist unquestionably accepts, a process that may be termed externalization or projection. The working scientist must be able to shift the conceptual framework from the private to the public level, where it can be shared generally by retrojection into disparate systems of individual scientists all over the world. The aim of the process is to arrive at statements that are invariant with respect to the individual observer—that is, insofar as possible, the same for each particular, purely personal framework into which it ultimately may be channeled. There is an analogy here with the method by which relativity physics selects statements and laws that are invariant with respect to transformation and are therefore generally applicable.

What is interesting is that on certain occasions, during the transformation of conceptions from the personal to the public realm, the scientist, perhaps unknowingly, smuggles the style, motivation, and commitment of his individual system and that of his society into his supposedly neutral, value-indifferent luggage. And it is at this point that the concept of projection will help us to understand how the style of contemporary personal and social thought introduces itself into scientific work.

Of the two main examples I wish now to investigate, the first refers to what is usually—and rather loosely—called anthropomorphic thinking in science. On this aspect of the mechanism of projection, a useful source is the work of Ernst Topitsch.[8] At the outset he notes that while the variety and number of conceptions by which attempts have been made to understand our environment and ourselves are enormous, the thought models that have played a major role fall into a quite limited number of categories. He reminds us that in the study of the psychology of development, Jean Piaget, among others, has stressed that the child conceives of the world as existing in analogical relation to his wishes and actions, his social connections and his handling of experience in general. In this sense phylogenetic and ontogenetic findings coincide; for until fairly recent times the scientist, too, has conceived of what is remote, unknown, or difficult to understand in terms of what is near, well-known, and self-evident in everyday terms. Social and artistic processes and productions have often served as explanations by analogy for the universe as a system—in short, by projecting outward into the universe conceptual images from the domain of social and productive action.

While it has by no means been widely recognized that this is an essential activity in the sciences, the same point has at least been noted in the social studies. Talcott Parsons, for example, has written:

> It is curious—and would merit investigation in terms of the sociology and psychology of knowledge—that the priority given to knowledge of the physical world in the development of modern philosophy reverses the priorities applying to the development of the human individual's knowledge of his own environment and, it seems, the formation of empirical knowledge in early cultural evolution. Since Freud, it has been known that the child's first structured orientation to his world occurs in the field of his *social* relationships. The "objects" involved in Freud's fundamental concept of object relations are "social" objects: persons in roles, particularly parents, and the collectivities of which they are parts and into which the child is socialized. This orientation includes an empirical cognitive component which is the foundation on which a child builds his later capacity for the scientific understanding of the empirical world. What is often interpreted as the child's "magical" thinking about the physical world probably evidences a lack of capacity to differentiate between physical and social objects.
>
> Similar things appear true of cultural evolution more generally, though . . . the parallels are far from exact. Perhaps the best single reference on the problems is the article *Primitive Classification* (1903) by Durkheim and Mauss . . . . This emphasizes, with special but not exclusive reference to the Australian aborigines, the priority of the social aspect of primitive categorization of the world, notable in the conception of spatial relations in terms of the arrangement of social units in the camp.[9]

Turning to the sciences more narrowly defined, it is not difficult to see that the hierarchical universe of Aristotle or of the medieval schoolmen was also abstracted from, a reflection of, the hierarchical class organization of society in which these thinkers lived. This still happens, of course, in our day: the chemist, the physicist, or the astronomer looks out and beholds a new world, one fitting to his time. For example, it is now a profoundly egalitarian rather than hierarchical universe, so much so that a whole theory of relativity (Milne's) has been built around the so-called cosmological principle, the principle that any observer anywhere in the universe interprets data in exactly the same way as any other observer elsewhere, making equivalent correlations between data and instants at which data are taken. It is a restless world, in which the parts are coupled by a complementaristic mutual engagement which is never a unidirectional action but always an interaction. It is, as it were, a class-unbounded world in which many old questions are meaningless but none is impious, and in which each of the few laws is presumed to have the widest possible scope.

Not only are certain conceptual images projected outward into the cosmos, but there is, as Topitsch stresses, also a projection *back,* a process by which the cosmos itself, in its anthropomorphic interpretation, may be *retrojected* into its original context, that of human action. He explains:

> The terrestrial state and terrestrial law must be assimilated to, or modeled upon, the cosmic state and law; the human ruler is the image, the son or deputy of the divine ruler of the world. Places of worship and cities are built according to the model of the supposed "world edifice" or "heavenly city," and music should be an echo of "the harmony of the spheres."
>
> Such conceptions were developed in the major cultures of the ancient East to become a mythology of great power and influence; in the Hellenistic age they fused with Greek thought; and they had their repercussions in Europe far on into the New Era. The conception of the ecclesiastical edifice as an image of the "Heavenly Jerusalem" or even of the cosmos was still familiar to the architects of the Gothic period and the Renaissance, so that an unbroken tradition leads from the Solar Kingdom of Egypt to that of Louis XIV. Moreover, astrology (which for thousands of years, far from being mere superstition, was a conception of the world equal in rank to philosophy) was founded on the same process of projecting conditions of immediate earthly reality into the cosmos (as in the naming of the stars), and then of retrojecting the "macrocosmos" so interpreted into the "microcosmos" of human existence.[10]

The interpretation of the cosmos and the individual by a projection and retrojection of social or technological modes of human behavior is a thematic tendency, that is, one not forced on us by contingent consideration. On this ground alone we expect that there exist also themes on the other side of the ledger, themes founded on the postulate of the antitype of human limitations and transitoriness, on the idea of a perfect entity. This entity, superior to all limitations and even above man's thought, is easily recognizable in scientific thought, from the beginning to this day, as the conception—a haunting and apparently irresistible one despite all evidence to the contrary—of the final, single, perfect object of knowledge to which the current state of science is widely thought to lead us, more or less asymptotically, but continually and inexorably.

Like the exemplification of this conception outside science—in the Supreme Being, or the millennial utopia—the final state of science is one that it is generally agreed cannot be defined with any degree of precision by means of concepts or the use of ordinary language. That would be incompatible with its perfection. It is seen as a conception far beyond those arising directly from an examination of the empirical world. Occasionally a scientist rashly dares to put this dream into words, and then it is likely to emerge that the best he can say is that the goal is already being achieved. The inadequacy of such a statement soon becomes apparent to everyone. Thus A. A. Michelson said in 1903:

> The more important fundamental laws and facts of physical science have all been discovered, and these are now so firmly established that the possibility of their ever being supplanted in consequence of new discoveries is exceedingly remote. . . . Our future discoveries must be looked for in the 6th place of decimals.[11]

Again, the physicist Robert B. Leighton wrote much more recently:

> . . . It is now believed that quantum electrodynamics provides an exact description of all physical phenomena which do not directly involve nuclear forces, the weak interactions, or gravitation: Nearly all of the data that appear in handbooks of physics and chemistry could, in principle, be calculated *from first principles* if sufficiently powerful mathematical techniques were known! With the rapid advances that are being made in particle physics, perhaps it is not too much to expect that in a few more decades *all* physical phenomena will be equally well understood.[12]

We may guess that there will very likely always be such expressions of hope in the imminent perfectibility of science to balance the sense of turmoil around us and of unexpected transitions beyond the horizon.

V

Returning to the methodological themes of anthropomorphic projection and retrojection in our own current scientific work, we would be fundamentally mistaken to regard these as an accidental rather than an essential and important element. I hold with Whorf and Sapir that a working language mirrors the internal metaphysics of the culture of which language is a part. This is true also for the language used in the scientific area of a culture. Niels Bohr thought that the same principle is applicable on a larger scale: in the essay *Natural Philosophy and Human Cultures,* he confesses his belief that "the traditional differences [of human cultures] in many ways resemble the different equivalent modes in which physical experience can be described."[13] More personally and specifically, Martin Deutsch, a nuclear physicist at the Massachusetts Institute of Technology, confessed in an article, *Evidence and Inference in Nuclear Research*: "In my own work I have been puzzled by the striking degree to which an experimenter's preconceived image of the process which he is investigating determines the outcome of his observations. The image to which I refer is the symbolic, anthropomorphic representation of the basically inconceivable atomic processes."[14]

One may go further than this. Not only are the atomic processes basically inconceivable once one leaves the level of common sense, but there is also considerable naïveté in accepting, even on the level of common sense, what we "see" in the laboratory. For most scientists, the creative scientific imagination, as Deutsch notes, can "function only by evoking potential or imagined sense impressions. . . . I have never met a physicist, at least not an experimental physicist, who does not think of the hydrogen atom by evoking a visual image of what he *would* see if the particular atomic model with which he is working existed literally on a [large] scale accessible to sense impressions. At the same time the physicist realizes that in fact the so-called internal structure of the hydrogen atom is *in principle* inaccessible to direct sensory perception."[15]

The more sophisticated science becomes, the more striking is this paradox. Even the simplest observation in any advanced science involves a formidable apparatus of theory. The valuable experimental observations in any modern laboratory, Deutsch recognizes, "seem virtually negligible in the totality of material involved in the experiment. . . . Almost all sense impressions concerning interpretation are irrelevant to the question investigated."[16] Thus, the energy, the size, the period of persistence of the phenomenon studied, all are minute compared to the other attendant data. To use the language of communication engineering, the ratio of signal to noise is extremely small in the laboratory.

In such situations the engineer knows that he must work with a very special kind of "receiver" in order to receive anything. A model is the maser (as used in radioastronomy) which in principle is a device that operates by re-emitting or releasing a signal at considerably larger energy than the incoming, noise-laden signal. This is achieved by having the emitter preloaded, as it were, to be triggered by the relatively low-energy, incoming signal. Similarly, it is possible to understand and use observations in physics today only if the scientist has, from the very beginning, a "well-structured image of the actual connections between the events taking place."

This is indeed far from the conventional idea that the scientist keeps a completely open mind. The more carefully we peer at the "faces" of our meters, therefore, the more we see the reflection of our own faces. Even in the most up-to-date physical concepts the anthropomorphic burden is very large. Particles still attract or repel one another, rather as do people; they "experience" forces, are captured or escape. They live and decay. Circuits "reject" some signals and "accept" others; and so forth. Deutsch notes: "An electron [or any particle of modern physics] is clearly not an object with the general properties of a ball which we would see if we had a sufficiently good microscope, or feel impinging on our hand if our nerves were a little more sensitive. We are not *forced* by direct sense perception to use this image. We have developed it because it allows us to reason from one experiment to the next by analogy; even in a mathematically sophisticated theory we must deal with formal thought processes designed to connect sensory impressions. It, too, must proceed by analogy with the connections established between such perceivable events."[17]

Here is only one of several reasons that some critics of science are so wrong in thinking of modern science as entirely depersonalized, cold and abstract, devoid of all personal concerns. If this were so, scientists would find their work lacking that secret source of excitement which cannot be easily analyzed but can be shared. More than that, without these tie-lines to personal styles and themes, the practitioners could not so successfully understand the content of public science itself.

Thus, even as in Babylon and Greece, where family relationships were projected into the very naming of constellations in the sky above, the nuclear physicist projects human relationships into his equipment and data. For example, for reasons that have become sound and even unavoidable through use and success, he prefers to "see" an experimental result, such as the bubble-chamber photograph in Figure 2, in terms of a life-cycle story. He will describe it as follows: a pion—an elementary particle whose track is marked π^- on the drawing that interprets the raw observations (Figure 3)—comes into the viewfield from the lower

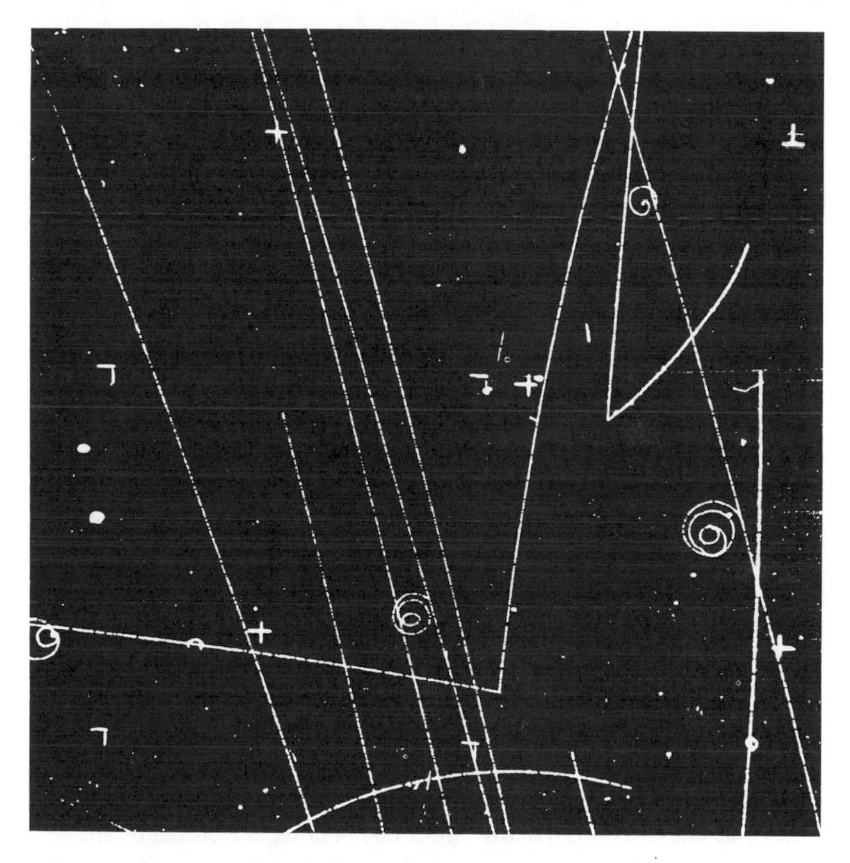

Figure 2. Associated production of neutral strange particles by π^- in liquid hydrogen bubble chamber.

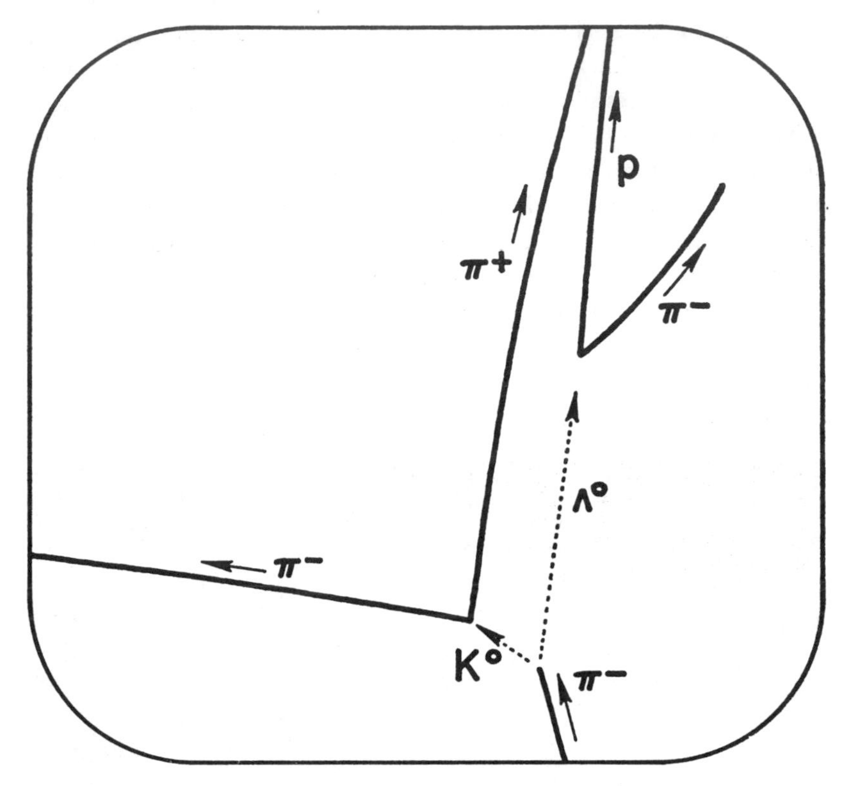

Figure 3. Schematic analysis of events in Figure 2.

center portion. It encounters a proton in the chamber and interacts with it to form two so-called strange particles (labeled K^0 and Λ^0—called "strange" because they were found to survive unexpectedly long, namely 10^{-10} sec.); these, being neutral, leave no trace to look at, until they decay. The product of each "strange" particle's decay is one negative and one positive particle, thereby producing in our viewfield, as it were, a third generation of particles, each again having its own characteristic lifetime.

This is a familiar, primordial type of drama or folk tale, one acted out in space and time. But it is one not forced on us by the "data." For it is at least conceivable that physicists might have chosen a quite different way of looking at "what happens" here; they might have started by seeing Figure 2 as a whole, as a calligraphic design (somewhat analogous to

Arman's products), without space-time development. Its meaning could have been sought in the structure of the apperception rather than in a story of evolution and devolution, of birth, adventure, and death.

Of course, one might object that this alternative would probably be quite barren in the framework of our science. But this recognition would simply reinforce the suspicion that the symbolic power of useful scientific concepts rests at least in some part on the fact that so many of these concepts have for so long been importing anthropomorphic projections from the world of the human drama.

VI

I turn finally to a process of projection in science which is not merely conceptual but temporal. It will help us to understand the fact that, at bottom, the work of major scientific "revolutionaries," like Einstein, is really not an act of iconoclastic destruction of the base of science and the rebuilding of a radically new state, a state possible only in the future. Rather, such advances can usually be seen to be projections back to an idealized, purified state of the past—and, in particular, a return to a state of imagined classical purity, to the ideal of an earlier science that supposedly had a small number and a low complexity of hypotheses.

To be sure, the dream of portraying with an earlier kind of simplicity a more highly developed state of science is, and always has been, in an important way doomed to failure. Simplicity in the new theory is always bought at the expense of rejecting regions of previously accessible speculation, or by considerably deepened mathematical sophistication, or by the invention of mechanisms that are at the time of announcement of the theory not at all amenable to experimental verification. Therefore, the call for simplicity has only a restricted validity. The motivation is, as it were, a longing to establish again an uncomplicated situation, a situation in which experience is dealt with in terms of one or a few large unities rather than detailed particulars—perhaps indeed an attempt to find a way back to a primitive or childlike state of reality.

In the case of Einstein, this projection to a paradisical past may have had two sources which reinforced each other. In his autobiographical account he spoke of his youthful belief in the reality of biblical events, a dream which was shattered at the age of twelve by contact with scientific works of the time. A quotation from his *Autobiographical Notes* seems to me particularly revealing because it shows a connection between the desire to return to a secularized paradise and the desire to escape to an extrapersonal level. Einstein wrote as follows:

> Thus I came—despite the fact that I was the son of entirely irreligious (Jewish) parents—to a deep religiosity, which, however, found an abrupt ending at the age of 12. Through the reading of popular scientific books I soon reached the conviction that much in the stories of the Bible could not be true. The consequence was a positively fanatic [orgy of] freethinking coupled with the impression that youth is intentionally being deceived by the state through lies; it was a crushing impression. Suspicion against every kind of authority grew out of this experience. . . . It is quite clear to me that the religious paradise of youth, which was thus lost, was a first attempt to free myself from the chains of the "merely personal." . . .The mental grasp of this extra-personal world within the frame of the given possibilities swam as highest aim half consciously and half unconsciously before my mind's eye.[18]

Einstein's attempt to restructure science, then, seems to me in several senses to be a return—first, to the childhood state of innocence by a secularization of the religious childhood paradise; second, to the early dream of a state or social environment greatly at variance with the harsh reality that he saw all around him—to a dream of a social environment which, in a word, characterizes the social childhood paradise; third, to an early state of science in which the purity of a few hypotheses supposedly was a primary characteristic.

It is perhaps not a mere coincidence that each of the physical scientists who by their work "completely revolutionized" science had a strong sense of history and an admiration for the ancients, and that both these traits are, as a matter of fact, rather lacking from the thoughts and vocabulary of most lesser scientists, those below the level of a Copernicus, a Newton, an Einstein, a Niels Bohr. This historic taste is, however, what one would expect of thinkers who at heart are purifiers and counter-revolutionaries.

Thus, Copernicus, directly and through Rheticus, is quite explicit about his having looked for a warrant for his heretical view in the writings of the ancients. He does not, of course, dare to mention Aristarchus in the final manuscript of DE REVOLUTIONIBUS, but he calls upon Hecetas of Syracuse, Heraclides, and Ecphantus the Pythagorean for support in the belief in the diurnal rotation of the earth, while Philolaus is called upon to support the doctrine of the annual orbit around the sun.

Newton was, of course, well versed in classical history and indeed preoccupied with biblical chronology. Although he was, with Leibniz, the discoverer of differential and integral calculus, Newton wrote the PRINCIPIA in the language of Apollonius and Archimedes. This incidentally accounts for the fact that soon the PRINCIPIA became almost entirely unreadable to most scientists. The geometrical style of proof has long since been supplanted by proofs involving the modern calculus. On this point Newton once explained, in the third person: "By the help of the new analysis Mr. Newton found out most of the propositions in his PRINCIPIA PHILOSOPHIAE. But because the ancients, for making things certain, admitted nothing in the geometry before it was demonstrated synthetically, he demonstrated the propositions synthetically, that the systems of the heavens might be founded upon good Geometry."[19]

The early, indeed the classically simple, state of science, then, is the true home of the "revolutionary" scientist's imagination. And how importantly the major scientific advances of which we speak are determined by a commitment to a style rather than "brute fact" is made clear by a very simple observation: each of the great advances plainly generated more strictly physical problems than it eliminated. They were *great advances* in the sense that they broke through to a new area of fruitful ignorance. Copernicus makes necessary the formulation of a completely new dynamics of celestial motion; he turns his back on Aristotelian physics, but he cannot put anything else in its place, and we have to wait until the Newtonian synthesis of mechanics for a new physics, a physics Copernicus would not have understood or supported. Newton, in turn, remains incomprehensible without an explanation of action at a distance, and that becomes one of the major and obsessive preoccupations of nineteenth-century physics—until Maxwell, Hertz, and Einstein give solutions which Newton would have found quite uncongenial.

Considerations of an aesthetic nature and the yearning for a simple state of science merge in the requirement that the number and types of hypotheses be severely restricted in any true theory. It is, I think, very significant that Cotes, Newton's disciple, eulogized his master in the preface to the second edition of the PRINCIPIA in these words: "He has so clearly laid open the most beautiful frame of the system of the world, that if King Alphonso were now alive he would not complain for want of the graces of simplicity or harmony." Simplicity, it should be recalled, is characteristic both of the state of theory in antiquity and of the desired state of theory in the future. The right path, these men seem to say, is, in science as in all mythically driven activities, from the past through the unfolding present into a regained state of the past.[20]

VII

I have presented here some speculative considerations on the various ways in which the state of affairs, the folklore and beliefs external to science (in the narrow sense of the word), affect the imagination of the acting scientist. Perhaps these thoughts will be of some use to those philosophers of science who have been struggling with the old question of what the source may be for the strong warrant for explanatory scientific principles—a warrant that cannot be found in experiment and logic alone.

But there is another, even more important reason for recognizing the existence of general stylistic commitments built into scientific work: in this way, we may hope to contribute a little to understanding the puzzling fact that science has indeed for so long been so successful and has remained so interesting. For without some such support for the imagination, coming to us from beyond the boundaries of science alone, without the help of all the best that has been thought and felt before us, how could we hope that the brief attention we can give to scientific problems during our short lives could even yield anything worthwhile? If at every turn we had to construct science anew out of science alone, without the guidance of style and knowledge in their widest sense, how could we hope to catch this complex and infinitely fascinating world with our minds at all?

NOTES

1. Ralph M. Blake, Curt J. Ducasse, and Edward H. Madden, THEORIES OF SCIENTIFIC METHOD (Seattle: University of Washington Press, 1960), p. 22. Some of these ideas have been elaborated by the author in another context in the essay *Über die Hypothesen, welche der Naturwissenschaft zu grunde liegen*, ERANOS JAHRBUCH 31 (Zurich: Rhein-Verlag, 1962), pp. 351–425.

2. Edith Hamilton and Huntington Cairns, eds., PLATO, THE COLLECTED DIALOGUES, Bollingen Series LXXI (New York: Pantheon Books, 1961), pp. 1511–1512, 12.966e–12.988b.

3. For example, see particularly *Republic*, 7.527–534a; *Epinomis*, 987–992; *Laws*, 820e–822d; and *Timaeus*, 38c–40d.

4. Peter Jones, *Arman and the Magic Power of Objects*, ART INTERNATIONAL

7, No. 3:41, 1963. See also my essay *On Style and Achievement in Physics: Further Contributions to Thematic Analysis*, ERANOS JAHRBUCH 33 (Zurich: Rhein-Verlag, 1964), pp. 319–363.

5. *Ibid.*, p. 42.

6. I have given a brief discussion of the distinction between thematic and phenomenic hypotheses in THE GRADUATE JOURNAL, 7:87–109, 1965, and reprinted here as Essay 1.

7. Gerald Holton and D. H. D. Roller, FOUNDATIONS OF MODERN PHYSICAL SCIENCE (Reading, Mass.: Addison-Wesley, 1958), p. 598.

8. Ernst Topitsch, *World Interpretation and Self Interpretation: Some Basic Patterns*, MYTH AND MYTHMAKING, ed. Henry A. Murray (New York: George Braziller, 1960), pp. 157–173. See also his book, THE ORIGIN AND END OF METAPHYSICS, Vienna: Springer-Verlag, 1958.

9. Talcott Parsons, *Unity and Diversity in the Modern Intellectual Disciplines: The Role of the Social Sciences*, DAEDALUS, 94:43, 1965.

10. Topitsch, *op. cit.*, p. 158.

11. Albert A. Michelson, LIGHT WAVES AND THEIR USES (Chicago: University of Chicago Press, 1903), pp. 23–24.

12. Robert B. Leighton, PRINCIPLES OF MODERN PHYSICS (New York: McGraw-Hill, 1959), pp. 678–679.

13. Niels Bohr, *Natural Philosophy and Human Cultures*, NATURE, 143: 268–272, 1939.

14. Martin Deutsch, *Evidence and Inference in Nuclear Research*, EVIDENCE AND INFERENCE, ed. Daniel Lerner (Glencoe, Illinois: The Free Press, 1959), pp. 96–100.

15. *Ibid.*, pp. 96–97.

16. *Ibid.*, p. 99.

17. *Ibid.*, p. 102.

18. Albert Einstein, *Autobiographical Notes*, ALBERT EINSTEIN: PHILOSOPHER-SCIENTIST, ed. P. A. Schilpp (New York: Harper & Brothers, 1959), pp. 3, 5.

19. Einstein, too, confessed the same admiration for ancient geometry. Exactly as was the case with Galileo, Einstein reports that it was the accidental encounter with Euclid that aroused his deep interest in the sciences as a child. Einstein says, "If Euclid failed to kindle your youthful enthusiasm, then you are not born to be a scientific thinker." Albert Einstein, *On the Method of Theoretical Physics*, IDEAS AND OPINIONS, Sonja Bargmann, trans. and ed. (London: Alvin Redman, 1954), p. 271.

20. Mircea Eliade, THE MYTH OF THE ETERNAL RETURN, Bollingen Series XLVI (New York: Pantheon Books, 1954), p. 95, has told us that "Archaic man . . . tends to set himself in opposition, by every means in his power, to history, regarded as a succession of events that are irreversible, unforseeable, possessed of autonomous value." This Eliade calls mastering the "terror of history." I take it he means history not as past but as the unfolding present. In that case, what he said of archaic man is also true of the scientist. Mastery over the unfolding present—in this is found the whole meaning of the words "mastery over nature."

It should be evident that the development of ideas traced in this article as applied to "revolutionary" scientists has a close analogy where "revolutionaries" transformed a field outside science—from the retrospective utopianism of the Founding Fathers of the American Republic to the neoclassicism of Stravinsky.

Figure 1 is reproduced from ART INTERNATIONAL, Volume 7, No. 3 (March 25, 1963) with permission of the publishers. The owner of the painting is M. Gunter Sachs, Lausanne, Switzerland.

Figures 2 and 3 are from INTRODUCTION TO THE DETECTION OF NUCLEAR PARTICLES IN A BUBBLE CHAMBER (prepared at the Lawrence Radiation Laboratory, The University of California at Berkeley), published by The Ealing Corporation, Cambridge, Massachusetts. ©1964, 1966. Reproduced by permission.

4 THE ROOTS OF COMPLEMENTARITY

Como, 1927

EACH AGE is formed by certain characteristic conceptions, those that give it its own unmistakable modernity. The renovation of quantum physics in the mid-1920's brought into public view just such a conception, one that marked a turning point in the road from which our view of the intellectual landscape, in science and in other fields, will forever be qualitatively different from that of earlier periods. It was in September 1927 in Como, Italy, during the International Congress of Physics held in commemoration of the one-hundredth anniversary of Alessandro Volta's death, that Niels Bohr for the first time introduced in a public lecture his formulation of complementarity.[1] It was reported that Bohr's audience contained most of the leading physicists of the world in this area of work, men such as Max Born, A. H. Compton, Peter Debye, Enrico Fermi, James Franck, Werner Heisenberg, Max von Laue, H. A. Lorentz, Robert Millikan, Wolfgang Pauli, Max Planck, Arnold Sommerfeld, and Otto Stern, among others. It was a veritable summit meeting. Only Einstein was conspicuously absent.

In the introduction to his lecture, Bohr said he would make use "only of simple considerations, and without going into any details of technical, mathematical character." Indeed, the essay contained only a few

This essay was originally published in DAEDALUS, Fall, 1970, pp. 1015–1055.

simple equations. Rather, its avowed purpose was a methodological one that, at least in this initial announcement, did not yet confess its ambitious scope. Bohr stressed only that he wanted to describe "a certain general point of view . . . which I hope will be helpful in order to harmonize the apparently conflicting views taken by different scientists."

He was referring to a profound and persistent difference between the classical description and the quantum description of physical phenomena. To review it, we can give four brief examples of the dichotomy:

1. In classical physics, for example in the description of the motion of planets or billiard balls or other objects which are large enough to be directly visible, the "state of the system" can (at least in principle) be observed, described, defined with arbitrarily small interference of the behavior of the object on the part of the observer, and with arbitrarily small uncertainty. In quantum description, on the other hand, the "state of the system" cannot be observed without significant influence upon the state, as for example when an attempt is made to ascertain the orbit of an electron in an atom, or to determine the direction of propagation of photons. The reason for this situation is simple: the atoms, either in the system to be observed or in the probe that is used in making the observation, are never arbitrarily fine in their response; the energy exchange on which their response depends is not any small quantity we please, but, according to the "quantum postulate" (Planck's fundamental law of quantum physics), can proceed only discontinuously, in discrete steps of finite size.

2. It follows that in cases where the classical description is adequate, a system can be considered closed although it is being observed, since the flow of energy into and out of the system during an observation (for example, of the reflection of light from moving balls) is negligible compared to the energy changes in the system during interaction of the parts of the system. On the other hand, in systems that require quantum description, one cannot neglect the interaction between the "system under observation," sometimes loosely called the "object," and the agency or devices used to make the observations (sometimes loosely called the "subject"). The best-known case of this sort is illustrated by Heisenberg's gamma-ray microscope, in which the progress of an electron is "watched" by scattering gamma rays from it, with the result that the electron itself is deflected from its original path.

3. In "classical" systems, those for which classical mechanics is ade-

quate, we have both conventional causality chains and ordinary space-time coordination, and both can exist at the same time. In quantum systems, on the other hand, there are no conventional causality chains; if left to itself, a system such as an atom or its radioactive nucleus undergoes changes (such as emission of a photon from the atom or a particle from the nucleus) in an intrinsically probabilistic manner. However, if we subject the "object" to space-time observations, it no longer undergoes its own probabilistic causality sequence. Both these mutually exclusive descriptions of manifestations of the quantum system must be regarded as equally relevant or "true," although both cannot be exhibited at one and the same time.

4. Finally, we can refer to Bohr's own illustration in the 1927 essay of "the much discussed question of the nature of light . . . [I]ts propagation in space and time is adequately expressed by the electromagnetic theory. Especially the interference phenomena *in vacuo* and the optical properties of material media are completely governed by the wave theory superposition principle. Nevertheless, the conservation of energy and momentum during the interaction between radiation and matter, as evident in the photoelectric and Compton effect, finds its adequate expression just in the light quantum idea put forward by Einstein."[2] Unhappiness with the wave-particle paradox, with being forced to use in different contexts two such antithetical theories of light as the classical wave theory and the quantum (photon) theory was widely felt. Einstein expressed it in April 1924 by writing: "We now have two theories of light, both indispensable, but, it must be admitted, without any logical connection between them, despite twenty years of colossal effort by theoretical physicists."[3]

The puzzle raised by the gulf between the classical description and quantum description was: could one hope that, as had happened so often before in physics, one of the two antithetical views would somehow be subsumed under or dissolved in the other (somewhat as Galileo and Newton had shown celestial physics to be no different from terrestrial physics)? Or would one have to settle for two so radically different modes of description of physical phenomena? Would the essential continuity that underlies classical description, where coordinates such as space, time, energy, and momentum can in principle be considered infinitely divisible, remain unyieldingly antithetical to the essential discontinuity and discreteness of atomic processes?

Considering the situation in 1927 in thematic terms, it was by that time clear that physics had inherited contrary themata from the "classical" period (before 1900) and from the quantum period (after 1900). A chief thema of the earlier period was continuity, although it existed side by side with the atomistic view of matter. A chief thema of the more recent period was discontinuity, although it existed side by side with the wave theory of electromagnetic propagation and of the more recent theories associated with de Broglie and Erwin Schrödinger.

In the older physics, also, classical causality was taken for granted, whereas in the new physics the concept of indeterminacy, statistical description, and probabilistic distribution as an inherent aspect of natural description were beginning to be accepted. In the older physics, the possibility of a sharp subject-object separation was not generally challenged; in the new physics it was seen that the subject-object coupling could be cut only in an arbitrary way. In Bohr's sense, a "phenomenon" is the description of that which is to be observed *and* of the apparatus used to obtain the observation.

Bohr's proposal of 1927 was essentially that we should attempt not to reconcile the dichotomies, but rather to realize the complementarity of representations of events in these two quite different languages. The separateness of the accounts is merely a token of the fact that, in the normal language available to us for communicating the results of our experiments, it is possible to express the wholeness of nature only through a complementary mode of descriptions.[4] The apparently paradoxical, contradictory accounts should not divert our attention from the essential wholeness. Bohr's favorite aphorism was Schiller's *Nur die Fülle führt zur Klarheit*. Unlike the situation in earlier periods, clarity does not reside in simplification and reduction to a single, directly comprehensible model, but in the exhaustive overlay of different descriptions that incorporate apparently contradictory notions.

Summarizing his Como talk, Bohr in 1949 stressed that the need to express one's reports ultimately in normal (classical) language dooms any attempt to impose a clear separation between an atomic "object" and the experimental equipment.

The new progress in atomic physics was commented upon from various sides at the International Physical Congress held in September 1927, at Como in commemoration of Volta. In a lecture on that occasion, I advocated a point of view conveniently termed "complementarity," suited to embrace the characteristic features of individuality of quantum phenomena, and at the same time to clarify the peculiar aspects of the observational problem in this field of experience. For this purpose, it is decisive to recognize that, *however far the phenomena transcend the scope of classical physical explanation, the account of all evidence must be expressed in classical terms.* The argument is simply that by the word "experiment" we refer to a situation where we can tell others what we have learned and that, therefore, the account of the experimental arrangement and of the results of the observations must be expressed in unambiguous language with suitable application of the terminology of classical physics.

This crucial point, which was to become a main theme of the discussions reported in the following, implies the *impossibility of any sharp separation between the behaviour of atomic objects and the interaction with the measuring instruments which serve to define the conditions under which the phenomena appear.* In fact, the individuality of the typical quantum effects finds its proper expression in the circumstance that any attempt of subdividing the phenomena will demand a change in the experimental arrangement, introducing new possibilities of interaction between objects and measuring instruments which in principle cannot be controlled. Consequently, evidence obtained under different experimental conditions cannot be comprehended within a single picture, but must be regarded as *complementary* in the sense that only the totality of the phenomena exhausts the possible information about the objects.[5]

What Bohr was pointing to in 1927 was the curious realization that in the atomic domain, the only way the observer (including his equipment) can be uninvolved is if he observes nothing at all. As soon as he sets up the observation tools on his workbench, the system he has chosen to put under observation and his measuring instruments for doing the job form one inseparable whole. Therefore the results depend heavily on the apparatus. In the well-known illustration involving a light beam, if the instrument of measurement contains a double pinhole through which the light passes, the result of observation will indicate that a wave phenomenon is involved; but if the "same" light beam is used when the measuring instrument contains a collection of recoiling scatterers, then the observation results will indicate that a stream of particles is involved. (Moreover, precisely the same two kinds of observations are obtained when, instead of the beam of light, one uses a beam of "particles" such as atoms or electrons or other subatomic particles.) One cannot construct an experiment which simultaneously exhibits the wave and particle aspects of atomic matter. A particular experiment will always show only one view or representation of objects at the atomic level.

The study of nature is a study of artifacts that appear during an engagement between the scientist and the world in which he finds himself. And these artifacts themselves are seen through the lens of theory. Thus, different experimental conditions give different views of "nature." To call light either a wave phenomenon or a particle phenomenon is impossible; in either case, too much is left out. To call light *both* a wave phenomenon and a particle phenomenon is to oversimplify matters. Our knowledge of light is contained in a number of statements that are seemingly contradictory, made on the basis of a variety of experiments under different conditions, and interpreted in the light of a complex of theories. When you ask, "What is light?" the answer is: the observer, his various pieces and types of equipment, his experiments, his theories and models of interpretation, *and* whatever it may be that fills an otherwise empty room when the lightbulb is allowed to keep on burning. All this, together, is light.

No objections seem to have been raised against Niels Bohr's paper at the Como meeting. On the other hand, at this first hearing the importance of the new point of view was not immediately appreciated. Apparently a typical comment overheard after Bohr's lecture was that it "will not induce any of us to change his own opinion about quantum mechanics."[6] A distinguished group of physicists, although a minority in the field, remained unconvinced by and indeed hostile to the complementarity point of view. Foremost among them was Einstein, who heard the first extensive exposition a month after the Como meeting, in October 1927, at the Solvay Congress in Brussels. Einstein had disliked even the earlier Göttingen-Copenhagen interpretations of atomic physics that were based on the themata of discontinuity and nonclassical causality. He had written to Paul Ehrenfest (August 28, 1926), "I stand before quantum mechanics with admiration and suspicion," and to Born (December 4, 1926) Einstein had said, "Quantum mechanics demands serious attention. But an inner voice tells me that this is not the true Jacob. The theory accomplishes a lot, but it does not bring us closer to the secrets of the Old One. In any case, I am convinced that He does not play dice."[7]

Almost a quarter of a century later Einstein was still in opposition, and added two objections to the complementarity principle: "to me it must seem a mistake to permit theoretical description to be directly dependent upon acts of empirical assertions, as it seems to be intended [for example] in Bohr's principle of complementarity, the sharp formulation of which, moreover, I have been unable to achieve despite much effort which I have expended on it."[8]

Bohr himself was aware from the beginning that the complementarity point of view was a program rather than a finished work; that is, it had to be extended and deepened by much subsequent work. It was to him "a most valuable incentive . . . to reexamine the various aspects of the situation as regards the description of atomic phenomena" and "a welcome stimulus to verify still further the rôle played by the measuring instruments."[9] However, as we shall see, over the years Bohr came to regard the complementarity principle as more and more important, extending far beyond the original context in which it had been announced. For his later, deep commitment to the conception, and for his awareness of the antiquity of some of its roots, we need cite here only an anecdotal piece of evidence. When Bohr was awarded the Danish Order of the Elephant in 1947, he had to supervise the design of a coat of arms for placement in the church of the Frederiksborg Castle at Hillerød. The device (see p. 122) presents the idea of complementarity: above the central insignia, the legend says *Contraria sunt complementa*, and at the center Bohr placed the symbol for Yin and Yang.

Lux versus Lumen

How did Bohr's complementarity point of view—so far from the older scientific tradition of strict separation between the observer and the observed—come to be developed? Finding the various roots of and the likely preparatory conditions for this transforming conception—those in physical theory and those in philosophical tradition—appears to me to be an interesting problem that is far from its unambiguous solution. However, there are already some useful results of the search, particularly insofar as they may have relevance for a better understanding of the mutual interaction of scientific and humanistic traditions.

The first direction to look is the development of the early ideas concerning the nature of light. That a modern thema was already inherent in the formulations that began in antiquity should not surprise us; we know from other studies that despite all change and progress of science,

Coat of arms chosen by Niels Bohr when he was awarded the Danish Order of the Elephant, 1947. From Stefan Rozental, ed., NIELS BOHR: HIS LIFE AND WORK AS SEEN BY HIS FRIENDS AND COLLEAGUES (New York: John Wiley & Sons, 1967).

the underlying, important themata are relatively few. In one guise or another they have been the mainstay of the imagination.

One of the favorite ancient ideas concerning the nature of light, originating in the Pythagorean school, postulated that rays are emitted by the eye to explore the world. Euclid spoke of the eye as if it were sending out visual rays whose ends probed the object, somewhat like the stick of a blind man tapping around himself. A somewhat more refined conception of this general sort is still found in Ptolemy in the second century A.D. in the ALMAGEST, and so was transmitted to a later period.

There is in these emission theories of light clearly an intimate interaction through contact between the observer and the observed. This is also true for the emanation tradition in another, less materialistic form.[10] Here, objects are thought to impress themselves upon our sight owing to a contact force similar to touch—action at a distance being ruled out in classical physics—and this touch reaches our souls by the action of the *eidola* or images or shadows which the emitting bodies send out. Plato held that as long as the eye is open it emits an inner light. For the eye to perceive, however, there must be outside the eye a "related other light," that of the sun or some other source that allows rays to come from the objects. Once more, a coupling between the outer and inner world is clearly attempted.

There were immense problems with emission theories. How, for example, can the eye pupil, only a few millimeters wide, admit the image that was emitted by a huge mountain? Nevertheless, the emanation theory was the take-off point for the optics developed in the seventeenth century. Here we find the modern idea that there is an infinite number of rays leaving from every point of an illuminated object in all directions. But the observer now stands off-stage, and he may or may not be the recipient of some of these ray bundles. The latter are no longer the *lux* of the ancients—*lux* being the word for light when it is regarded as a subjective phenomenon—but rather the *lumen*, a kind of stream of light "objects."

The modern period started effectively with Kepler, who in his writing on Witelo in 1604 and later in the DIOPTRICS of 1611 described how light is refracted by a sphere, for example in a spherical bottle filled with water; he applied his findings to the pupil of the eye. Here was the basic new idea in the optics of vision: the eyeball, and the lens in front of it, focus the ray bundles that come through the pupil, and at the focus the sensorium is stimulated in some way—*which is simply not discussed as part of optics.* In the DIOPTRICS Kepler showed for the first time how lenses really work. Significantly, most images that can be constructed diagrammatically by ray optics can, in fact, not be seen at all by an eye placed at the instrument. Gone are the eidola and species, the "recognition" of soul by soul in Neoplatonist discussion of optics—but gone also is the close coupling of the observer and the observed. The *lumen* had won over the *lux*.

We see how the science of optics became "modern": by an act of breaking the bonding that was self-evident for the ancients, by disengaging the conceptions of what goes on "out there, objectively speaking" on the one hand, and what the eye does with light on the other hand. At some point someone had to do what Kepler, in preparing for Newton, finally did, namely to get interested in bundles of light rays coming together on a screen outside an eye—or, what is for the physics of light significantly exactly the same thing, on the retina or screen in back of the eye—and to stop thinking about the sense impressions produced at such a focus at the same time. As Müller's influential LEHRBUCH DER PHYSIK said in 1926, just one year before Bohr's formulation of the idea of complementarity, the first task of physical optics "is the sharp separation between the objective ray of light and the sensory impression of light. The subject of discussion of physical optics is the ray of light, whereas the inner processes between eye and brain"—says the LEHRBUCH, dismissing the matter—"are in the domain of physiology, and perhaps also psychology."

We see here an attempt at precisely the same separation of primary and secondary qualities, between the numerical and affective aspects of nature, that, as it had turned out three centuries before, was the key with which Galileo and others at that time managed to go from the mechanics of antiquity to modern mechanics. We recall that it was Galileo who did for particles, such as falling stones, what Kepler did for light—namely, to remove the language of volition and teleology, and to fortify the notion of "impersonal," causal laws of motion. The Newtonian science of light has no primary place for the observer and his sense impression. In this manner, the important, basic properties of light could be discovered: the finite propagation speed, the existence of light rays outside the range to which the eye is sensitive, the analogy between light rays and other radiation such as X-rays, and so forth.

The decoupling between *lux* and *lumen*, between subject and object, observer and the observed, and with it the destruction of the earlier, holistic physics, was a painful and lengthy process. The reason why it was ultimately victorious is the reason why the same process in all other parts of science worked: once the separation was made, there ensued a dazzling enrichment of our intellectual and material world. By 1927, a reader of physics texts was bound to feel that the modern theory of light, from electromagnetic theory to the design of optical instruments, devoted its attention entirely to *lumen*, and was a field just as deanthropomorphized as all other parts of the developed physical sciences.

But the seed of a new view of light was present, carried in the early historic development which we have sketched, in the prescientific, commonsense notions that everyone begins with—and in the operational meaning of some of the main concepts of optics. Thus we turn to a second main line of ideas leading to the complementarity point of view.

Operational Meanings

One of the oldest and most elementary building blocks of optics is: light travels in any homogeneous medium in straight lines. But let us consider for a moment why we believe that this statement is true.

We can check it most directly in an experimental way by inserting a screen or scatterer, such as chalk dust, in different parts of the same beam. If we consider this closely, we notice that such a method destroys the light beam that we wanted to examine. The insertion of the apparatus interferes with the phenomenon.

This situation is typical on the atomic scale. There are no comparable problems when one wishes to check, say, Newton's First Law of Motion for ordinary physical objects, for example, by watching or photographing a ball rolling on a flat table. We can verify that a material object in a force-free medium will travel in straight lines without drastically interrupting the object's path. The small effects of the apparatus can be removed by calculation. The fact that the observer and the "object" must share between them at least one indivisible quantum is here negligible, that is, can be made an arbitrarily small part of the phenomenon. From past observations we can therefore extrapolate with certainty the paths the object will take in the future. Space-time descriptions and classical causality apply without difficulty. Not so for beams of light and of other particles on the atomic scale. The more certainly we have ascertained their past, the less certainly we can follow their subsequent progress; the effects of the perturbing interaction with the apparatus cannot be taken out by calculation, but are intrinsically probabilistic. In fact, owing to the uncertainty principle, it is not even possible to define precisely the initial state of the system in the sense required by the classical view of causality.

If we do not wish to intercept the whole beam, we can try to discover whether a beam goes in a straight line by another method: by placing a number of slits at some distance from one another, but all along the same axis, then checking if light penetrates this whole set of collimators. But there are now two problems. First, how do we know whether the slits are indeed arranged in a straight line? We might check it with a straight edge—but we know the straight edge is straight because we can sight directly along it and see no curves or protrusions. Clearly, this process of sighting, or anything equally effective, relies on using a light beam to sight along the ruler. And that, of course, is circular reasoning, assuming, in setting up the instrument, what the experiment is designed to prove.

The paradox is not inescapable, there are other, although cumbersome, methods for lining up the slits without assuming anything about light. But again, we run into trouble. The more closely we wish to define the line along which the beam is to travel, and consequently the narrower we make the slit, the more we find that the beam's energy is spread out into the "shadow," turning, as it were, a corner on going through the slit. This is the phenomenon of diffraction. It is exceedingly easy to demonstrate with the crudest equipment, even just by letting light from a candle pass through the narrow space between two fingers held closely to the eye.

We are dealing here with an instrumental coupling between observer (equipment) and the entity to be observed. As soon as we try to give an operational meaning to the phrase "light travels in any homogeneous medium in straight lines," we see what a poor statement it is.

As a result, a physicist is likely to prefer another statement, more general but which can be reduced in the limit to the one above. It is Fermat's Principle of Least Time, derived from a statement that dates from about 1650. Between any two points, light will go along that path in which the time spent in transit is less than the time that would be spent in any other path. This view explains why a light beam appears to go in a straight line in a homogeneous medium, and also how a beam is reflected or refracted at the interfaces of two media. But the statement harbors the curious idea that light is "exploring" to find the quickest path, as if light were scouting around in the apparatus. We get here a hint of instrumental coupling of the most intimate sort. The suspicion arises that the properties we assign to light are to some degree the properties of the boxes through which light has to find its way.

This becomes quite obvious and unmistakable when we turn to another well-known experiment. When light is sent through a double slit, an interference pattern characteristic of the geometry of the arrangement is obtained on a screen. If one of the two slits is blocked off, a rather different pattern of interference results. All this can be easily understood with elementary constructions from the classical theory of light. However, if a very weak beam of light is used with the double slit experiment, so that at any given time it is exceedingly unlikely that more than a single photon travels through the apparatus, a remarkable thing will be observed: even though one cannot help using classical language and thinking that a single photon will have to go through either one or the other of the two slits at a given time, it will be found that as long as both slits are kept open, the interference pattern accumulating in due course on a photographic plate placed at the screen has exactly the same characteristics as that for the earlier double slit experiment, when the beam was so strong that at any given moment some photons were passing through one of the slits and some photons were passing through the other. Equally remarkable, if one now closes one of the slits toward which the very weak beam of photons is being sent, the interference pattern accumulated over a period changes to the pattern characteristic of a strong light beam passing through a single slit. The fact that for a weak light beam the interference pattern depends on the number of slits available—even though there is no evident way in which the single photon can "know" if the other slit is open—is an indication that the experimental observations of light yield characteristics of the box and its slits as much as of light itself. In short, the experiments are made on the entity light + box. Here, then, in the operational examination of laws of light propagation, is a second path leading to the complementarity idea.

From Correspondence to Complementarity

Yet another primary influence on Bohr was, of course, the achievement and failures of physics in his own work from about 1912 to about 1925. Bohr's model of the hydrogen atom of 1912–1913 is now usually remembered best for the magnificent accomplishment of predicting the frequencies of the emission spectrum. To do this, Bohr essentially tried to reconcile the two apparently antithetical notions about light, both of which had had their successes—the electromagnetic theory of Maxwell, according to which light propagates as a wavelike disturbance characterized by continuity, and, on the other hand, Einstein's theory that light energy is characterized by discreteness and discontinuity. As Einstein had put it in his 1905 paper presenting a "heuristic" point of view concerning the interaction of light and matter: "The energy in the light propagated in rays from a point is not smeared out continuously over larger and larger volumes, but rather consists of a finite number of energy quanta localized at space points, which move without breaking up, and which can be absorbed or emitted only as wholes."

By 1912 the indisputable evidence for Einstein's outrageous notion was not yet at hand, but some experiments on the photoelectric effect, including those with X-rays, began to make it plausible.[11] Indeed, it was not until Millikan's experiment, published in 1916, and A. H. Compton's experiment of 1922, that the quantum theory of light was seen everywhere to be unavoidable.

It is therefore, in retrospect, even more remarkable how courageous Niels Bohr's work of 1912–1913 was. Let us recall his model of the hydrogen atom in its initial form, even though it was soon made more accurate though more complex. Bohr's hydrogen atom had the nucleus at the center (where Ernest Rutherford, in whose Manchester laboratory Bohr was a guest, had just then discovered it to be) and the electron orbiting at some fixed distance around the nucleus. When the sample is heated or the atoms are otherwise excited by being given extra energy, the electron of the excited atom will not be in the normal, innermost orbit, or ground state, but will be traveling in a more distant orbit. At some point the electron will jump from the outer orbit to one of the allowed inner orbits, and in so doing will give up the energy difference between these orbits, or stationary states, in the form of a photon of energy $h\nu$. This corresponds to the emission of light at the observed frequency ν or the corresponding wave length $\lambda = c/\nu$ (where c is the speed of light). The various observed frequencies emitted from an excited sample of hydrogen atoms were therefore interpreted to be a stream of photons, each photon having the energy corresponding to the allowed transition between stationary states.

The success of the model in explaining all known spectrum lines of hydrogen, in predicting other series that were also found, and in giving a solid foothold on the explanation of chemical properties, could not hide the realization, fully apparent to Niels Bohr himself, that the model carried with it a number of grave problems. First of all, it used simultaneously two separate notions which were clearly conflicting: the classical notion of an identifiable electron moving in an identifiable orbit like a miniature planetary system, and the quantum notion that such an electron is in a stationary state rather than continually giving up energy while orbiting (as it should do on the basis of Maxwell's theory, amply tested for charges circulating in structures of large size). Bohr's postulate that the electron would not lose energy by radiation while in an orbit, but only on transition from one orbit to the other, was necessary to "save" the atom from gradually collapsing with the emission of a spectrum line of continuously changing frequency. Also, contrary to all previous ideas, the frequency of the emitted photon was not equal to the frequency of the model's orbiting electron, either in its initial or in its final stationary state.

Looking back later on the situation of about 1913, Merle A. Tuve noted that the Bohr atom was "quite irrational and absurd from the viewpoint of classical Newtonian mechanics and Maxwellian electrodynamics . . . . Various mathematical formalisms were devised which simply 'described' atomic states and transitions, but the same arbitrary avoidance of detailed processes, for example, descriptions of the actual *process* of transition, were inherent in all these formulations."[12]

Niels Bohr himself took pains to stress these conflicts from the beginning. In fact, the explanation of the spectral lines, which were the most widely hailed achievement, more or less constituted an afterthought in his own work. His interest was precisely to examine the area of conflict between the conceptions of ordinary electrodynamics and classical mechanics on the one hand and quantum physics on the other. As Jammer pointed out, "Not only did Bohr fully recognize the profound chasm in the conceptual scheme of his theory, but he was convinced that progress in quantum theory could not be obtained unless the antithesis between quantum-theoretic and classical conceptions was brought to the forefront of theoretical analysis. He therefore attempted to trace the roots of this antithesis as deeply as he could. It was in this search for fundamentals that he introduced the revolutionary conception of 'stationary' states, 'indicating thereby that they form some kind of waitting places between which occurs the emission of the energy corresponding to the various spectral lines,' [as Bohr put it in an address of December 20, 1913, to the Physical Society in Copenhagen.]"[13] At the end of his address, Bohr said, "I hope I have expressed myself sufficiently clearly so that you appreciate the extent to which these considerations conflict with the admirably coherent group of conceptions which have been rightly termed the classical theory of electrodynamics. On the other hand, by emphasizing this conflict, I have tried to convey to you the impression that it may also be possible in the course of time to discover a certain coherence in the new ideas."

This methodological strategy of *emphasizing conceptual conflict as a necessary preparation for its resolution* culminated, fourteen years later, in the announcement of the complementarity principle. In the meantime, Bohr formulated a proposal that turned out to be a moderately successful half-way house toward the reconciliation between classical and quantum mechanics, a conception which, from about 1918 on, became known as the correspondence principle.

In essence, Bohr still hoped for the resolution between opposites by attending to an area where they overlap, namely the extreme cases where quantum theory and classical mechanics yield to each other. For example, for very large orbits of the hydrogen atom's electron, the neighboring, allowed stationary states in Bohr's model come to be very close together. It is easily shown that a transition between such orbits, on the basis of quantum notions, yields a radiation of just the same frequency expected on classical grounds for a charged particle orbiting as part of a current in a circular antenna—and, moreover, the frequency of radiation would be equal to the frequency of revolution in the orbit. Thus for sufficiently large "atoms," and conversely for sufficiently small "circuits" scaled down from the normal size of ordinary electric experiments, a coincidence, or correspondence, of predictions is obtained from the two theories.

In this manner, classical physics becomes the limiting case of the more complex quantum physics: our more ordinary, large-scale experiments fail to show their inherently quantal character only because the transitions involved are between states characterized by high quantum numbers. In this situation the quantum of action relative to the energies involved in the system is effectively zero rather than having a finite value, and the discreteness of individual events is dissolved, owing to the large number of events, in an experienced continuum.

The correspondence principle came to be developed in the hands of Bohr and his collaborators into a sophisticated tool. The basic hope behind it was explained by Bohr in a letter to A. A. Michelson on February 7, 1924:

It may perhaps interest you to hear that it appears to be possible for a believer in the essential reality of the quantum theory to take a view which may harmonize with the essential reality of the wave-theory conception even more closely than the views I expressed during our conversation. In fact on the basis of the correspondence principle it seems possible to connect the discontinuous processes occurring in atoms with the continuous character of the radiation field in a somewhat more adequate way than hitherto perceived. . . . I hope soon to send you a paper about these problems written in cooperation with Drs. Kramers and Slater.[14]

But, shortly after the publication in 1924 of the paper by Bohr, Kramers, and Slater,[15] experiments were initiated by Walter Bothe and Hans Geiger and by A. H. Compton and A. W. Simon—with unambiguously disconfirming results. The correspondence principle, it appeared now clearly, had been a useful patch over the fissure, but it was not a profound solution.

Even before the discovery, major problems known to be inherent in the Bohr atom included the following: the fact that the antithetical notions of the wave (implied in the frequency or wave length of light emitted) and of the particle (implied in the then current idea of the electron) were by no means resolved, but on the contrary persisted unchanged in the model of the atom; so did the conflict between the antithetical notions of classical causality on the one hand (as in the presumed motion of the electrons in their orbits) and of probabilistic features on the other (as for the transitions between allowed orbits); and even the notion of the "identity" of the atom had to be revised, for it was no longer even in principle observable and explorable as a separate entity without interfering with its state. Each different type of experiment produces its own change of state, so that different experiments produce different "identities."

Such questions remained at the center of discussion among the most concerned physicists. Schrödinger and de Broglie, for example, hoped to deal with the glaring contrast between the themata of continuity and discontinuity by providing a wave-mechanical explanation for phenomena that previously had been thought to demand a language of quantization. As Schrödinger wrote in his first paper on the subject,[16] "It is hardly necessary to point out how much more gratifying it would be to conceive a quantum transition as an energy change from one vibrational mode to another than to regard it as a jumping of electrons. The variation of vibrational modes may be treated as a process continuous in space and time and enduring as long as the emission process persists." Thus, space-time description and classical causality would be preserved.

The reception accorded to Schrödinger's beautiful papers was interesting. Heisenberg had obtained essentially the same results in a quite different way through his matrix mechanics; as Jammer notes, "it was an *algebraic* approach which, proceeding from the observed discreteness of spectral lines, emphasized the element of *discontinuity*; in spite of its renunciation of classical description in space and time it was ultimately the theory whose basic conception was the *corpuscle*. Schrödinger's, in contrast, was based on the familiar apparatus of differential equations, akin to the classical mechanics of fluids and suggestive of an easily visualizable representation; it was an *analytical* approach which, proceeding from a generalization of the classical laws of motion, stressed the element of *continuity*."[17] "Those who in their yearning for continuity hated to renounce the classical maxim *natura non facit saltus* acclaimed Schrödinger as the herald of a new dawn. In fact, within a few brief months, Schrödinger's theory 'captivated the world of physics' because it seemed to promise 'a fulfillment of that long-baffled and insuppressible desire' [in the words of K. K. Darrow, The Bell System Technical Journal, 6 (1927)]....Planck reportedly declared 'I am reading it as a child reads a puzzle,' and Sommerfeld was exultant."[18] So, of course, was Einstein, who as early as 1920 had written to Born, "that one has to solve the quanta by giving up the continuum, I do not believe."

We are, of course, dealing here with the kind of intellectual commitment, or "insuppressible desire," that characterizes a true thematic attachment. Rarely has there been a more obvious fight between different themata vying for allegiance, or a conflict between the aesthetic criteria of scientific choice in the face of the same set of experimental data. And nothing is more revealing of the true and passionate motivation of scientists than their responses to each other's antithetical constructs. In a letter to Pauli, Heisenberg wrote: "The more I ponder about the physical part of Schrödinger's theory, the more disgusting [*desto abscheulicher*] it appears to me." Schrödinger, on his side, freely published his response to Heisenberg's theory: "I was discouraged [*abgeschreckt*] if not repelled [*abgestossen*]."[19]

Different aspects of thematic analysis and thematic conflict were the subject of previous articles.[20] In these studies I pointed out a number of other theme-antitheme couples, which may be symbolized by $(\theta, \bar{\theta})$. What Bohr had done in 1927, shortly after the Heisenberg-Schrödinger debates, was to develop a point of view which would allow him to *accept both members of the $(\theta, \bar{\theta})$ couple as valid pictures of nature*, accepting the continuity-discontinuity (or wave-particle) duality as an irreducible fact, instead of attempting to dissolve one member of the pair in the other as he had essentially tried to do in the development of the correspondence-principle point of view. Secondly, Bohr saw that the $(\theta, \bar{\theta})$ couple involving discrete atomism on the one hand and continuity on the other is related to other $(\theta, \bar{\theta})$ dichotomies that had obstinately refused to yield to bridging or mutual absorption (for example, the subject-object separation versus subject-object coupling; classical causality versus probabilistic causality). The consequence Bohr drew from these recognitions was of a kind rare in the history of thought: he introduced explicitly a new thema, or at least identified a thema that had not yet been consciously a part of contemporary physics. Specifically, Bohr asked that physicists accept both θ and $\bar{\theta}$—though both would not be found in the same plane of focus at any given time. Nor are θ and $\bar{\theta}$ to be transformed into some new entity. Rather, they both exist in the form *Either θ Or $\bar{\theta}$* the choice depending on the theoretical or experimental questions which you may decide to ask. We see at once why all parties concerned, both those identified with θ and those identified with $\bar{\theta}$, would not easily accept a new thema which saw a basic truth in the existence of a paradox that the others were trying to remove.

Poul Martin Møller and William James

Another root of the complementarity conception can be discerned in Niels Bohr's work when we carefully read and reread his own statements of the complementarity point of view. For it is at first curious and then undeniably significant that from the very beginning in 1927, Niels Bohr cited experiences of daily life to make apparent the difficulty of distinguishing between object and subject, and, as Oskar Klein wrote in a retrospective essay, "to facilitate understanding of the new situation in physics, where his view appeared too radical or mysterious even to many physicists."[21] In this connection, according to Klein, Bohr chose a particularly simple and vivid example: the use one may make of a stick when trying to find one's way in a dark room. The man, the stick, and the room form one entity. The dividing line between subject and object is not fixed. For example, the dividing line is at the end of the stick when the stick is grasped firmly. But when it is loosely held, the stick appears to be an object being explored by the hand. It is a striking reminder of the situation described in the classical emanation theory of light in which we first noted the problem of coupling between observer and observed.

On studying Bohr's writings one realizes by and by that his uses of apparently "extraneous" examples or analogies of this sort are more than mere pedagogic devices. In his September 1927 talk, the final sentence was "I hope, however, that the idea of complementarity is suited to characterize the situation, which bears a deep-going analogy to the general difficulty in the formation of human ideas, inherent in the distinction between subject and object." Similar and increasingly more confident remarks continued to characterize Bohr's later discussions of complementarity. Thus in his essay *Quantum Physics and Philosophy* (1958), the lead essay in the second collection of Bohr's essays under the title ESSAYS 1958–1962 ON ATOMIC PHYSICS AND HUMAN KNOWLEDGE,[22] Bohr concluded, "It is significant that . . . in other fields of knowledge, we are confronted with situations reminding us of the situation in quantum physics. Thus, the integrity of living organisms, and the characteristics of conscious individuals and human cultures present features of wholeness, the account of which implies a typical complementarity mode of description . . . . *We are not dealing with more or less vague analogies, but with clear examples of logical relations which, in different contexts, are met with in wider fields.*" It will be important for our analysis to try to discern clearly what Bohr means in such passages.

Some illumination is provided by a story which Niels Bohr loved to tell in order to illustrate and make more understandable the complementarity point of view. Léon Rosenfeld, a long-term associate of Niels Bohr, who has also been concerned with the origins of complementarity, told how seriously Bohr took his task of repeatedly telling the story. "Everyone of those who came into closer contact with Bohr at the Institute, as soon as he showed himself sufficiently proficient in the Danish language, was acquainted with the little book: it was part of his initiation."[23]

The "little book" which Bohr used was a work of the nineteenth-century poet and philosopher, Poul Martin Møller. In that light story, THE ADVENTURES OF A DANISH STUDENT, Bohr found what he called a "vivid and suggestive account of the interplay between the various aspects of our position." A student is trying to explain why he cannot use the opportunity for finding a practical job, and reports the difficulties he is experiencing with his own thought process:

> My endless enquiries make it impossible for me to achieve anything. Furthermore, I get to think about my own thoughts of the situation in which I find myself. I even think that I think of it, and divide myself into an infinite retrogressive sequence of "I"s who consider each other. I do not know at which "I" to stop as the actual one, and in the moment I stop at one, there is indeed again an "I" which stops at it. I become confused and feel a dizziness, as if I were looking down into a bottomless abyss, and my ponderings result finally in a terrible headache.

Further, the student remarks:

> The mind cannot proceed without moving along a certain line; but before following this line, it must already have thought it. Therefore one has already thought every thought before one thinks it. Thus every thought, which seems the work of a minute, presupposes an eternity. This could almost drive me to madness. How could then any thought arise, since it must have existed before it is produced? . . . The insight into the impossibility of thinking contains itself an impossibility, the recognition of which again implies an inexplicable contradiction.[24]

Bohr used the situation in the story not as a distant, vague analogy; rather, it is one of those cases which, "in different contexts, are met with in wider fields." Moreover, the story seems appropriate for two other reasons. Bohr reports that conditions of analysis and synthesis of psychological experiences "have always been an important problem in philosophy. It is evident that words like thoughts and sentiments, referring to mutually exclusive experiences, have been used in a typical complementary manner since the very origin of language."[25] Also, the humane setting of the Danish story, and the fact that it renders a situation in words rather than scientific symbols, should not mislead us into thinking that it is thereby qualitatively different from the information supplied in scientific discourse. On the contrary: Bohr said, in defending the complementarity principle, "The aim of our argumentation is to emphasize that all experience, whether in science, philosophy, or art, which may be helpful to mankind, must be capable of being communicated by human means of expression, and it is on this basis that we shall approach the question of unity of knowledge."[26] We shall come back to this important statement presently.

Now one must confess that it is on first encounter curious, and at least for a professional physicist perhaps a little shocking, to find that the father of the complementarity principle, in these passages and others, should frequently have gone so far afield, by the standards of the scientific profession, in illustrating and extending what he took to be the full power of the complementarity point of view. In looking for the roots of the complementarity principle, we might grant more readily the three avenues shown so far, namely through the history of the concept of light, the operational definition of light behavior, and through Bohr's own work in physics. But in pursuing this new avenue, we seem to be leaving science entirely.

I imagine that many of Bohr's students and associates listened to his remarks with polite tolerance, perhaps agreeing that there might be a certain pedagogic benefit, but not a key to the "unity of knowledge." To the typical scientist, the student in Møller's story who becomes dizzy when he tries to think about his own thoughts, because precise "thought" and "thought *about* thought" are complementary with respect to each other and so mutually exclusive at the same time, would seem somehow to have a problem different from that of the experimenter who cannot simultaneously show both the wave characteristics and the particle characteristics of a light beam. Similarly, the intrusion of the student as introspective observer upon his own thought processes seems to have after all only a thin connection with the intrusion of the macroscopic laboratory upon the submicroscopic quantum events being studied.

It was therefore surprising and revealing when it was found recently, almost by accident, that one of the roots of the modern complementarity point of view in Niels Bohr's own experience was probably just this wider, more humanistic context shown in the previous quotations. The discovery I speak of came about in a dramatic way. A few years ago, the American Physical Society and the American Philosophical Society engaged in a joint project to assemble the sources for the scholarly study of the history of quantum mechanics. This project, under the general directorship of Thomas S. Kuhn, spanned a number of years, and one of its functions was to obtain interviews with major figures on the origins of their contributions to quantum physics. An appointment for a number of interviews was granted by Niels Bohr, and the fifth interview was conducted on November 17, 1962, by Kuhn and Aage Petersen. In the course of the interview, Petersen, who was Niels Bohr's long-time assistant, raised the question of the relevance of the study of philosophy in Bohr's early thoughts. The following interchange occurred, according to the transcript:

AaP: How did you look upon the history of philosophy? What kind of contributions did you think people like Spinoza, Hume, and Kant had made?

NB: That is difficult to answer, but I felt that these various questions were treated in an irrelevant manner [in my studies].

AaP: Also Berkeley?

NB: No, I knew what views Berkeley had. I had seen a little in Høffding's writings, but it was not what one wanted.

TSK: Did you read the works of any of these philosophers?

NB: I read some, but that was an interest by [and here Bohr suddenly stopped and exclaimed]—oh, the whole thing is coming [back to me]! I was a close friend of Rubin [a fellow student, later psychologist], and, therefore, I read actually the work of William James. William James is really wonderful in the way he makes it clear—I think I read the book, or a paragraph, called . . . No, what is that called? it is called "The Stream of Thoughts," where he in a most clear manner shows that it is quite impossible to analyze things in terms of—I don't know what to call it, not atoms, I mean simply, if you have some things . . . they are so connected that if you try to separate them from each other, it just has nothing to do with the actual situation. I think that we shall really go into these things, and I know something about William James. That is coming first up now. And that was because I spoke to people about other things, and then Rubin advised me to read something of William James, and I thought he was most wonderful.

TSK: When was this that you read William James?

NB: That may be a little later, I don't know. I got so much to do, and it may be at the time I was working with surface tension [1905], or it may be just a little later. I don't know.

TSK: But it would be before Manchester [1912]?

NB: Oh yes, it was many years before.[27]

Niels Bohr clearly was interested in pursuing this further—"we shall really go into these things." But alas, the next day Bohr suddenly died.

There are enough leads to permit plausible speculations on this subject. K. T. Meyer-Abich reports in his interesting book, KORRESPONDENZ, INDIVIDUALITÄT UND KOMPLEMENTARITÄT (Wiesbaden, 1965) that among German scientists it was remembered that Bohr used to cite William James and only a few other western philosophers. Moreover, Niels Bohr himself, in an article in 1929[28] makes lengthy excursions into psychology in order to use analogies that, in Meyer-Abich's opinion, could well refer directly to William James's chapter on the "Stream of Thought" in James's book, THE PRINCIPLES OF PSYCHOLOGY (1890). On the other hand, doubts have been raised about the timing. Léon Rosenfeld[29] has expressed his strong belief that the work of William James was not known to Niels Bohr until about 1932. He recalls that in or about 1932, Bohr showed Rosenfeld a copy of James's PRINCIPLES OF PHYCHOLOGY. Rosenfeld believes that a few days earlier Bohr had had a conversation with Rubin, the psychologist and Bohr's former fellow student. Rubin may have sent the book to Bohr after their conversation. Bohr showed excited interest in the book, and especially pointed out to Rosenfeld the passages on the "stream of consciousness." During the next few days, Bohr shared the same excitement with several visitors, and Rosenfeld retained the definite impression that this was Bohr's first acquaintance with William James's work. In Rosenfeld's opinion, more relevant than speculation concerning an early influence of James was a remark made by Bohr: after discussing his "early philosophical meditations and his pioneering work of 1912–1913, he told me [Rosenfeld] in an unusually solemn tone of voice, 'and you must not forget that I was quite alone in working out these ideas, and had no help from anybody.' "[30]

In view of remarkable analogies or similarities between the ideas of

James and of Bohr, to be shown below, one can choose either to believe, with Meyer-Abich and Jammer, that Bohr had read James early enough to be directly influenced, or to believe with Rosenfeld that Bohr had independently arrived at the analogous thoughts (perhaps brought to them by other forces such as those we have already cited, or additional ones such as contemplation of the concepts of multiform function and Riemann surfaces).[31] In some ways the second alternative is the more interesting though difficult one, for it hints that here may be a place to attack the haunting old question why and by what mechanisms the same themata attain prominence in different fields in nearly the same periods. Still, no matter which view one chooses to take at this time, reading William James's chapter "Stream of Thought" in the light of Bohr's remark in the interview of November 1962 comes as a surprise to a physicist familiar with Bohr's contributions to atomic physics.[32]

James first insists that thought can exist only in association with a specific "owner" of the thought. Thought and thinker, subject and object, are tightly coupled. The objectivization of thought itself is impossible. Hence one must not neglect the circumstances under which thought becomes the subject of contemplation. "Our mental reaction to every given thing is really a resultant of our experience in the whole world up to that date. From one year to another we see things in new lights . . . . The young girls that brought an aura of infinity—at present hardly distinguishable existences; the pictures—so empty; and as for the books, what was there to find so mysteriously significant in Goethe?" One can here imagine the sympathetic response of Bohr, who wrote, "for objective description and harmonious comprehension it is necessary in almost every field of knowledge to pay attention to the circumstances under which evidence is obtained."

There is another sense in which consciousness cannot be concretized and atomized. James writes, "Consciousness does not appear to itself chopped up in bits; it flows. Let us call it the stream of thought, of consciousness, or of subjective life." Yet there does exist a discontinuous aspect: the "changes, from one moment to another, in the quality of the consciousness." If we use the vocabulary of quantum theory, James here proposes a sequence of individual changes between stationary states, with short periods of rest in these states—a metaphor that brings to mind Bohr's notion of 1912–1913 of the behavior of the electron in the hydrogen atom. To quote James, "Like a bird's life, [thought] seems to be made of an alternation of flights and perchings. The rhythm of language expresses this, where every thought is expressed in a sentence and every sentence closed by a period. . . . Let us call the resting places the 'substantive parts,' and the places of flight the 'transitive parts,' of the stream of thought."

But here enters a difficulty; in fact, the same one that plagued the student in Møller's story. The difficulty is, in James's words, "introspectively, to see the transitive parts for what they really are. If they are but flights to conclusions, stopping them to look at them before a conclusion is reached is really annihilating them." However, if one waits until one's consciousness is again in a stationary state, then the moment is over. James says, "Let anyone try to cut a thought across in the middle and get a look at its section, and he will see how difficult the introspective observation of the transitive tract is . . . . Or if our purpose is nimble enough and we do arrest it, it ceases forthwith to be itself . . . The attempt at introspective analysis in these cases is in fact like . . . trying to turn up the light quickly enough to see how the darkness looks." Letting thoughts flow, and making thoughts the subject of introspective analysis are, as it were, two mutually exclusive experimental situations.

It is from such a vantage point that one may attempt to interpret some of the novel features of Bohr's 1927 paper on complementarity to have been influenced either by a reading of James, or by thinking independently on parallel lines—and thereby understand better the final passage in Bohr's paper: "I hope, however, that the idea of complementarity is suited to characterize the situation, which bears a deep-going analogy to the general difficulty in the formation of human ideas, inherent in the distinction between subject and object."[33]

At this point, one might well ask where the term "complementarity" itself, which Bohr introduced into physics in 1927, may have come from. There are a number of fields from which the term may have been adapted, including geometry or topology. But both Meyer-Abich and Jammer point to a more provocative possibility, namely the chapter on *The Relations of Minds to Other Things*, in William James's PRINCIPLES OF PSYCHOLOGY (1890), just one chapter prior to that on *The Stream of Thought*. In the subsection " 'Unconsciousness' in Hysterics," James relates cases of hysterical anaesthesia (loss of the natural perception of sight, hearing, touch, and so on), and notes that P. Janet and A. Binet "have shown that during the times of anaesthesia, and coexisting with it, *sensibility to the anaesthetic parts is also there, in the form of a secondary consciousness* entirely cut off from the primary or normal one, but susceptible of being *tapped* and made to testify to its existence in various odd ways."[34]

The chief method for tapping was Janet's method of "distraction." If Janet put himself behind hysteric patients who were "plunged in conversation with a third party, and addressed them in a whisper telling them to raise their hand or perform other simple acts [including writing out answers to whispered questions] they would obey the order given, although their *talking* intelligence was quite unconscious of receiving it."[35] If interrogated in this way, hysterics responded perfectly normally when, for example, their sensibility to touch was examined on areas of skin that had been shown previously to be entirely anaesthetic when examined through their primary consciousness.

In addition, some hysterics could deal with certain sensations only in either one consciousness *or* the other, but not in both at the same time. Here James cites a famous experiment in a striking passage:

> M. Janet has proved this beautifully in his subject Lucie. The following experiment will serve as the type of the rest: In her trance he covered her lap with cards, each bearing a number. He then told her that on waking she should *not see* any card whose number was a multiple of three. This is the ordinary so-called "post-hypnotic suggestion," now well known, and for which Lucie was a well-adapted subject. Accordingly, when she was awakened and asked about the papers on her lap, she counted and said she saw those only whose number was not a multiple of 3. To the 12, 18, 9 etc., she was blind. But the *hand*, when the sub-conscious self was interrogated by the usual method of engrossing the upper self in another conversation, wrote that the only cards in Lucie's lap were those numbered 12, 18, 9, etc., and on being asked to pick up all the cards which were there, picked up these and let the others lie. Similarly when the sight of certain things was suggested to the subconscious Lucie, the normal Lucie suddenly became partially or totally blind. "What is the matter? I can't see!" the normal personage suddenly cried out in the midst of her conversation, when M. Janet whispered to the secondary personage to make use of her eyes.[36]

James gives these and other examples to support a conclusion in which he defines the concept of complementarity in psychological research:

> It must be admitted, therefore, that in *certain persons*, at least, *the total possible consciousness may be split into parts which coexist but mutually ignore*

each other, and share the objects of knowledge between them. More remarkable still, they are *complementary*. Give an object to one of the consciousnesses, and by that fact you remove it from the other or others. Barring a certain common fund of information, like the command of language, etc., what the upper self knows the under self is ignorant of, and *vice versa*.[37]

The analogy with Bohr's concept of complementarity in physics is striking, quite apart from the question of the genetic connection between these two uses of the same word.

Christian Bohr and Harald Høffding

Bohr's affinity for ideas analogous to those of William James was preceded by a philosophical and personal preparation that goes back to his childhood. In his essay, *Glimpses of Niels Bohr as a Scientist and Thinker*, Oskar Klein, one of Bohr's earliest collaborators, provides a revealing picture of the young man.

Niels Bohr himself and his brother Harald, a brilliant mathematician, liked to give examples of the innocently credulous—and at the same time resolute—way in which as a child he accepted what he saw and heard. They also spoke of geometrical intuition he developed so early . . . . The first feature appeared for instance in believing literally what he learned from the lessons on religion at school. For a long time this made the sensitive boy unhappy on account of his parents' lack of faith. When later, as a young man, he began to doubt, he did so also with unusual resolution and thereby developed a deep philosophical bent similar to that which seems to have characterized the early Greek natural philosophers.[38]

Christian Bohr, Niels Bohr's father, was professor of physiology at the University of Copenhagen. His work involved him in one of the important philosophical debates of the last part of the nineteenth century, the differences between and relative merits of the "vitalistic" theories and the mechanistic conceptions of life processes. In several ways, Christian Bohr's interests shaped his son's ideas and preoccupations. We know that as a youth, Niels Bohr was allowed to work in the laboratory of his father, and to meet the scholars interested in philosophy with whom Christian Bohr kept close contact, such as Harald Høffding, professor of philosophy at the University of Copenhagen. Høffding often visited the Bohr household, and Niels Bohr attested to the profound influence he received from early childhood by being permitted to stay and listen during meetings of an informal club made up of his father, Høffding, the physicist Christian Christiansen, and the philologist Hans Thomsen. Høffding, in turn, described Christian Bohr as a scientist who recognized "strict application of physical and chemical methods of physiology" in the laboratory, but who, outside the laboratory, "was a keen worshipper of Goethe. When he spoke of practical situations or of views of life, he liked to do so in a dialectic manner."[39]

We may understand the implications of this description best through Oskar Klein, who remembers a characterization which Niels Bohr gave him: "He mentioned his father's idea that teleology, when we want to describe the behavior of living beings, may be a point of view on a par with that of causality. The idea was later to play an essential role in Bohr's attempt to throw light on the relation between the biologist's and the physicist's way of describing nature."[40]

Niels Bohr entered the university in 1903, and soon took Høffding's course in the history of philosophy and logic. He also belonged to a student's club in which the questions raised in Høffding's lectures on philosophy were discussed. (Another member was Rubin.) While Bohr, as indicated in his last interview, felt no great attraction to philosophical systems (such as those of "Spinoza, Hume, and Kant"), there is little doubt about the lasting impression Høffding made on Bohr—perhaps most of all because of Høffding's active interest in the applicability to philosophy of the work of what he called *philosophierende Naturforscher*, from Copernicus to Newton and from Maxwell to Mach. For example, the latter two are discussed at some length in Høffding's MODERNE PHILOSOPHEN which appeared in 1904 in Danish (1905 in German) as successor to his monumental HISTORY OF MODERN PHILOSOPHY.

There also appears to have been a personal sympathy between the older and the younger man. While still Høffding's student, Bohr pointed out some error in Høffding's exposition, and Høffding, in turn, allowed Bohr to help him correct proofs of the offending passage. A warm friendship developed eventually that was freely acknowledged on both sides, as indicated, for example, by Niels Bohr's acknowledgment of Harald Høffding's influence on him, on the occasion of Høffding's eighty-fifth birthday,[41] and conversely in letters of Høffding to Emile Myerson in 1926 and 1928.[42] The first of these letters, incidentally, is dated December 13, 1926, shortly before Bohr's vacation trip to Norway in early 1927 during which, according to Heisenberg and others, Bohr's ideas on complementarity were developed in the form he announced later in 1927. Another letter was written half a year after the presentation of the complementarity principle at Como. In it, Høffding writes to Meyerson (March 13, 1928): "Bohr declares that he has found in my books ideas which have helped the scientists in the 'understanding' of their work, and thereby they have been of real help. This is great satisfaction for me, who feels so often the insufficience of my special preparation with respect to the natural sciences."[43]

Among all the philosophers and scientists discussed by Høffding, it is unlikely that any interested student of Høffding's will have failed to encounter some aspects of William James's work. An admirer, like James, of G. T. Fechner (the father of psychophysics), Høffding devoted his first book to psychology (Danish edition, 1882). At about the time Bohr took his philosophy course, Høffding used the occasion of the St. Louis meeting of 1904 to visit James in the United States. James, in turn, supplied an appreciative preface for the English translation (of 1905) of Høffding's PROBLEMS OF PHILOSOPHY—a book which Høffding reported later to have originated in his university lectures in 1902.[44] And in the same year of Høffding's visit to James, Høffding expressed in his MODERNE PHILOSOPHEN his admiration for James's work, to whom the concluding chapter is devoted, with such comments as "James belongs to the most outstanding contemporary thinkers .... The most important of his writings is THE PRINCIPLES OF PSYCHOLOGY."

Kierkegaard

In Høffding's own life, a crucial and early influence was the work of Kierkegaard, as he freely confessed.[45] Høffding reported that in a youthful crisis, in which he was near "despair," he had found solace and new strength through Kierkegaard's writings, and he mentions particularly Kierkegaard's work now known as STAGES ON LIFE'S WAY. Høffding became known as one of the prominent exponents and followers of Kierkegaard; indeed, the second major work Høffding published was the book, KIERKEGAARD ALS PHILOSOPH.[46]

Whether Niels Bohr caught some of his own interest in Kierkegaard while a student of Høffding is not known, but the fact of this early interest is well documented. Thus it is remembered that in 1909 Niels sent his brother Harald as a birthday gift Kierkegaard's book, STAGES ON LIFE'S WAY, with a letter saying, "It is the only thing I have to send; but I do not believe that it would be very easy to find anything better. In any case I have had very much pleasure in reading it, I even think that it is one of the most delightful things I have ever read."[47] Then he added that he did not fully agree with all of Kierkegaard's views. One can well imagine that Niels Bohr could enjoy the aesthetic experience and the moral passion, without having to agree also with the antiscientific attitude of much of the work.

Bohr's remarks about Kierkegaard bring us to the last of the various possible avenues that prepared for the complementarity notion. While this is not the proper place for a searching examination of those elements in Kierkegaard's works for which analogous elements have been noted in Bohr's work,[48] it will be of interest to remind ourselves of one or two chief features that characterized the writing of both Kierkegaard and his chief interpreter in Denmark, Høffding.

Kierkegaard's existentialism was rooted in German Romanticism, upholding the individual and the momentary life situation in which he finds himself against the rationality and objective abstraction championed by eighteenth-century Enlightenment. The denial of the subjective, Kierkegaard argued, leads to self-contradictions, for even the most abstract proposition remains the creation of human beings. In a reaction to Hegel and to some aspects of Kant, Kierkegaard wrote about science in his journal: "Let it deal with plants and animals and stars, but to deal with the human spirit in that way is blasphemy, which only weakens ethical and religious passions." Truth cannot be found without incorporating the subjective, particularly in the essentially irrational, discontinuous stages of recognitions leading to the achievement of insight. As John Passmore writes, "each major step on the way to truth is a free decision. Our progress, according to Kierkegaard, from the aesthetic to the scientific point of view, and then again from the scientific to the ethical and from the ethical to the religious, cannot be rationalized into an orderly, formally justifiable, step from premise to conclusions: It is in each case a leap to a quite new way of looking at things."[49]

What is perhaps of greatest interest to us is the accentuation of the role of discontinuity in Kierkegaard's work. Here we can do no better than cite at some length the section on Kierkegaard in Høffding's own chief work, A HISTORY OF MODERN PHILOSOPHY:

[Kierkegaard's] leading idea was that the different possible conceptions of life are so sharply opposed to one another that we must make a choice between them, hence his catchword *either—or*; moreover, it must be a choice which each particular person must make for himself, hence his second catchword, *the individual.* He himself designated his thought "qualitative dialectic," by which he meant to bring out its opposition to the doctrine taught by Romantic speculation of continuous development by means of necessary inner transitions. Kierkegaard regarded this doctrine as pure fantasticalness—a fantasticalness, to be sure, to which he himself had felt attracted.[50]

What is essential for us to notice is that a main feature of Kierkegaard's "qualitative dialectic" is an acceptance of thesis and antithesis, *without* proceeding to another stage at which the tension is resolved in a synthesis. Thus he draws a line between thought and reality which must not be allowed to disappear. Høffding writes: "Even if thought should attain coherency it does not therefore follow that this coherency can be preserved in the practice of life. . . . Such great differences and oppositions exist side by side that there is no thought which can embrace them all in a 'higher unity.' "[51] "Kierkegaard came more and more to regard the capability of embracing great contrasts and of enduring the suffering which this involves as the criterion of the sublimity and value of a conception of life."[52]

Kierkegaard's stress on discontinuity between incompatibles, on the "leap" rather than the gradual transition, on the inclusion of the individual, and on inherent dichotomy was as "nonclassical" in philosophy as the elements of the Copenhagen doctrine—quantum jumps, probabilistic causality, observer-dependent description, and duality—were to be in physics.

Now it would be as absurd as it is unnecessary to try to demonstrate that Kierkegaard's conceptions were directly and in detail translated by Bohr from their theological and philosophical context to a physical context. Of course, they were not. All one should do is permit oneself the open-minded experience of reading Høffding and Kierkegaard through the eyes of a person who is primarily a physicist—struggling, as Bohr was, first with his 1912–1913 work on atomic models, and again in 1927, to "discover a certain coherence in the new ideas" while pondering the conflicting, paradoxical, unresolvable demands of classical physics and quantum physics which were the near-despair of most physicists of the time. It is in this frame of mind that one can best appreciate, for example, Høffding's discussion of Kierkegaard's indeterministic notion of the "leap":

> In Kierkegaard's ethics the qualitative dialectic appears partly in his conception of choice, of the decision of the will, partly in his doctrine of stages. He emphatically denies that there is any analogy between spiritual and organic development. No gradual development takes place within the spiritual sphere, such as might explain the transition from deliberation to decision, or from one conception of life (or "stadium") to another. Continuity would be broken in every such transition. As regards the choice, psychology is only able to point out possibilities and approximations, motives and preparations. The choice itself comes with a jerk, with a leap, in which something quite new (a new quality) is posited. Only in the world of possibilities is there continuity; in the world of reality decision always comes through a breach of continuity. But, it might be asked, cannot this jerk or this leap itself be made an object of psychological observation? Kierkegaard's answer is not clear. He explains that the leap takes place between two moments, between two states, one of which is the last state in the world of possibilities, the other the first state in the world of reality. It would almost seem to follow from this that the leap itself cannot be observed. But then it would also follow that it takes place unconsciously—and the possibility of the unconscious continuity underlying conscious antithesis is not excluded.[53]

It is at this point that the writings of Høffding and Kierkegaard most evidently overlap with the teachings of William James. In fact, there are two specific periods where the overlapping conceptions of Kierkegaard, Høffding, and James can plausibly have been influential for Bohr in the sense of providing sympathetic preparation or support: one came in Bohr's work during the early period, from 1912 through the correspondence point of view (that is, in the analogy between Bohr's nonclassical transitions of the electron between stationary states on one hand, and Kierkegaard's "leaps" or James's transient flights and "transitive parts" on the other hand). The other came in the period from about 1926, when Bohr's complementarity point of view was being developed; and here we have already pointed to possible sources or antecedents for Bohr's analogies in passages such as the conclusion of his September 1927 address ("the idea of complementarity is suited to characterize the situation, which bears a profound analogy to the general difficulty in the formation of human ideas, inherent in the distinction between subject and object"), as well as passages in a paper of 1929 ("Strictly speaking, the conscious analysis of any concept stands in a relation of exclusion to its immediate application"; "The necessity of taking recourse to a complementary, or reciprocal mode of description is perhaps familiar to us from psychological problems"; "In particular, the apparent contrast between the continuous onward flow of associative thinking and the preservation of the unity of the personality exhibits a suggestive analogy with the relation between the wave description of the motions of material particles, governed by the superposition principle, and their indestructible individuality").[54]

One characteristic trait of Bohr should not be overlooked in this discussion, for without it the necessary predisposition for reaching the complementarity point of view would have been missing. I refer to Bohr's well-known dialectic style of thinking and of working. One of those who worked with him longest, Léon Rosenfeld, attests that Bohr's "turn of mind was essentially dialectical, rather than reflective. . . . He needed the stimulus of some form of dialogue to start off his thinking."[55] Rosenfeld also records a well-known dictum of Bohr: "Every sentence I say must be understood not as an affirmation, but as a question." Bohr's habit of work was frequently to develop a paper during dictation, walking up and down the room and arguing both with himself and a fellow physicist whom he had persuaded to be his sounding-board, transcriber, and critic—and whom he was likely to leave in an exhausted state at the end. As Einstein, Heisenberg, Schrödinger, and many others had to experience, it seemed as if Bohr looked for and fastened with greatest energy on a contradiction, heating it to its utmost before he could crystallize the pure metal out of the dispute. Bohr's method of argument shared with the complementarity principle itself the ability to exploit the clash between antithetical positions. We have given earlier only the first line of a couplet from Schiller, reported to have been one of Bohr's favorite sayings: after the line "Only wholeness leads to clarity" there follows "And truth lies in the abyss":

Nur die Fülle führt zur Klarheit,
Und im Abgrund wohnt die Wahrheit.

Of Niels Bohr stories there are legions, but none more illuminating than that told by his son Hans concerning the fundamentally dialectic definition of truth. Hans reports that one of the favorite maxims of his father was the distinction between two sorts of truth: trivialities, where opposites are obviously absurd, and profound truths, recognized by the fact that the opposite is also a profound truth.[56] Along the same line, there has been a persistent story that Bohr had been impressed by an example or analogue for the complementarity concept in the mutually exclusive demands of justice and of love. Jerome S. Bruner has kindly given me a first-hand report of a conversation on this point that took place when he happened to meet Niels Bohr in 1943 or early 1944 for the first time. "The talk turned entirely on the complementarity between affect and thought, and between perception and reflection. [Bohr] told me that he had become aware of the psychological depths of the concept of complementarity when one of his children had done something inexcusable for which he found himself incapable of appropriate punishment: 'You cannot know somebody at the same time in the light of love and in the light of justice!' I think that those were almost exactly the words he used. He also . . . talked about the manner in which introspection as an act dispelled the very emotion that one strove to describe."[57]

Complementarity Beyond Physics

We can now ask: what was Bohr's real ambition for the complementarity conception? It certainly went far beyond dealing with the paradoxes in the physics of the 1920's. Not only were some of the roots of the complementarity principle outside physics, but so also was its intended range of application. Let me remind you of Bohr's statement: "The integrity of living organisms and the characteristics of conscious individuals and human cultures present features of wholeness, the account of which implies a typically complementary mode of description . . . . We are not dealing with more or less vague analogies, but with clear examples of logical relations which, in different contexts, are met with in wider fields."[58] The complementarity principle is a manifestation of a thema in a sense which I have previously developed[59]—one thema in the relatively small pool of themata from which the imagination draws for all fields of endeavor. When we devote attention to a particular thema in physics or some other science, whether it be complementarity, or atomism, or continuity, we must not forget that each special statement of the thema is an aspect of a general conception which, in the work of a physicist or biologist or other scientist, is exemplified merely in a specific form. Thus a general thema, θ, would take on a specific form in physics that might be symbolized by θ_{ϕ}, in psychological investigation by θ_ψ, in folklore by θ_μ, and so on. The general thema of discontinuity or discreteness thus appears in physics as the θ_ϕ of atomism, whereas in psychological studies it appears as the thema θ_ψ of individualized identity. One may express a given θ as the sum of its specific exemplifications, as symbolized (without straining for precision) by the expression:

$$\theta = \sum_{n=\alpha}^{n=\omega} \theta_n$$

From this point of view we realize that Bohr's proposal of the complementarity principle was nothing less than an attempt to make it the cornerstone of a new epistemology. When "in general philosophical perspective . . . we are confronted with situations reminding us of the situation in quantum physics,"[60] it is not that those situations are in some way pale reflections or "vague analogies" of a principle that is basic only in quantum physics; rather, the situation in quantum physics is only one reflection of an all-pervasive principle. Whatever the most prominent factors were which contributed to Bohr's formulation of the complementarity point of view in physics—whether his physical research or thoughts on psychology, or reading in philosophical problems, or controversy between rival schools in biology, or the complementary demands of love and justice in everyday dealings—it was the *universal* significance of the role of complementarity which Bohr came to emphasize.

Moreover, this universality explains how it was possible for Bohr to gain insight for his work in physics from considerations of complementary situations in other fields. For as Léon Rosenfeld accurately remarks, "As his insight into the role of complementarity in physics deepened in the course of these creative years, he was able to point to situations in psychology and biology that also present complementary aspects; and the considerations of such analogies in epistemological respect in its turn threw light on the unfamiliar physical problems."[61] "Bohr devoted a considerable amount of hard work to exploring the possibilities of application of complementarity to other domains of knowledge; he attached no less importance to this task than to his purely physical investigations, and he derived no less satisfaction from its accomplishment."[62]

During the last thirty years of his life, Bohr took many opportunities to consider the application of the complementarity concept in fields outside of physics. Rosenfeld reports that the first important opportunity of this kind offered itself when Bohr was invited to address a biological congress in Copenhagen in 1932.[63] Starting from the idea of complementarity as used for understanding the dual aspects of light, Bohr then proceeded to point to the application of complementarity relations in biology. Rosenfeld's account of the talk is worth citing in detail:

> This had a special appeal to him: he had been deeply impressed by his father's views on the subject, and he was visibly happy at being now able to take them up and give them a more adequate formulation. [His father], in the work of the reaction against mechanistic materialism at the beginning of the century, had put up a vigorous advocacy of the teleological point of view in the study of physiology: without the previous knowledge of the function of an organ, he argued, there is no hope of unravelling its structure or the physiological processes of which it is the seat. At the same time, he stressed, with all the authority of a life devoted to the analysis of the physical and chemical aspects of such processes, the equally imperious necessity of pushing this analysis to the extreme limit which the technical means of investigation would permit us to reach . . . .
>
> Such reflections came as near as one would expect at the time to establishing a relation of complementarity between the physico-chemical side of the vital processes, governed by the kind of causality we are accustomed to herald as the truly scientific one, and the properly functional aspect of these processes, dominated by teleological or finalistic causality. In the past, the two points of view, under varying forms, have always been put in sharp opposition to each other, the general opinion being that one of them had to prevail to the exclusion of the other, that there was no room for both in the science of life. Niels Bohr could now point out that this last belief was only the result of a conception of logic which the physicists had recognized as too narrow, and that the wider frame of complementarity seemed particularly well-suited to accommodate the two standpoints, and make it possible without any contradiction to take advantage of both of them, quite in the spirit of his father's ideas. Thus, an age-long sterile conflict would be eliminated and replaced by a full utilization of all the resources of scientific analysis.[64]

One need not be tempted into imagining Bohr in a Hamletlike striving to establish his father's ideas; but one also need not remain untouched by the closing of the circle. For surely one of the paths leading to complementarity had opened while Niels Bohr was in his father's laboratory and shop club.

In the years following the Congress of 1932 Bohr took his point of view before an even wider audience; in addition to his written and spoken contributions before physical scientists, he presented himself at such meetings as the Second International Congress for the Unity of Science in Copenhagen (June 1936) in a discussion on "Causality and Complementarity"; the International Congress for Physics and Biology (October 1937) on "Biology and Atomic Physics"; the International Congress for Anthropology and Ethnology in Copenhagen (1938) on "Natural Philosophy and Human Cultures"; and on many later occasions of a similar sort.[65]

In each of these lectures Bohr provided a new set of illustrations of the common theme. Thus in his address before the anthropologists in 1938, on the eve of World War II, Bohr stressed complementary features of human societies. He also returned to the problem posed by the student (*licentiate*) in Møller's story. As Rosenfeld writes:

> He could now look back at the duality of aspects of psychical experience with all the mastery he had acquired over the nature of complementarity relations, and point out that this duality corresponded to different ways of drawing a separation between the psychical process which was chosen as the object of observation and the observing subject: drawing such a separation is precisely what we mean when we speak of fixing our attention on a definite aspect of the process; according as we draw the line, we may experience an emotion as part of our subjective feeling, or analyze it as part of the observed process. The realization that these two situations are complementary solves the riddle of the licentiate's egos observing each other, and is in fact the only salvation from his qualms.[66]

Speaking before the Congress of the Fondation Européenne de la Culture in Copenhagen on October 21, 1960, in an address entitled "The Unity of Human Knowledge," Bohr returned again to the need to search, within the great diversity of cultural developments, "for those features in all civilizations which have their roots in the common human situation." He developed these ideas in sociological and political context, particularly since he was increasingly more preoccupied with helping to "promote mutual understanding between nations with very different cultural backgrounds."[67] Deeply concerned about the dangers of the Cold War, Bohr spent a good part of his later years on political and social questions, including work on plans for peaceful uses of nuclear energy and for arms control. In these and other articles on this topic, one can discern Bohr's dissatisfaction with his own state of understanding; the problems posed by national antagonisms did not seem to be fully understandable in the same terms that had seemed to him successful in physics and psychology. As he confessed at the end of his lecture before the Royal Danish Academy of Sciences in 1955, "The fact that human cultures, developed under different conditions of living, exhibit such contrasts with respect to established traditions and social patterns allows one, in a certain sense, to call such cultures complementary. However, we are here in no way dealing with definite, mutually exclusive features, such as those we meet in the objective description of general problems of physics and psychology, but the differences in attitude which can be appreciated or ameliorated by an expanded intercourse between peoples."[68]

Bohr returned to the same theme repeatedly. For example, in the essay quoted earlier, *The Unity of Human Knowledge*, Bohr reexamined the requirement that even the most abstract principles of quantum physics, for example, must be capable of being rendered in commonsense, classical language. "The aim of our argumentation," Bohr wrote, "is to emphasize that all experience, whether in science, philosophy, or art, which may be helpful to mankind, must be capable of being communicated by human means of expression, and it is on this basis that we shall approach the question of unity of knowledge."[69]

The last phrase, used in the title of the essay, suddenly puts into perspective for us that Bohr's manifold and largely successful ambitions place him in the tradition typified by another "philosophizing scientist," one who belonged to the generation before Bohr—a man whom Bohr, like many others, had read early, and whose views Høffding had described in a sympathetic way in his MODERNE PHILOSOPHEN and in PROBLEMS OF PHILOSOPHY. It is Ernst Mach.

Bohr seems to have mapped out for himself the same grand, interdisciplinary task—in his forceful and innovative influence on physics and on epistemology, in his deep interest in the sciences far beyond physics itself, even in his active and liberal views on social-political questions. And as physicist, physiologist, psychologist, and philosopher, Ernst Mach had also wanted to find a principal point of view from which research in any field could be more meaningfully pursued. This point of view Mach thought to have found by going back to that which is given before all scientific research, namely the world of sensations. On this basis,

Mach had established himself as the patriarch of the Unity of Science movement. In his turn, Niels Bohr, starting from the profound reexamination of the problem of sensation and particularly of object-subject interaction, also hoped he had found (in the complementarity point of view) a new platform from which to evaluate and solve the basic problems in a variety of fields, whether in physics, psychology, physiology, or philosophy.

Bohr's achievement, from 1927 on, of attaining such a principal point of view was not an accidental development. On the contrary, it was the fulfillment of an early ambition. A biographer of Bohr records that "as a young student, fired with the ideas Høffding was opening to him, Bohr had dreamed of 'great inter-relationships' between all areas of knowledge. He had even considered writing a book on the theory of knowledge. . . . But physics had drawn him irresistibly."[70] In the end, Bohr's attempt to understand the unity of knowledge (a topic on which he wrote nearly two dozen papers) on the basis of complementarity could be seen as precisely the fulfillment of the desire to discover the "great inter-relationships among all areas of knowledge."

Bohr's aim has a grandeur which one must admire. But while his point of view is accepted by the large majority in physics itself, it would not be accurate to say that it is being widely understood and used in other fields; still less has it swept over philosophy the way Mach's views did among the generation of scientists brought up before the theory of relativity and quantum mechanics. Even those who in their professional work in physics have experienced the success of the complementarity point of view at first hand find it hard or uncongenial to transfer to other areas of thought and action, as a fundamental thematic attitude, the habit of accepting basic dualities without straining for their mutual dissolution or reduction. Indeed, we tend to be first of all reductionists, perhaps partly because our early intellectual heroes have been men in the tradition of Mach and Freud, rather than Kierkegaard and James.

Perhaps, also, it is just a matter of time—more time needed to assimilate a new thema widely enough; to sort out the merely seductive and the solid applications; and to learn to perceive the kind of grandeur in the scope of the new notion which Robert Oppenheimer delineated:

> An understanding of the complementary nature of conscious life and its physical interpretation appears to me a lasting element in human understanding

and a proper formulation of the historic views called psychophysical parallelism.

For within conscious life, and in its relations with the description of the physical world, there are again many examples. There is the relation between the cognitive and the affective sides of our lives, between knowledge or analysis, and emotion or feeling. There is the relation between the aesthetic and the heroic, between feeling and that precursor and definer of action, the ethical commitment; there is the classical relation between the analysis of one's self, the determination of one's motives and purposes, and that freedom of choice, that freedom of decision and action, which are complementary to it....

To be touched with awe, or humor, to be moved by beauty, to make a commitment or a determination, to understand some truth—these are complementary modes of the human spirit. All of them are part of man's spiritual life. None can replace the others, and where one is called for, the others are in abeyance....

The wealth and variety of physics itself, the greater wealth and variety of the natural sciences taken as a whole, the more familiar, yet still strange and far wider wealth of the life of the human spirit, enriched by complementary, not at once compatible ways, irreducible one to the other, have a greater harmony. They are the elements of man's sorrow and his splendor, his frailty and his power, his death, his passing, and his undying deeds.[71]

NOTES

1. After much further work, Bohr published the lecture in 1928 under the title *The Quantum Postulate and the Recent Development of Atomic Theory*; it has been reprinted in several places, for example, as one of the four essays in the collection by Niels Bohr, ATOMTHEORIE UND NATURBESCHREIBUNG, Berlin: Springer Verlag, 1931; also published as ATOMIC THEORY AND THE DESCRIPTION OF NATURE, Cambridge: University Press; New York: Macmillan Co., 1934.

2. Bohr, *The Quantum Postulate*, ATOMIC THEORY AND THE DESCRIPTION OF NATURE, p. 55.

3. Albert Einstein, *Das Comptonsche Experiment,* BERLINER TAGEBLATT (April 20, 1924), supplement, p. 1; cited by M. J. Klein, *A Twentieth-century Challenge to Energy Conservation* (forthcoming).

4. Bohr, *The Quantum Postulate*, ATOMIC THEORY AND THE DESCRIPTION OF NATURE, pp. 54–55. Bohr introduced the need for working out a "complementarity theory" in the following, rather overburdened sentence: "The

very nature of the quantum theory thus forces us to regard the space-time co-ordination and the claim of causality, the union of which characterizes the classical theory, as complementary but exclusive features of the description, symbolizing the idealization of observation and definition respectively." Max Jammer, to whose book THE CONCEPTUAL DEVELOPMENT OF QUANTUM MECHANICS (New York: McGraw-Hill, 1966, pp. 351–352) we shall frequently refer, adds: "This statement, in which the term 'complementary' appeared for the first time and in which spatiotemporal description is referred to as complementary to causal description, contained the essence of what later became known as the 'Copenhagen interpretation' of quantum mechanics."

Heisenberg's uncertainty principle, formulated early in 1927, had given a first indication of complementary relations between physical concepts, though in a restricted sense. The uncertainty principle tells us that if we attempt to localize a particle in space (or time), we shall, during the measurement process, impart to the particle momentum (or energy) within a range of values that increases as we decrease the size of the space-time region on which we wish to focus attention. Position and momentum are not mutually exclusive notions since both are needed to specify the state of a system and both can be measured in the same experiment. But they are complementary in the restricted sense that they cannot both at the same time be ascertained with arbitrarily high precision; that is, the more precision is obtained in one measurement, the less it is possible to have in the other. In contrast, the wave-particle aspects of matter are complementary *and* mutually exclusive; an atomic entity cannot exhibit both its particle and its wave properties simultaneously. It is for this reason that textbooks often say that Bohr's statement of complementarity at Como transcended the Heisenberg uncertainty principle.

5. Niels Bohr, *Discussion with Einstein on Epistemological Problems in Atomic Physics*, in P. A. Schilpp, ed., ALBERT EINSTEIN: PHILOSOPHER-SCIENTIST (Evanston, Ill.: The Library of Living Philosophers, 1949), pp. 209–210; italics in original.

6. Jammer, *op. cit.*, p. 354.

7. *Ibid.*, p. 358 (Footnote 128, translation by G. H.).

8. Albert Einstein, *Remarks Concerning the Essays Brought Together in this Co-operative Volume*, in Schilpp, *op. cit.*, p. 674.

9. Bohr, *Discussion with Einstein on Epistemological Problems in Atomic Physics*, in Schilpp, *op. cit.*, p. 218.

10. For some aspects of the early history of the theories of light, see the interesting book by Vasco Ronchi, OPTICS, THE SCIENCE OF VISION, trans. Edward Rosen, New York: New York University Press, 1957; or Johann Müller, LEHRBUCH DER PHYSIK, Braunschweig: Friedrich Vieweg & Son, 1926. I have relied on both extensively.

11. For example, see W. H. Bragg, *The Consequences of the Corpuscular Hypothesis of the γ and X Rays and the Range of β Rays*, PHILOSOPHICAL MAGAZINE AND JOURNAL OF SCIENCE, 20, No. 117:385–416, September, 1910.

12. Merle A. Tuve, *Physics and the Humanities — The Verification of Complementarity*, THE SEARCH FOR UNDERSTANDING, ed. Caryl P. Haskins (Washington, D.C.: Carnegie Institution, 1967), p. 46.

13. Jammer, *op. cit.*, p. 87.

14. Quoted, with permission, from a letter of Niels Bohr in the American Philosophical Society Library, Philadelphia. I thank Dorothy M. Livingston for having drawn this letter to my attention.

15. Niels Bohr, H. A. Kramers, and J. C. Slater, *The Quantum Theory of Radiation,* PHILOSOPHICAL MAGAZINE AND JOURNAL OF SCIENCE, 47, No. 281: 785–802, May, 1924. The German version is *Über die Quantentheorie der Strahlung*, ZEITSCHRIFT FÜR PHYSIK, 24:69–87, June, 1924.

16. Erwin Schrödinger, *Quantisierung als Eigenwertproblem*, ANNALEN DER PHYSIK, 79:375, July, 1926.

17. Jammer, *op. cit.*, pp. 271–272; italics in original.

18. *Ibid.*, p. 271.

19. As quoted, *ibid.*, p. 272.

20. Gerald Holton, *On the Thematic Analysis of Science: The Case of Poincaré and Relativity*, in MÉLANGES ALEXANDRE KOYRÉ (Paris: Hermann, 1964), Volume II, pp. 257–268, and reprinted here as Essay 6; *Presupposition in the Construction of Theories*, THE GRADUATE JOURNAL, 7, No. 1:87–109, 1965–66, and reprinted here as Essay 1; and *Science and New Styles of Thought*, THE GRADUATE JOURNAL, 7, No. 2:339–442, 1967, and reprinted here as Essay 3.

21. Oskar Klein, *Glimpses of Niels Bohr as Scientist and Thinker*, in NIELS BOHR: HIS LIFE AND WORK AS SEEN BY HIS FRIENDS AND COLLEAGUES, ed. Stefan Rozental (New York: John Wiley & Sons, 1967), p. 93.

22. Niels Bohr, *Quantum Physics and Philosophy*, ESSAYS 1958–1962 ON ATOMIC PHYSICS AND HUMAN KNOWLEDGE (New York: Interscience Publishers, 1963), p. 7: italics supplied.

23. Léon Rosenfeld, *Niels Bohr in the Thirties: Consolidation and Extension of the Conception of Complementarity*, in Rozental, ed., *op. cit.*, p. 121.

24. Cited in Bohr, *The Unity of Human Knowledge*, ESSAYS 1958–1962 ON ATOMIC PHYSICS AND HUMAN KNOWLEDGE, p. 13, and Léon Rosenfeld, *Niels Bohr's Contribution to Epistemology*, PHYSICS TODAY, 16, No. 10:47–54, October, 1963. In this article and elsewhere, Rosenfeld has insisted on the importance of the story for Bohr; moreover, Rosenfeld believes that the struggle of the student with his many egos was "the only object lesson in dialectical

thinking that Bohr ever received and the only link between his highly original reflection and philosophical tradition" (p. 48).

25. Bohr, *The Unity of Human Knowledge*, Essays 1958–1962, p. 12.

26. *Ibid.*, p. 14.

27. The permission granted by the estate of Niels Bohr and by the American Philosophical Society to reproduce this section of the interview is gratefully acknowledged.

28. Bohr, *The Quantum of Action and the Description of Nature*, Atomic Theory and the Description of Nature, pp. 92–101.

29. Letter to the author, February 28, 1968.

30. *Ibid.* In an interview conducted with Werner Heisenberg by T. S. Kuhn for the History of Quantum Physics project on February 11, 1963, Heisenberg volunteered that James was one of Bohr's favorite philosophers; the chapter on the "stream of thought" seemed to have made a profound impression on Bohr. Heisenberg placed these discussions somewhere between 1926 and 1929, most probably around 1927. When told of doubts about the timing, Heisenberg responded that he could not "guarantee" that these discussions with Bohr had not been after 1932.

31. Rosenfeld, *Niels Bohr's Contribution to Epistemology*, p. 49. See also Léon Rosenfeld, Niels Bohr (Amsterdam: North-Holland Publishing Co., 1945, 1961), pp.12–13.

32. We follow here the sequence given in K. M. Meyer-Abich, Korrespondenz, Individualität, und Komplimentarität: Eine Studie zur Geistesgeschichte der Quantentheorie in den Beiträgen Niels Bohrs (Wiesbaden: Franz Steiner Verlag, 1965), p. 133 ff.

33. Bohr, *The Quantum Postulate and the Recent Development of Atomic Theory*, Atomic Theory and the Description of Nature, p. 91.

34. William James, The Principles of Psychology (New York: Dover Publications, 1950), Volume I, p. 203; italics in original in all passages quoted from Principles of Psychology.

35. *Ibid.*, p. 204.

36. *Ibid.*, pp. 206–207.

37. *Ibid.*, p. 206.

38. Klein, *Glimpses of Niels Bohr*, in Rozental, ed., *op. cit.*, p. 74. One notices here the remarkable similarity of Bohr's experience with that recorded in Einstein's *Autobiographical Notes*—the same early religious acceptance in contrast to his parents' beliefs, followed by a loss or rejection of "the religious paradise of youth," as Einstein called it.

39. *Childhood and Youth*, in Rozental, ed., *op. cit.*, p. 13.

40. Klein, *Glimpses of Niels Bohr*, in Rozental, ed., p. 76.

41. See Jammer, *op. cit.*, p. 349.

42. *Ibid.*, pp. 347, 349.

43. Ruth Moore, in her book, NIELS BOHR: THE MAN, HIS SCIENCE, AND THE WORLD THEY CHANGED (New York: Alfred A. Knopf, 1966), p. 432, records that on one wall of Bohr's house in Carlsberg, there "were portraits of those nearest to Bohr, grouped reverentially together": Bohr's father and mother, his brother Harald, his grandfather Adler, and "Bohr's teacher Høffding. If any doubts existed of Høffding's influence on Bohr's life, it was settled by the placement of his portrait."

44. Harald Høffding in Raymond Schmidt, ed., DIE PHILOSOPHIE DER GEGENWART IN SELBSTDARSTELLUNGEN (Leipzig: Felix Meiner, 1922), p. 86.

45. For example, *ibid.*, p. 75.

46. Danish edition, 1892; German edition, 1896.

47. *Childhood and Youth*, in Rozental, *op. cit.*, p. 27. See also the account of J. Rud Nielsen, *Memories of Niels Bohr*, PHYSICS TODAY, 16, No. 10: 27–28, October, 1963. Referring to a visit from Bohr in 1933, Nielsen wrote: "Knowing Bohr's interest in Kierkegaard, I mentioned to him the translation made by Prof. Hollander of the University of Texas, and Bohr began to talk about Kierkegaard: 'He made a powerful impression upon me when I wrote my dissertation in a parsonage in Funen, and I read his works night and day,' he told me. 'His honesty and willingness to think the problems through to their very limit is what is great. And his language is wonderful, often sublime. There is, of course, much in Kierkegaard that I cannot accept. I ascribe that to the time in which he lived. But I admire his intensity and perseverance, his analysis to the utmost limit, and the fact that through these qualities he turned misfortune and suffering into something good.' "

48. A preliminary treatment of the subject has been made in the section *The Philosophical Background of Nonclassical Interpretations*, Jammer, *op. cit.*, pp. 166–180.

49. John A. Passmore, A HUNDRED YEARS OF PHILOSOPHY, rev. ed. (London: Gerald Duckworth, 1957), pp. 462–463.

50. Harald Høffding, A HISTORY OF MODERN PHILOSOPHY, ed. and trans. B. E. Meyer (New York: Dover Publications, 1955), Volume II, p. 286. The work was originally issued in 1893 and intended to cover the ground to 1880. The English translation was published in 1900. Høffding also explored the role of discontinuity in other contexts, e.g. in MODERNE PHILOSOPHEN (1904), where he contrasts at length the older *Kontinuitätsphilosophie* (as in Taine, Fouillée, Wundt, Ardigò) with the more recent *Diskontinuitätsphilosophie* (e.g. Renouvier, "der Nestor der Philosophie der Gegenwart," and Boutroux).

51. *Ibid.*, pp. 286–287.

52. *Ibid.*, pp. 288–289.

53. *Ibid.*, pp. 287–288.

54. Bohr, *The Quantum of Action and the Description of Nature*, ATOMIC THEORY AND THE DESCRIPTION OF NATURE, pp. 96, 99–100.

55. Rosenfeld, *Niels Bohr in the Thirties*, in Rozental, ed., *op. cit.*, p. 117.

56. Hans Bohr, *My Father*, in Rozental, ed., *op. cit.*, p. 328.

57. Jerome S. Bruner, private communication to the author, December 25, 1967. Bruner added a comment which will become relevant to us in what follows below: "I knew Bohr for years afterwards and again spent several hours with him when he was at the Institute for Advanced Study at Princeton, and he came to visit. He had an extraordinary sensitivity for psychological problems, and indeed he once repeated Mach's famous remark about basically our only two sciences: one treats sensation as external and is physics, the other treats it as internal and is psychology. He did not cite this old saw of Mach's approvingly, but urged that there was a grain of truth in it."

58. Bohr, *Quantum Physics and Philosophy*, ESSAYS 1958–1962, p. 7.

59. See note 20.

60. Bohr, *Quantum Physics and Philosophy*, ESSAYS 1958–1962, p. 7.

61. Rosenfeld, *Niels Bohr in the Thirties*, in Rozental, ed., *op. cit.*, p. 116.

62. *Ibid.*, p. 120.

63. Niels Bohr, *Light and Life*, address at the Second International Congress for Light Therapy, Copenhagen, August 1932, NATURE, 131, No. 3308: 421–423, March, 1933; 131, No. 3309: 457–459, April, 1933.

64. Rosenfeld, *Niels Bohr in the Thirties*, in Rozental, ed., *op. cit.*, pp. 132–133.

65. For a partial bibliography of Bohr's writings, see Meyer-Abich, *op. cit.*, pp. 191–199.

66. Rosenfeld, *Niels Bohr in the Thirties*, in Rozental, ed., *op. cit.*, pp. 135–136. A useful summary of Bohr's views concerning the application of the complementarity conception to physics, biology, psychology, and social anthropology is given in Niels Bohr, *On Atoms and Human Knowledge,* DAEDALUS, Spring, 1958, pp. 164–175.

67. Bohr, *The Unity of Human Knowledge*, ESSAYS 1958–1962, pp. 14–15.

68. Bohr, *On Atoms and Human Knowledge*, pp. 174–175.

69. Bohr, *The Unity of Human Knowledge*, ESSAYS 1958–1962, p. 14.

70. Moore, *op. cit.*, pp. 406–407. There is a great deal of evidence of the large scale of Bohr's later hopes along these lines. In his 1933 discussion, J. Rud Nielsen (*Memories of Niels Bohr*, p. 27) reports: "Bohr talked a good deal about his plans for future publications. 'I believe that I have come to a certain stage of completion in my work,' he said, 'I believe that my conclusions have wide application also outside of physics. . . . I would like to write a book that could be used as a text. I would show that it is possible to reach all important results with very little mathematics. In fact, in this manner one would

in some respects achieve greater clarity.' This book, which Bohr referred to as his testament, was never written."

Similarly, Rosenfeld (*Niels Bohr's Contributions to Epistemology*, p. 54) writes: "Bohr had great expectations about the future role of complementarity. He upheld them with unshakable optimism, never discouraged by the scant response he got from our unphilosophical age. . . . Bohr declared, with intense animation, that he saw the day when complementarity would be taught in the schools and become part of general education."

71. J. Robert Oppenheimer, SCIENCE AND THE COMMON UNDERSTANDING (New York: Simon & Schuster, 1953), pp. 81–82.

An early draft of this essay was presented at the *Tagung* of *Eranos* (August 1968). I have profited from discussions with students in my seminar, particularly Bernard Lo and Kellogg Stelle, and with Dr. Arthur I. Miller.

II *On Relativity Theory*

5 ON THE ORIGINS OF THE SPECIAL THEORY OF RELATIVITY

WHEN I was asked to discuss a problem of theory construction and the logic of discovery, I noted particularly the request to bring out the historical-sociological aspects. This directive was a pleasant surprise, for I recalled that Hans Reichenbach had flatly declared himself for the opposite view when he said, "The philosopher of science is not much interested in the thought processes which lead to scientific discoveries . . . that is, he is not interested in the context of discovery, but in the context of justification."[1] If, therefore, I shall make some remarks on the origins of Einstein's special theory of relativity, I will be disobeying the Reichenbachian dictum. However, I draw further strength for this resolution from Einstein, who himself declared for the value of the historical treatment of the rise of key theories in science.

I speak of Einstein's work because his case is both typical and special. The rise of relativity theory shares many features with the rise of other important scientific theories in our time, and in addition it is of course very much more: to find another work that illuminates as richly the relationship between physics, mathematics, and epistemology, or between experiment and theory, or one with the same range of scientific,

Originally presented as a lecture at a joint meeting of the American Association for the Advancement of Science, the American Philosophical Association, and the Philosophy of Science Association, this essay was published in AMERICAN JOURNAL OF PHYSICS, Volume XXVIII, No. 7, pp. 627–636, 1960.

philosophical, and general intellectual implications, one would have to go back to Newton's PRINCIPIA. The theory of relativity was a key development, both in physical science itself and also in modern philosophy of science. The reason for its dual significance is that Einstein's work provided not only a new principle of physics, but, as A. N. Whitehead said, "a principle, a procedure, and an explanation." Accordingly, the commentaries on the historical origins of the theory of relativity have tended to fall into two classes, each having distinguished proponents: the one views it as a mutant, a sharp break with respect to the work of the immediate predecessors of Einstein; the other regards it as an elaboration of then current work, e.g., by Lorentz and Poincaré.

To my mind, the Einsteinian innovation is understood best by superposition of both views, by seeing the discontinuity of methodological orientation within an historically continuous scientific development.[2] Before we come to discuss this, and if we take seriously my point of view, we should first be ready to investigate a number of real problems of the historical or even "historical-sociological" kind: What are the sources for a study of the origins of the special theory of relativity (*RT*) and what is their probable reliability? What was the state of science around 1905, what were the contributions which prepared the field for the *RT*, and what did Einstein know about them? What were the steps by which Einstein reached the conclusions he published in 1905? To what extent was this work a link in a continuous chain, having as its immediate predecessors Lorentz and Poincaré? What was the role of experiment in the genesis of the *RT*, and what the role of the existence of contradictory hypotheses? What part did epistemological analysis play in Einstein's thought? What was the early reception of the *RT* among scientists? In particular, what was Einstein's relation with Mach, Lorentz, and Planck? What may we say about the style of Einstein's work and his personal orientations? What, if anything, in the origins and content of the *RT* is typical of other theories with great impact on science? And even, what methodological principles for the study of the history of science emerge from this study?

We would find that the existing literature is not always of help in studying such questions. The literature on the *RT* is, of course, vast. LeCat[3] listed over 3400 scientific papers in the field up to 1922, with an approximately exponential growth giving a sevenfold increase in seven years. Biographically or philosophically oriented analyses are also fairly numerous (for example, by Schlick, Reichenbach, Frank, Meyerson, Cassirer, Whitehead, Wenzel, Grünbaum, Polanyi, Margenau, Lenzen, Bridgman, and Northrop.) It may be remarked there has so far been no full-scale historical study. A number of valuable essays exist in this direction (for example, by Born, Dugas, Kuznetsov, von Laue, Pauli, Straneo, and Whittaker); these are generally concerned with the chronological development of physics and typically constitute a portion of a longer work having a purpose different from that of a primarily historical-philosophical study. For the latter, the best source is at present indeed Einstein's own set of papers.

Continuity in Einstein's Work

To these papers we must turn to discover, for example, the elements of continuity linking Einstein's first publication on the *RT* with his other work at the time and with the older tradition itself. After the paper of 1905,[4] Einstein returned to the exposition of the *RT* several times, and each restatement is of interest. For instance, in his book ÜBER DIE SPEZIELLE UND DIE ALLGEMEINE RELATIVITÄTSTHEORIE he emphasized in his introduction that "the author has made the greatest effort to present the main ideas . . . on the whole in the sequence and in such context as they in fact arose."[5] It is not surprising that the sequence given there is not in accord with the sequence of steps in the 1905 paper itself, but the historian of science finds an interesting problem in the fact that neither of these is in accord with other autobiographical or biographical accounts.

When one studies the relativity papers in the larger contextual setting of Einstein's other scientific papers, particularly those on the quantum theory of light and on Brownian motion which also were written and published in 1905, one notices two crucial points. While the three epochal papers of 1905—sent to the ANNALEN DER PHYSIK at intervals of less than eight weeks—seem to be in entirely different fields, closer study shows that they arose in fact from the same general problem, namely, the fluctuations in the pressure of radiation. In 1905, as Einstein later wrote to von Laue,[6] he had already known that Maxwell's theory leads to the wrong prediction of the motion of a delicately suspended mirror "in a Planckian radiation cavity." This connects on the one hand with the consideration of Brownian motion as well as to the quantum structure of radiation, and on the other hand with Einstein's more general reconsideration of "the electromagnetic foundations of physics" itself.[7]

One also finds that the style of the three papers is essentially the same and reveals what is typical of Einstein's work at that time. Each begins with the statement of formal asymmetries or other incongruities of a predominantly aesthetic nature (rather than, for example, a puzzle posed by unexplained experimental facts), then proposes a principle—preferably one of the generality of, say, the Second Law of Thermodynamics, to cite Einstein's repeated analogy—which removes the asymmetries as one of the deduced consequences, and at the end produces one or more experimentally verifiable predictions.

Specifically, Einstein's first paper on the quantum theory of light opens in a typical manner:

> There exists a radical formal difference between the theoretical representations which physicists have constructed for themselves concerning gases and other ponderable bodies on the one hand, and Maxwell's theory of electromagnetic processes in so-called empty space on the other hand.[8]

The significant starting point is a formalistic difference between theoretical representations in two fields of physics which, to most physicists, were so widely separated that no such comparison would have invited itself and therefore no such discrepancy would be noted. The discrepancy Einstein points out is between the discontinuous or discrete character of particles and of their energy on one hand, and the continuous nature of functions referring to electromagnetic events and of the energy per unit area in an expanding wave front on the other hand. The discussion of the photoelectric effect, for which this paper is mostly remembered, occurs toward the end, in a little over two pages out of the total sixteen. The prescription for obtaining an experimental verification of his point of view is given in a single, typically succinct Einsteinian sentence (straight-line relation with constant slope between frequency of light and stopping potential for all electrode materials).

In his second paper published in 1905,[9] Einstein points out in the second paragraph that the range of application of classical thermodynamics may be discontinuous even in volumes large enough to be microscopically observable. He ends with the equation giving Avogadro's number in terms of observables in the study of particle motion, and with the one-sentence exhortation: "May some investigator soon succeed in deciding the question which has been raised here, and which is important for the theory of heat!" Significantly, Einstein reported the following year[10] that only after the publication of this paper was his attention drawn to the experimental identification, as long ago as 1888, of Brownian motion with the effect whose existence he had deduced as a necessity from the kinetic-molecular theory. In his *Autobiographical Notes* he repeats that he did the work of 1905 "without knowing that observations concerning Brownian motion were already long familiar."[11]

The third paper of 1905[12] is, of course, Einstein's first paper on the *RT*. He begins again by drawing attention to a formal asymmetry, i.e., in the description of currents generated during relative motion between magnets and conductors. The paper does not invoke explicitly any of the several well-known experimental difficulties—and the Michelson and Michelson-Morley experiments are not even mentioned when the opportunity arises to show in what manner the *RT* accounts for them. At the end, Einstein briefly mentions here, too, specific predictions of possible experiments (giving the equation "according to which the electron must move in conformity with the theory presented here").[13]

Return to a Classic Restriction on Hypotheses

The recognition of these common elements in the three papers prepares us for the essential realization that the fundamental postulates appearing in each of the three papers are *heuristic*. The heuristic nature of the postulate of relativity was from the beginning apparent to Einstein (as he asserted in 1907 and later) because of the restriction of the *RT* to translational motions and to gravitation-free space.[14]

The study of the three papers together reveals also the extent to which Einstein's *RT* represents an attempt to restrict hypotheses to the most *general kind* and the *smallest number* possible—a goal on which Einstein often insisted.[15] In the 1905 paper on *RT*, he makes, in addition to the two "conjectures" raised to "postulates" (i.e., of relativity and of the constancy of light velocity) only four other assumptions: one of the isotropy and homogeneity of space, the others concerning three logical properties of the definition of synchronization of watches. In contrast, H. A. Lorentz's great paper[16] which appeared a year before Einstein's publication and typified the best work in physics of its time—a paper which Lorentz declared to be based on "fundamental assumptions" rather than on "special hypotheses"—contained in fact eleven ad hoc hypotheses: restriction to small ratios of velocities v to light velocity c, postulation a priori of the transformation equations (rather than their derivation from other postulates), assumption of a stationary ether, assumption that the stationary electron is round, that its charge is uniformly distributed, that all mass is electromagnetic, that the moving electron changes one of its dimensions precisely in the ratio of $(1-v^2/c^2)^{1/2}$ to one, that forces between uncharged particles and between a charged and uncharged particle have the same transformation properties as electrostatic forces in the electrostatic system, that all charges in atoms are in a certain number of separate "electrons," that each of these is acted on only by others in the same atom, and that atoms in motion as a whole deform as electrons themselves do. It is for these reasons that Einstein later maintained that the *RT* grew out of the Maxwell-Lorentz theory of electrodynamics "as an amazingly simple summary and generalization of hypotheses which previously have been independent of one another. . . ."[17]

If one has studied the development of scientific theories, one notes here a familiar theme: *the so-called scientific "revolution" turns out to be at bottom an effort to return to a classical purity.* This is not only a key to a new evaluation of Einstein's contribution, but indicates a fairly general characteristic of great scientific "revolutions." Indeed, while it is usually stressed that Einstein challenged Newtonian physics in fundamental ways, the equally correct but neglected point is the number of methodological correspondences with earlier classics, for example, with the PRINCIPIA.

Here a listing of some main parallels between the two works must suffice: the early postulation of general principles which in themselves do not spring directly from experience; the limitation to a few basic hypotheses;[18] the exceptional attention to epistemological rules in the body of a scientific work; the philosophical eclecticism of the author; his ability to dispense with mechanistic models in a science which in each case was dominated at the time by such models;[19] the small number of specific experimental predictions; and the fact that the most gripping effect of the work is its exhibition of a new point of view.

The central problem, moreover, is the same in both works: the nature of space and time and what follows from it for physics. Here, the basic attitudes have in both cases more in common than appears at first reading. That Newton's absolute space and absolute time were not meaningful concepts in the sense of laboratory operations was, of course, not the original discovery of Mach; rather, it was freely acknowledged by Newton himself. But Einstein was also quite explicit that in replacing absolute Newtonian space and time with an infinite ensemble of rigid meter sticks and ideal clocks he was not proposing a laboratory-operational definition. He stated it could be realized only to some degree, "not even with arbitrary approximation," and that the fundamental role of the whole conception, both on factual and on logical grounds, "can be attacked with a certain right."[20] Thus the *RT* merely shifted the locus of space time from the sensorium of Newton's God to the sensorium of Einstein's abstract *Gedanken*experimenter—as it were, the final secularization of physics.

In his tribute on the occasion of the 200th anniversary of Newton's death, Einstein wrote: "I must emphasize that Newton himself was better aware of the weaknesses inherent in his intellectual edifice than the generations of learned scientists which followed him. This fact has always aroused my deep admiration. . . ."[21] He then immediately draws attention to the fact that "Newton's endeavors to represent his system as necessarily conditioned by experience and to introduce the smallest number of concepts not directly referable to empirical objects is everywhere evident." He recalls that Newton regarded the law of gravitational interaction as a heuristic device, "not supposed to be a final explanation, but a rule derived by induction from experience." When the essay ends with Einstein clearly associating himself with a view of causality which he characterizes as "Newtonian," he could well have widened the context of that remark.

Time-Dependence in Source Materials

I cannot avoid a word of warning on the use of sources such as Einstein's writings, particularly to an audience not professionally engaged in the study of the history of science. This has to do with the fact that in many important particulars the writings of one man do not by any means necessarily overlap. I am not speaking merely of the fact that Einstein regarded the discoverer, and particularly himself, as a very poor source of information concerning the genesis of his own ideas, and suggested rather that this study was one of the most interesting tasks for the historian of science. No, I have in mind the simple, yet often neglected fact that Einstein as a person with a single, unchanging identity, in a real sense never existed, just as there never was a single unchangeable entity called Galileo or Newton or Dalton. Einstein himself saw this clearly when he wrote at the start of his *Autobiographical Notes*:

> The exposition of that which is worthy of communication does nonetheless not come easy; today's person of 67 is by no means the same as was the one of 50, of 30, or of 20. Every reminiscence is colored by today's being what it is, and therefore by a deceptive point of view. . . .

And it is not only growth or change—it is also the difference between experience lived and experience reported.

> In this case it is well possible that such an individual in retrospect sees a uniformly systematic development, whereas the actual experience takes place in kaleidoscopic particular situations.[22]

These two effects, coupled with Einstein's large output of writings of both a scientific and a popular kind, explain why everyone—from the extreme positivist to the critical realist—can find some part of Einstein's work to nail to his mast as a battle flag against the others.

There are two ways of dealing with this problem in historically oriented work. The first is to be explicitly careful in the evaluation of all sources, including autobiographical statements, to allow a time-dependent weighting factor. This has always been true, but is particularly pertinent in modern physics where changes per unit time are far larger than before. Revealing examples, and very worthwhile topics of study, are Einstein's attitude toward the ether problem, or his relation to Ernst Mach, or his more general epistemological position. Concerning the first of these, for instance, Einstein underwent a profound change of orientation between the statement near the beginning of his fundamental 1905 paper: "The introduction of a 'luminiferous ether' will prove to be superfluous inasmuch as the view here to be developed will not require an 'absolutely stationary space' provided with special properties"—a provocative remark on which Dugas astutely comments "such a declaration, made on the threshold of his theory, could only alienate him from the physicists imbued with the classical representation"[23]—to his Leiden speech of 1920 on "Äther und Relativitätstheorie" in which he says near the close:

> Recapitulating, we may say that according to the general theory of relativity space is endowed with physical qualities; in this sense, therefore, there exists an ether. According to the general theory of relativity, space without ether is unthinkable; for in such a space there not only would be no propagation of light, but also no possibility of existence for standards of space and time (measuring rods and clocks), nor therefore any space-time intervals in the physical sense.[24]

To the student of the nature of scientific theories, a sequence of individual documents on a particular topic from one pen represents therefore, as it were, a sequence of cross sections in space-time, from which he is challenged to reconstruct the progress or worldline of the topic. Particularly in recent and contemporary physics, no single segment of this worldline may be safely extrapolated; a quick turn is always likely. This enhances the interest: the reconstruction of the changing course of opinion on a topic becomes doubly important, and these changes in one topic may often be correlated with changes in another topic. In the case of Einstein, for example, the attitudes toward the ether, toward Mach, toward epistemology and metaphysics generally, and toward religion, all show closely correlated changes in time. This itself poses new and valuable problems, both to the historian and to the philosopher of science.

The Complementarity of Source Materials

There is a second problem involving divergent or contradictory views concerning a scientist's work. It is generated not by internal changes or conflicts, but by external ones. I can discuss this in the briefest way by pointing to the question of what one is to do with biographical works which are not in agreement.

Such biographies are a precious set of sources for the study of the origins of the relativity theory. Among the principal ones that appeared in Einstein's own time are, in order of publication, those by Moszkowski, Reiser, Reichinstein, Marianoff and Wayne, Seelig, Frank, Infeld, and Vallentin.[25] Each has interest in its own right, but naturally enough they differ vastly in their points of view as well as on factual matters. One can begin to discern the Vivianis and Stukleys now, the sources of future myths and the sources of reliable references. It was therefore important to discover the unpublicized fact that one of these was written under a pseudonym by a relative of Einstein and checked by him for factual accuracy, that another was publicly disowned by Einstein, that he made an attempt in a third case to persuade the author—whom he did not trust to be fair or accurate—to forego publication, that he was pleased with the material in another of these books, and so forth.

The uncommonly large amount and variety of material emphasizes the problems the historian of science must face. The different points of view from which two or more honest biographies are written yield, of course, different interpretations. On some matters of "fact" (as, for example, dates and places) one certainly can ask for agreement or accuracy in some absolute sense. But on larger and more qualitative questions (for example, the acceptance of the theory) one can profitably adopt the attitude that evidence obtained by biographical research under different points of view cannot be comprehended within a single picture, but must be regarded as complementary in the sense that *only the totality of the presentations exhausts the possible information about the subject*. This will be recognized as closely analogous to one part of the complete statement of the complementarity principle in physics.[26] To look for an "independent" view in qualitative matters in any other way is likely to lead one merely to take a position equidistant from all others, or from the "isms" that motivate them.

The complementarity principle tells the physicist also that it is not possible to make a sharp separation between the behavior of atomic objects and the interaction with the measuring instruments which serve to define the conditions under which the phenomena appear. This statement, too, has a close parallel in the study of the history of philosophy of science, and one must therefore be aware that the scholar and the subject of his study together form one system in which it is not meaningful to try to achieve a complete separation of one part from the other. It is in this spirit that one must understand, and use, the picture of Einstein as a revolutionary, which is painted by a revolutionary, and that of Einstein as a positivist, as presented by a positivist. To one who is committed to the existence of a real medium to explain the transmission of light through space, the *RT* is important primarily insofar as it adds to or subtracts from this position. Only with this explicit recognition can one use a number of accounts together, each of which would otherwise appear to present a strikingly different person or work.

Whittaker's Accounts of the Origins of Einstein's Work

To illustrate this point concretely I wish to turn to a question on which a dispute has been active: namely, to what extent Einstein's work was original rather than anticipated by, or specifically based on, other published work. Particularly interesting is the essay on Einstein by Sir Edmund Whittaker.[27] Whittaker's commitment to the nineteenth-century tradition of physics and to the ether theory is illustrated in his well-known book, A HISTORY OF THE THEORIES OF AETHER AND ELECTRICITY,[28] and also by his excellent contributions in the field of classical mechanics. Moreover, in the second volume of the HISTORY, completed in 1953, which carries the story to 1926, Whittaker had largely dismissed Einstein's paper of 1905 on the *RT* as one "which set forth the relativity theory of Poincaré and Lorentz with some amplifications, and which attracted much attention."[29]

This presentation evoked considerable criticism, some of which I know to have reached Whittaker while his book was still in manuscript, and some of which reached him by the time he composed the biographical memoir after Einstein's death in 1955. It is therefore noteworthy that in his 1955 necrology for Einstein, Whittaker did not change his earlier evaluation. For example, he repeated that Poincaré in a speech in St. Louis, in September, 1904[30] had coined the phrase "principle of relativity." Whittaker asks how physics could have been reformulated in accordance with "Poincaré's principle of relativity," and he reports that with respect to the laws of the electromagnetic field this "discovery was made in 1903 by Lorentz," citing a paper by Lorentz in the PROCEEDINGS of the Amsterdam Academy of Sciences, for the year 1903.[31] Whittaker shows that "the fundamental equations of the aether in empty space" are invariant under suitably chosen (i.e., Lorentz) transformations, and he concludes with the remarkable sentence: "Einstein [in the *RT* paper of 1905] adopted Poincaré's principle of relativity (using Poincaré's name for it) as a new basis for physics and showed that the group of Lorentz transformations provided a new analysis connecting the physics of bodies in motion relative to each other."[32]

Since Whittaker's analysis has been and is likely to continue to be given considerable weight, it is necessary to examine it closely. It turns out to be an excellent example of the proposition that no such analysis can be considered meaningful except insofar as it deals both with the material it purports to cover *and* with the prior commitments and prejudices of the scholar himself. Here is a brief summary of main findings when Whittaker's analysis is considered in this light.

(1) Einstein's *RT* paper of 1905 was indeed one of a number of contributions by many different authors in the general field of the electrodynamics of moving bodies. In the ANNALEN DER PHYSIK alone there are eight papers from 1902 up to 1905 concerned with this general problem. Einstein himself always insisted on this aspect of continuity. The earliest evidence is in a letter written in the spring of 1905 to his friend Conrad Habicht, describing his various investigations. In one sentence he describes the developing *RT* paper: "The fourth work lies at hand in drafts [*liegt im Konzept vor*] and is an electrodynamics of moving bodies making use of a *modification* of the theory of space and time; you will surely be interested in the purely kinematic part of this work."[33] Seelig also quotes a later remark of Einstein which gives in one sentence his often repeated attitude: "With respect to the theory of relativity it is not at all a question of a revolutionary act, but of a natural development of a line which can be pursued through centuries."[34]

On the other hand, to say that Einstein's paper "attracted much attention" is correct only if one neglects the first few years after publication. For the early period, a more characteristic reaction was, in fact, either total silence, or the response to be found in the first paper in the ANNALEN DER PHYSIK that mentioned Einstein's work on the *RT*. It was the claim of a categorical experimental disproof of Einstein's theory.[35]

(2) The paper by Poincaré of 1904 which Whittaker cites turns out not to enunciate a new relativity principle, but is rather a very acute and penetrating though qualitative summary of the difficulties which contemporary physics was then making for six classical laws or principles, including what is in effect the Galilean-Newtonian principle of relativity. The list given by Poincaré is as follows: the Law of Conservation of Energy; the Second Law of Thermodynamics; the Third Law of Newton; "the principle of relativity, according to which the laws of physical phenomena should be the same whether for an observer fixed or for an observer carried along in a uniform movement or translation . . ."; the

principle of conservation of mass; and the principle of least action.[36] Of the principle of relativity Poincaré complains that it "is battered" by current developments in electromagnetic theory, although, he says, it "is confirmed by daily experience" and "imposed in an irresistible way upon one's good sense." Poincaré's main point is to show the need for a new development, the outlines of which he suggests in these words: "Perhaps likewise we should construct a whole new mechanics, that we only succeed in catching a glimpse of, where inertia increasing with the velocity, the velocity of light would become an impassable limit."[37] Thus he illustrates both the power of his intuition and the qualitative nature of the suggestion.

(3) It is more difficult to discuss the 1903 paper of Lorentz which Whittaker, both in his book and in his Memoir, cited specifically as the work that spelled out most of the basic details of Einstein's *RT* of 1905. In the first place, this paper does not exist. What Whittaker clearly wished to refer to is the paper Lorentz published a year later, in 1904.[16] Since Whittaker was otherwise very careful with the voluminous citations of references, this repeated slip, which doubles the time interval between the work of Lorentz and of Einstein, is not merely a mistake. It is at least a symbolic mistake, symbolic of the way a biographer's preconceptions interact with his material.

(4) Whittaker clearly implied that Einstein used Lorentz's transformation equation published in 1904. He therefore chose to neglect that both Einstein and those close to him have repeatedly said that Einstein had not read Lorentz's 1904 paper.[38]

(5) Even if one does not wish to rely on the word of Einstein and other prominent physicists of his time in this matter, there are four items of internal evidence in Einstein's 1905 paper which indicate that he had not read Lorentz's of 1904. Einstein does write the transformation equations in a form equivalent to those of Lorentz (or, for that matter, of Voigt's of 1887); but whereas Lorentz had assumed these equations a priori in order to obtain the covariance of Maxwell's equations in free space, Einstein *derived* them from the two fundamental postulates of the *RT*. He therefore did not need to know of Lorentz's paper of 1904.[39]

Secondly, as Einstein's first two major papers of 1905 show, he was in the habit of giving credit in footnotes to the work of others which he might be using; the absence of a specific reference to the 1904 paper of

Lorentz may therefore be taken at its face value, the more so since Einstein twice in the text of this same paper refers to Lorentz by name in citing the then current electromagnetic theory in the form Lorentz had given it in his book of 1895.[40] Parenthetically, one may also say that it is rather preposterous to suggest that a young man of Einstein's temperament and painful honesty, and one who, as the letters to Lorentz soon thereafter show, revered Lorentz deeply, should knowingly be using, without acknowledgment, an important new finding in the recent work of the foremost theoretical physicist in this field.[41]

Next, in the second paragraph of his paper, Einstein recalls that the "laws of electrodynamics and optics" have been found to "be valid for all frames of reference for which the equations of mechanics hold good" to the first order of the quantity v/c. But one of the main points of the 1904 paper of Lorentz was his claim to have extended the theory to the *second* order of v/c. And a fourth internal evidence is Einstein's choice of convention in the expression for force and mass in the dynamics of charged particles; this choice[42] is far less suitable than Lorentz's, forcing Planck to point this out in 1906.

(6) Quite apart from the question whether Einstein's 1905 paper was written independently of Lorentz's is the equally significant fact that in a crucial sense Lorentz's paper was of course not on the relativity theory as we understand the term since Einstein. Lorentz's fundamental assumptions are not relativistic; as Born says, "he never claimed to be the author of the principle of relativity,"[43] and, on the contrary, referred to it as "Einstein's Relativitätsprinzip" in his lectures of 1910. In Lorentz's essay on "The Principle of Relativity of Uniform Translation," published in 1922,[44] six years before Lorentz's death, he still asked that space be considered to have "a certain substantiality; and if so, one may, in all modesty, call true time the time measured by clocks which are fixed in this medium, and consider simultaneity as a primary concept."[45] In his 1904 paper he had postulated the nonrelativistic addition theorem for velocities, $v = V + u$, and even in the 1922 book he did not consider the velocity of light as inherently the highest attainable velocity of material bodies.

(7) Finally, we note another set of important differences between Lorentz's accomplishment of 1904 and what Whittaker implies. Strictly speaking, the Lorentz theory of 1904 applies only to small values of v/c, since the constant l which is taken to be 1 for small values of v/c

enters in the first power in the transformation equations for x and t. Also, Maxwell's equations in the presence of charges are not completely invariant in Lorentz's treatment even at small speeds v, since in the primed (moving) system, a term is left over in the expression for div$'D'$, namely, div$'D' = [1 - (vu_x'/c^2)]\rho'$, as compared to div$D = \rho$.[46] We have already noted the number of ad hoc hypotheses which Lorentz was forced to introduce, and which robbed the theory of electromagnetic phenomena of the generality typical of fundamental conceptions.

In closing, I return to my initial remarks: the detailed study of the historical situation is, to my mind, an important first step in those discussions which try to base epistemological considerations on "real" cases. This is not always done easily; but it is through the dispassionate examination of historically valid cases that we can best become aware of the preconceptions which underlie all philosophical study.

NOTES

1. Hans Reichenbach, *The Philosophical Significance of the Theory of Relativity*, in Paul Arthur Schilpp, ed., ALBERT EINSTEIN: PHILOSOPHER-SCIENTIST (Evanston, Ill.: Library of Living Philosophers, 1949), p. 292.

2. Gerald Holton, *Continuity and Originality in Einstein's Special Relativity Theory*, IX CONGRESO INTERNACIONAL DE HISTORIA DE LAS CIÉNCIAS, GUIONES DE LAS COMMUNICACIONES (Barcelona-Madrid, 1959), Volume II, p. 41.

3. Maurice LeCat, BIBLIOGRAPHIE DE LA RELATIVITÉ, Brussels: Maurice Lamertin, 1924.

4. Albert Einstein, *Zur Elektrodynamik bewegter Körper,* ANNALEN DER PHYSIK 17:891–921, 1905.

5. Albert Einstein, ÜBER DIE SPEZIELLE UND DIE ALLGEMEINE RELATIVITÄTSTHEORIE, Braunschweig: Friedrich Vieweg & Son, 1917.

6. Letter of January 17, 1952 (unpublished). See also Max Born, *Physics and Relativity* in André Mercier and Michel Kervaire, eds., FÜNFZIG JAHRE RELATIVITÄTSTHEORIE (Bern: Birkhäuser, 1956), pp. 248–249.

7. Albert Einstein, *Autobiographical Notes*, in Schilpp, *op. cit.*, p. 47.

8. Albert Einstein, *Über einen die Erzeugung und Verwandlung des Lichtes betreffenden heuristischen Gesichtspunkt*, ANNALEN DER PHYSIK, 17:132, 1905.

9. Albert Einstein, *Über die von der molekularkinetischen Theorie der Wärme geforderte Bewegung von in ruhenden Flüssigkeiten suspendierten Teilchen*, ANNALEN DER PHYSIK 17:560, 1905.

10. Albert Einstein, *Zur Theorie der Brownschen Bewegung*, ANNALEN DER PHYSIK, 19:371–381, 1906.

11. Einstein, *Autobiographical Notes, loc. cit.* See also Leopold Infeld, ALBERT EINSTEIN, HIS WORK AND ITS INFLUENCE ON OUR WORLD, rev. ed. (New York: Charles Scribner's Sons, 1950), pp. 97–98.

12. See note 4, p. 891.

13. *Ibid.*, p. 921.

14. On a few occasions, although not in the original paper, Einstein made this point. E.g., in *Bemerkungen zu der Notiz von Hrn. Paul Ehrenfest: Die Translation deformierbarer Elektronen und der Flächensatz*, ANNALEN DER PHYSIK, 23:206, 1907. "The relativity principle [is to be regarded] . . . solely as a heuristic principle, which, considered by itself contains only assertions about rigid bodies, clocks, and light signals."

15. Cf. Albert Einstein, *The Problem of Space, Ether and the Field in Physics*, IDEAS AND OPINIONS, trans. and rev. Sonja Bargmann (New York: Crown Publishers, 1954), p. 282. "The theory of relativity is a fine example of the fundamental character of the modern development of theoretical science. The initial hypotheses become steadily more abstract and remote from experience. On the other hand, it gets nearer to the grand aim of all science, which is to cover the greatest possible number of empirical facts by logical deductions from the smallest possible number of hypotheses or axioms."

16. H. A. Lorentz, *Electromagnetic Phenomena in a System Moving with Any Velocity Smaller Than That of Light*, KON. AKADEMIE VAN WETENSHAPPEN AMSTERDAM, PROCEEDINGS OF THE SECTION OF SCIENCES, 6:809–831, 1904 (English-language edition). This paper, originally presented as part of the proceedings of the meeting of April 23, 1904, was first published in June, 1904, in the Dutch-language edition of the PROCEEDINGS (12:896–1009, 1904).

17. See note 5, p. 28. See also Albert Einstein, *Zum Relativitätsproblem*, SCIENTIA, 15:337, 1914.

18. Wolfgang Pauli, in RELATIVITÄTSTHEORIE, Volume 19 of ENCYCLOPÄDIE DER MATHEMATISCHEN WISSENSCHAFTEN, Leipzig: B. G. Teubner, 1921, trans. Gerard Field as THEORY OF RELATIVITY (New York: Pergamon Press, 1958), p. 5, unwittingly draws forceful attention to this particular point when summarizing his analysis of the *RT* in the following words: "The postulate of relativity implies that a uniform motion of the center of mass of the universe relative to a closed system will be without influence on the phenomena in such a system." Note the correspondence with the main hypothesis in the last edition of the PRINCIPIA.

19. Cf. Max von Laue, *Einstein und die Relativitätstheorie*, NATURWISSENSCHAFTEN, 43:1–8, 1956.

20. Albert Einstein, LES PRIX NOBEL EN 1921–1922 (Stockholm, 1923), p. 2. See also Albert Einstein, *Dialog über Einwände gegen die Relativitätstheorie*, NATURWISSENSCHAFTEN, 6:697–702, 1918.

21. Albert Einstein, *Newtons Mechanik und ihr Einfluss auf die Gestaltung der theoretischen Physik*, NATURWISSENSCHAFTEN, 15:273–276, 1927; reprinted as *The Mechanics of Newton and Their Influence on the Development of Theoretical Physics*, IDEAS AND OPINIONS, pp. 257–258.

22. See note 7, pp. 3–7.

23. René Dugas, A HISTORY OF MECHANICS, trans. J. R. Maddox (New York: Central Book Co., 1955), p. 490.

24. Albert Einstein, *Ether and Relativity*, in SIDELIGHTS ON RELATIVITY (London: Methuen & Co., 1922), p. 23. The next sentences reaffirm the difference between this and other ether models. "But this ether may not be thought of as endowed with the quality characteristic of ponderable media, as consisting of parts which may be tracked through time. The idea of motion may not be applied to it."

In this connection, see also the essay *Relativity and the Problem of Space*, which Einstein added in the 15th edition (as Appendix V) of RELATIVITY, THE SPECIAL AND THE GENERAL THEORY, trans. Robert W. Lawson (London: Methuen & Co., 1954), pp. 135–157. Commenting on it to Carl Seelig in ALBERT EINSTEIN (Zurich: Europa Verlag, 1954), p. 291, Einstein wrote: "In particular, it is shown that the development has a close connection with Descartes' argument for the non-existence of 'empty space.' "

25. Philipp Frank, EINSTEIN, HIS LIFE AND TIMES, trans. George Rosen, ed. Suichi Kusaka, New York: Alfred A. Knopf, 1947.

Leopold Infeld, ALBERT EINSTEIN, HIS WORK AND ITS INFLUENCE ON OUR WORLD, rev. ed., New York: Charles Scribner's Sons, 1950.

Dimitri Marianoff and Palma Wayne, EINSTEIN, AN INTIMATE STUDY OF A GREAT MAN, Garden City, New York: Doubleday, Doran & Co., 1944.

Alexander Moszkowski, EINSTEIN, THE SEARCHER; HIS WORK EXPLAINED FROM DIALOGUES WITH EINSTEIN, trans. Henry L. Brose, New York: E. P. Dutton & Co., 1921.

David Reichinstein, ALBERT EINSTEIN, SEIN LEBENSBILD UND SEINE WELTANSCHAUUNG, Prague: Ernst Ganz, 1935.

Anton Reiser, ALBERT EINSTEIN, A BIOGRAPHICAL PORTRAIT, New York: A. & C. Boni, 1930.

Carl Seelig, ALBERT EINSTEIN, A DOCUMENTARY BIOGRAPHY, trans. Mervyn Savill, London: Staples Press, 1956.

Antonia Vallentin, EINSTEIN, A BIOGRAPHY, trans. Moura Budberg, London: Weidenfeld and Nicolson, 1954.

26. I employ it here as a suggestive, though not prescriptive, analogy.

27. Sir Edmund Whittaker, *Albert Einstein*, BIOGRAPHICAL MEMOIRS OF FELLOWS OF THE ROYAL SOCIETY (London: Royal Society, 1955), pp. 37–67.

28. Sir Edmund Whittaker, A HISTORY OF THE THEORIES OF AETHER AND ELECTRICITY: FROM THE AGE OF DESCARTES TO THE CLOSE OF THE NINETEENTH CENTURY, London and New York: Longmans, Green & Co., 1910; rev. and enlarged as A HISTORY OF THE THEORIES OF AETHER AND ELECTRICITY: THE CLASSICAL THEORIES, London and New York: Nelson & Sons, 1951.

29. Sir Edmund Whittaker, A HISTORY OF THE THEORIES OF AETHER AND ELECTRICITY: THE MODERN THEORIES, 1900–1926 (London: Nelson & Sons, 1953), p. 40.

30. J. H. Poincaré, *L'Etat actuel et l'avenir de la physique mathématique*, BULLETIN DES SCIENCES MATHÉMATIQUES (1904), Première Partie, 302–324; the English translation is *The Principles of Mathematics—Empirical Physics*, MONIST, 15, No. 1:1–24, 1905.

31. The citation given is "PROC. ACAD. SCI. AMST. (English ed.) (1903), 6, 809."

32. See note 27, p. 42.

33. Seelig, *op. cit.*, p. 89. Emphasis added.

34. *Ibid.*, p. 97.

35. Walter Kaufmann, *Über die Konstitution des Elektrons*, ANNALEN DER PHYSIK 19:495, 1905.

36. Poincaré, *op. cit.*, p. 5.

37. *Ibid.*, p. 23.

38. See the footnote on this point by Arnold Sommerfeld in the reprints and translations of Einstein's 1905 paper in the Teubner and Methuen editions of the collection of essays on the *RT* [e. g., THE PRINCIPLE OF RELATIVITY (London: Methuen & Co., 1923)], or Pauli, *op. cit.*, p. 3, or Einstein's letter to Carl Seelig: "As for me, I knew only Lorentz's important work of 1895 . . . but not Lorentz's later works and also not the inquiries of Poincaré connected with them. In this sense my work of 1905 was independent." TECHNISCHE RUNDSCHAU, 47: Bern, May 6, 1955, cited in Born, *op. cit.*, p. 248.

39. This is by no means the only such case in Einstein's early scientific career. In fact, his work on thermodynamics and fluctuation phenomena in the period 1902–1905 was to a large extent a repetition of available material;

as Einstein said later, "Not acquainted with the earlier investigations of Boltzmann and Gibbs, which had appeared earlier and actually exhausted the subject, I developed the statistical mechanics and the molecular-kinetic theory of thermodynamics which was based on the former." (See note 7, p. 47.) Einstein's unawareness in 1905 of the earlier identification of Brownian motion has been referred to previously. Anton Reiser (*op. cit.*, p. 52), provides the report that at his university Einstein planned to build a device for measuring the ether drift, not knowing of Michelson's apparatus; although this earliest example is quite understandable in terms of the incompleteness of Einstein's training at that point, it illustrates a remark made often about him by his friends: that he read little, but thought much.

40. H. A. Lorentz, VERSUCH EINER THEORIE DER ELEKTRISCHEN UND OPTISCHEN ERSCHEINUNGEN IN BEWEGTEN KÖRPERN, Leiden: E. J. Brill, 1895.

41. Einstein later accurately reported that "At the turn of the century, H. A. Lorentz was regarded by theoretical physicists of all nations as the leading spirit; and this with fullest justification." Albert Einstein, *H. A. Lorentz, His Creative Genius and His Personality*, in H. A. LORENTZ, IMPRESSIONS OF HIS LIFE AND WORK, ed. G. L. de Haas-Lorentz (Amsterdam: North-Holland Publishing Co., 1957), p. 5.

42. As most recently remarked by Max von Laue, *Einstein und die Relativitätstheorie*, NATURWISSENSCHAFTEN, 43:1–8, 1956, in documenting his belief that Einstein did not know of Lorentz's 1904 paper.

43. Born, *op. cit.*, p. 248.

44. A. D. Fokker, ed.; translated as part 2 of Volume III of LECTURES ON THEORETICAL PHYSICS, London: Macmillan & Co., 1931.

45. *Ibid.*, p. 211.

46. Whittaker, (See note 29, p. 31), says Lorentz "obtained a transformation in a form which is exact to all orders of the small quantity v/c," although strictly speaking this is correct only for free space and relatively small values of v.

6 POINCARÉ AND RELATIVITY

IN EACH of the books and essays from the hand of Alexandre Koyré that illuminate the area of contemporary scholarship in the history of science, there are two kinds of passages that have been especially memorable to me. One is the category of those brilliant summaries of new recognitions and those *Wegweiser* to previously unexplored territory which have become part of all our thinking. The other category is made up of those sudden openings through which we can glimpse a reminder of the philosopher's intention behind the historian's treatise. One of the latter passages contains a challenge that has by no means been met. Koyré writes:

Pour moi qui ne crois pas à l'interprétation positiviste de la science — ni même à celle de Newton — l'histoire racontée de façon si brillante par M. Crombie contient une leçon bien différente : l'empirisme pur — et même la ' philosophie expérimentale ' — ne conduisent nulle part; et ce n'est pas en renonçant au but apparemment inaccessible et inutile de la connaissance du réel, mais au contraire en le poursuivant avec hardiesse, que la science progresse sur la voie sans fin qui la conduit à la vérité . . . Les grandes révolutions scientifiques du xxe siècle — autant que celles du xviie ou du xixe — bien que

Based on a paper delivered at the 10th International Congress of the History of Science, Cornell University, August, 1960, this essay was originally published in MÉLANGES ALEXANDRE KOYRÉ, PUBLIÉS À L'OCCASION DE SON SOIXANTE DIXIÈME ANNIVERSAIRE (Paris: Hermann, 1964), pp. 257–268.

fondées naturellement sur la découverte de faits nouveaux — ou sur l'impossibilité de les vérifier — sont fondamentalement des révolutions *théoriques* dont le résultat ne fût pas de mieux relier entre elles les ' données de l'expérience', mais d'acquérir une nouvelle conception de la réalité profonde qui sous-tend ces 'données'.[1]

". . . ne conduisent nulle part." And a little earlier in the same article: "Le positivisme est fils de l'échec et du renoncement."[2] Even if one does not align oneself with the conception of *réalité profonde,* and of truth as the unattainable end of the search, one immediately recognizes the strength of this view which in every age has animated the work of the scientists themselves, whether they expressed it or not. And one is forced to ask oneself: what are the directions in which neo-positivism might be susceptible of growth and transformation? For it is clear that such further growth, starting from its undoubted strength, is both possible and inevitable.

I

We may best indicate one of these possible directions by means of a case study belonging to the group of examples where a prominent contributor to science seemed to hold back or draw back from the full understanding or full exploitation of a great advance growing out of his own work, or when he turns in an apparently unreasonable manner against the work of others whom we now know really to have been supporters of his own cause. A well-remembered example of the first kind is Galileo with respect to the law of inertia, or Planck with respect to the quantum theory in its early years. Examples of the second kind are Galileo versus Kepler's ellipsi, Dalton versus Gay-Lussac's work, or Planck versus Einstein's photon. Henri Poincaré's position with respect to relativity theory in general, and to Einstein's contribution in particular, is an example fitting both types.

How shall we understand such dilemmas? One way is to study each case separately to discern its own complex of reasons and unreasons. This has been done in several instances, sometimes with striking persuasiveness. But one can go beyond the individual case and construct a schema which explains such strange impasses in more general terms.

Let us recall the case of Poincaré: around 1905, Poincaré was simultaneously Professor of Mathematical Physics, of Astronomy, and of Celestial Mechanics at the University of Paris. At 51 years of age, he was at the peak of his immense powers and illustrious career. In the report on the Bolyai Prize of the Hungarian Academy of Sciences (awarded in 1905), Poincaré was called "incontestablement le premier et le plus puissant chercheur du temps présent dans le domaine des Mathématiques et de la Physique mathématique."[3] And the more closely one studies his writings, the more remarkable it seems that he should have stopped short of full presentation of relativity theory as we know it, and, more than that, remained unshakably against Einstein's interpretation to the end. One could cite again and again passages that are very much what Einstein would have written, or even more severe—e.g., on H. A. Lorentz's work: if he has succeeded, "ce n'est qu'en accumulant les hypothèses."[4]

It is not surprising that Einstein and some of his friends of that period have repeatedly testified that their reading of Poincaré's book SCIENCE AND HYPOTHESIS (1902) was an experience of considerable influence on Einstein.[5] Thus we recall Poincaré's denial of absolute space,[6] his objection to absolute movement,[7] his reference to the Principle of Relative Motion,[8] and to a "Principle of Relativity,"[9] and his search for invariant forms of physical laws under transformations. In particular, one should read the publication of which Louis de Broglie has said: "Dans un remarquable mémoire écrit avant les travaux d'Einstein et paru dans les RENDICONTI DEL CIRCOLO MATEMATICO DI PALERMO, où il a étudié la dynamique de l'électron d'une façon approfondie, il a donné les formules de la cinématique relativiste."[10]

Poincaré went further than Lorentz and denied absolute time and the "intuition" of simultaneous events at two different places[11]—although he did not advance to a discussion of simultaneity for events observed from differently moving systems. But the existence of the ether is rarely doubted, for, like Lorentz, Poincaré explained by compensation of effects the apparent validity of absolute laws in moving inertial systems, and maintained the privileged position of the ether.[12] Nevertheless, he was also forced to make uncomfortable qualifications, such as those in LA SCIENCE ET L'HYPOTHÈSE: "C'est pour échapper à cette dérogation aux lois générales de la mécanique que nous avons inventé l'éther . . . . Il faudrait bien, si on ne voulait changer toute la mécanique, introduire l'éther, pour que cette action, que la matière paraîtrait subir, fût contrebalancée par la réaction de la matière sur quelque chose . . . ." And "on croit toucher l'éther du doigt."[13]

There, as in other places, one notes a tentative, indeterminate method of argument showing through the beautiful prose at precisely those places where Poincaré must in fact have felt in greatest trouble. A significant example is his discussion of time and simultaneity, where he toys with defining a uniform time by going back to a Newtonian Sensorium of "une intelligence infinie," "une sorte de grande conscience qui verrait tout, et qui classerait tout *dans son temps*."[14] He rejects it because this infinite intelligence, "si même elle existait, serait impénétrable pour nous" —but he does not know what else to put in its place. In the end, Poincaré remains suspended between relativism and absolutism. Why?

Such a question has an answer in two parts, one specific for the case in hand, the other, general. As to the former, Poincaré was by nature and talent a gradualist. In one of the last lectures (11 April 1912), he summarized his preferred program (while confessing that it is not always a possible one) in these words:

Les théories anciennes reposent sur un grand nombre de coïncidences numériques qui ne peuvent être attribuées au hasard; nous ne pouvons donc disjoindre ce qu'elles ont réuni; nous ne pouvons plus briser les cadres, nous devons chercher à les plier . . . .[15]

And in another paper, eight years earlier (in 1904), he gave a program for "a more satisfactory theory of the electrodynamics of bodies in motion": "Let us take, therefore, the theory of Lorentz, turn it in all senses, modify it little by little, and perhaps everything will arrange itself."[16]

Moreover, Poincaré's gradualism was appropriate and successful, entirely in keeping with the style of work that characterizes one of the two main types of major contributors. For on the one hand there are those who, like Einstein, point out and open up large areas of ignorance and fruitful new work. And on the other hand there are those who succeed in bringing long-standing problems to a higher stage of completion. Poincaré's work in physics is of the second kind. His strength is again and again to rescue the physics of Newton, of Maxwell, of Lorentz.

Thus his first great prize, in 1889, came to him for work on the three-body problem. In it he answered the long-standing question, "Can the law [of gravitation] of Newton by itself explain all the astronomical phenomena?" with "Yes, to a very high probability."[17] His work included improvement of the solution concerning the stability of Saturn's rings. It was Poincaré who had brought Maxwell's work to France (1888) and who saw his own main contribution to relativity theory explicitly to be the perfection of Lorentz's theory. He had shown in 1900 how the introduction of an electromagnetic momentum, transported by the ether, would rescue both the Law of Conservation of Momentum and Lorentz's theory, not to speak of giving a purpose to the ether. Later, the introduction of the ether's pressure on electrons (to account for their stability and deformation) provided for it still another task.

Poincaré was, in short, the brilliant conservator of his day.[18] In his formulation, most of the bothersome questions of the relativity of time and simultaneity did not have to be raised. Classical physics remained the safe rock upon which necessary modifications were to be built.

II

A second part of the answer has lately been sought in Poincaré's conventionalist epistemology. In an interesting paper, M. Théo Kahan blames (somewhat as deBroglie did in SAVANTS ET DÉCOUVERTES[19]) "sa philosophie conventionaliste qui accordait aux lois de la géométrie et de la physique tout au plus un caractère de conventions utiles, sans signification réelle profonde."[20] While I do not disagree that this played a role, I am struck as forcefully by an apparently quite unrelated, different train in Poincaré, namely, that he insisted repeatedly that the relativity principle is simply an *experimental fact.*

Thus, in the essay *L'Espace et le temps,* he wrote "le principe de la relativité physique, nous l'avons dit, est un fait expérimental, au même titre que les propriétés des solides naturels; comme tel, il est susceptible d'une incessante révision. . . ."[21] And when Poincaré came to write the essay *La Mécanique et l'optique*, he made a very significant confession. The news reached him that the results obtained by the great experimentalist Walter Kaufmann (1906) disproved the relativity theory of Lorentz (as well as that of Einstein). To Poincaré, the principle of relativity, on which he had based his own great work published that same year in the RENDICONTI, was immediately suspect, and he now wrote: "Le Principe de la Relativité n'aurait donc pas la valeur rigoureuse qu'on était tenté de lui attribuer."[22]

Einstein, on his side, neither accepted nor disproved Kaufmann's claim. With the characteristic certainty of a man for whom the fundamental hypothesis is *not* contingent either on experiment or on heuristic (conventionalistic) choice, Einstein waited for others to show, over the next years, that Kaufmann's experiments had not been decisive.[23] For the crucial difference between Einstein and Poincaré was that Einstein had fully embraced relativity, by elevating what he called his *Vermutung* [conjecture] namely, the Principle of Relativity, to the status of a *Voraussetzung* [postulate]. I refer to such a postulate as a thematic proposition or thematic hypothesis. And it is precisely such nonverifiable and nonfalsifiable (and not even quite arbitrary) thematic hypotheses which are most difficult to advance or to accept. It is they which are at the heart of major changes or disputes, and whose growth, reign, and decay are much-neglected indicators of the most significant developments in the history of science. With these last statements we have been led by a specific example to the threshold of a discussion that develops this position in detail. In the limits of space made available I can only put forward the most essential terminology [by summarizing here some portions of Section *III* of Essay 1].

III

All modern analyses of science agree that two types of propositions are scientifically *not* meaningless, namely, propositions concerning empirical matters of "fact," and propositions concerning the calculus of logic and mathematics that helps us to structure and analyze.

We can use these two generalized coordinates to define an x-y plane. It is the plane in which scientific discourse usually proceeds. A concept such as force may be considered as a point in the x-y plane. The projection on the x, or phenomenic, dimension corresponds to the empirical meaning of "force," i.e., its detection and measurement through, say, the distortion experienced by standard objects. The projection of the concept "force" on the y- dimension is its analytic meaning (vector property, e.g., parallelogram law of composition).

In the same manner, we analyze a statement, e.g., a hypothesis or the law of universal gravitational attraction, in terms of its phenomenic and heuristic-analytic components. Such an analysis is a "contingency analysis," because the value of a statement in the x-y plane is contingent on the possibility of (1) checking the phenomenic component (e.g., whether two masses do move closer in a Cavendish experiment) and (2) checking the heuristic-analytic component (e.g., whether the analysis in terms of vectors in Euclidean space is more appropriate than, say, in terms of scalars). The x-y plane is thus the "contingent plane," where scientific concepts and propositions have both empirical and analytical relevance.

Now it has been the claim of modern positivism that statements are scientifically meaningful only insofar as they have components in the contingent plane. This attitude has also been the ruling one in the younger sciences such as psychology, and also in history, particularly the history of science. From Bacon, Kepler, and Newton on, all who have claimed not to feign hypotheses have been concerned with keeping the hypotheses they must use in the contingent plane. And this is one reason why science has grown so rapidly since the early part of the seventeenth century.

The fact, however, is that this aim is not and never can be fully achieved, and the analysis of historic cases in science should more widely begin to take into account that concepts and hypotheses used in science are meaningful not only in the contingent plane. The contingent components are merely two of three components, the results of a projection from an x-y-z space to the x-y plane. A concept such as force, for example, has also a *thematic* component, which is directly coupled neither to phenomena nor to tautological, analytic statements, but to the persisting theme of an active potency principle that stands behind the whole sequence of concepts from which our idea of force has developed: *energeia, anima, vis, Kraft*. Similarly, a hypothesis or a law such as the Law of Conservation of Momentum or of Energy can be more fully understood as a line element or configuration in three-dimensional "proposition space" with a projection on the third axis, that of themata corresponding to the delineation of the persistent motif or thema of "constancy" or "conservation."

Poincaré himself clearly expressed the role of themata in a passage in LA SCIENCE ET L'HYPOTHÈSE:

Comme nous ne pouvons pas donner de l'énergie une définition générale, le principe de la conservation de l'énergie signifie simplement qu'il y a *quelque chose* qui demeure constant. Eh bien, quelles que soient les notions nouvelles que les expériences futures nous donneront sur le monde, nous sommes sûrs d'avance qu'il y aura quelque chose qui demeurera constant et que nous pourrons appeler *énergie* [to which we now add: even when we used to call it only mass.][24] (Emphasis in original.)

A thematic position, or methodological thema, is a guiding theme in the pursuit of scientific work, such as the thema of expressing laws of constancy, of extremum, or of impotency, or quantification, or Rules of Reasoning. A thematic proposition, or thematic hypothesis, is a statement that is directly neither verifiable nor falsifiable, e.g., Einstein's principle of relativity in its modern sense and of the constancy of light velocity in free space—and Poincaré's basic belief in the ether.

The recognition of such thematic differences may help us understand the widespread feeling of paradox and outrage when a new thema is proposed in opposition to the prevalent ones—as was, of course, the case with relativity theory, so much so that Poincaré, to the end of his life in 1912, referred to Einstein's theory of relativity never once in print (and to Einstein, as far as I could discover, only once, on the subject of the photon, and in a derogatory way).

Finally, more can be said concerning the fact that it is the mark of a certain type of genius, in particular of the nonconservatory mind, to be "themata-prone"—at least for a time. On this point, nothing is more revealing than the transcript of the Solvay Congress of 1911, where Einstein and Poincaré met. The relativity theory was no longer an issue. The new topic was the quantum theory. It raised new, grave problems; for example, how could one understand probabilistic behavior on the atomic scale?

And right there and then, it was Einstein's turn to begin resisting a new theme, that of inherently probabilistic behavior. It now was he who began to warn that one should not feign a hypothesis located so far from the contingent plane. The next chapter in the history of science, from the point of view of thematic analysis, had opened.

NOTES

1. Alexandre Koyré, *Les Origines de la science moderne*, Diogène, 16:40–41, 1956.

2. *Ibid.*, p. 36.

3. Gustave Rados, *Rapport sur le Prix Bolyai*, Bulletin des Sciences Mathématiques, 30:105, 1906.

4. Henri Poincaré, La Valeur de la science (Paris: Ernest Flammarion, [1905]), p. 187. Poincaré had objected previously to the introduction of new ad hoc hypotheses in the then current theory of electric and optical phenomena (e.g., Henri Poincaré, Rapports du Congrès de Physique de 1900, pp. 22–23). Lorentz took the objections seriously enough to draw attention to them in his paper *Electromagnetic Phenomena in a System Moving at Any Velocity Smaller Than That of Light*, Kon. Akademie van Wetenschappen, Amsterdam, Proceedings of the Section of Sciences, English-language edition, 6:809, 1904. There he agreed that "surely this course of inventing special hypotheses for each new experimental result is somewhat artificial. It would be more satisfactory if it were possible to show by means of certain fundamental assumptions and without neglecting terms of one order of magnitude or another, that many electromagnetic actions are entirely independent of the motion of the system. Some years ago, I already sought to frame a theory of this kind. I believe it is now possible to treat the subject with better results." The fact was, however, that the objection against the proliferation of hypotheses remained in force against the new paper also: see Gerald Holton, *On the Origins of the Special Theory of Relativity*, American Journal of Physics, 28:627–636, 1960; reprinted here as Essay 5.

5. E.g., see Albert Einstein, Lettres à Maurice Solovine (Paris: Gauthier-Villars, 1956), p. viii; unpublished letter of Einstein to M. Besso, 6.3. 1952; Carl Seelig, Albert Einstein (Zurich: Europa Verlag, 1954), p. 69.

6. E.g., Henri Poincaré, La Science et l'hypothèse (Paris: Ernest Flammarion, 1902), p. 111; La Valeur de la science, pp. 272–273.

7. See citations in Augustin Sesmat, Systèmes de référence et mouvements, No. 486 of Actualités scientifiques et industrielles (Paris: Hermann, 1937), pp. 38–40.

8. Poincaré, La Science et l'hypothèse, Chapter 7.

9. Poincaré, *The Principles of Mathematical Physics*, an address at the International Congress of Arts and Sciences, St. Louis, 1904; republished in The Monist, 15:1–24, 1905.

10. Louis de Broglie, Savants et découvertes (Paris: Éditions Albin-Michel, 1951), p. 50. The reference is to the article *Sur la dynamique de l'électron*, Rendiconti del Circolo Matematico di Palermo, 21:129–176, 1906, which was presented by Poincaré on July 23, 1905. A short abstract appeared earlier in Comptes Rendus, 140:1504–1508, 1905. Einstein's first paper on relativity theory appeared in the issue of Annalen der Physik dated 26 September 1905, and bears the legend "Bern, Juni 1905 (Eingegangen am 30. Juni 1905)."

11. Poincaré, LA SCIENCE ET L'HYPOTHÈSE, p. 111.

12. SESMAT, *op. cit.*, p. 40.

13. Cf. also Poincaré, *Notice sur ses travaux scientifiques* (1902), quoted in Vito Volterra, *et al.*, HENRI POINCARÉ, L'OEUVRE SCIENTIFIQUE, L'OEUVRE PHILOSOPHIQUE (Paris: Felix Alcan, 1914), pp. 221–222. "Les hypothèses relatives à ce que je viens d'appeler la forme ne peuvent pas être vraies ou fausses, elles ne peuvent être que commodes ou incommodes. Par exemple, l'existence de l'éther, celle même des objets extérieurs, ne sont que des hypothèses commodes. . . ." Here Pierre Boutroux adds the query: "Ces conclusions sont-elles exagérées?" To answer this, we point to the rarity with which Poincaré expressed a conventionalistic attitude to the existence of the ether, and the seriousness and success with which he did regard the ether and its functions. Thus Poincaré was even more willing to grant the conventionalistic status of Euclidean geometry, but he also showed that it was not just a matter of convenience, e.g., in LA SCIENCE ET L'HYPOTHÈSE, p. 93: "La géométrie euclidienne n'a donc rien à craindre des expériences nouvelles." Similarly, in DERNIÈRES PENSÉES (Paris: Ernest Flammarion, 1913), p. 196: ". . . les atomes ne sont plus une fiction commode. . . ." The ether, also, was to Poincaré not merely "une fiction commode."

14. Poincaré, LA VALEUR DE LA SCIENCE, p. 47.

15. Henri Poincaré, *Les Rapports de la matière et de l'éther*, JOURNAL DE PHYSIQUE, 2:360, 1912; and DERNIÈRES PENSÉES, pp. 219–220.

16. Poincaré, *The Principles of Mathematical Physics*, p. 19.

17. Jacques Hadamard, *Le Problème des trois corps*, in Volterra, *et al.*, *op. cit.*, p. 88.

18. Charles Nordmann, *Henri Poincaré, son oeuvre scientifique—sa philosophie*, REVUE DES DEUX MONDES, Series 6, 11:347, 1912. In this eulogy, Nordmann accurately points out that Poincaré's work could be considered "comme le couronnement de trois siècles de recherches," and he adds that posterity would place Poincaré's work on celestial mechanics "à côté des immortels PRINCIPES de Newton. . . ." Paul Langevin hit astutely and more clearly the same point when he wrote two years later, in *Le Physicien*, in Volterra, *et al.*, *op. cit.*, p. 170: "[Poincaré in 1904] voyait avec un peu d'inquiétude ébranler, grâce aux instruments forgés par lui-même, le vieil édifice de la dynamique newtonienne qu'il avait récemment encore couronné par ses admirables travaux sur le problème des trois corps et la forme d'équilibre des corps célestes."

19. de Broglie, *op. cit.*, p. 51.

20. Théo Kahan, *Sur les Origines de la théorie de la relativité restreinte*, REVUE D'HISTOIRE DES SCIENCES ET DE LEURS APPLICATIONS, 12:162, 1959.

21. Poincaré, *L'Espace et le temps*, DERNIÈRES PENSÉES, p. 51. Cf. also RENDICONTI DE CIRCOLO MATEMATICO DI PALERMO, p. 129: "Que ce postulat [de relativité] jusqu'ici d'accord avec l'expérience, doive être confirmé ou infirmé plus tard par des expériences plus précises, il est en tout cas intéressant de voir quelles en peuvent être les conséquences."

22. Henri Poincaré, *La Mécanique et l'optique*; Book 3, Chapter 2 of SCIENCE ET MÉTHODE (Paris: Ernest Flammarion, 1908), p. 248.

23. A full discussion of *Experimental Investigations on the Mass of the Electron* is given in Chapter 7 of H. A. Lorentz, LECTURES ON THEORETICAL PHYSICS (London: Macmillan & Co., 1931), Volume III. Summarizing Kaufmann's experiments (which were not definitely disproved until the work of Neumann, 1914; and Guye and Lavanchy, 1916), Lorentz says: ". . . a number of sources of experimental error can be pointed out. Thus, e.g., the vacuum was not high enough. In fact, now and then a spark passed between the plates of the condenser, which shows that there was always some ionisation current left between these plates, and that, therefore, the homogeneity of the electric field was not above doubt. In fine, no definite verdict can be based upon Kaufmann's experiments in favour of either theory" (p. 274). [Einstein's reaction to Kaufmann's "disproof" is of considerable importance and will be discussed in Essay 8 (*Mach, Einstein, and the Search for Reality*).]

24. Poincaré, LA SCIENCE ET L'HYPOTHÈSE, p. 195.

7 INFLUENCES ON EINSTEIN'S EARLY WORK

N March 11, 1952, Albert Einstein wrote to Carl Seelig:

> Between the conception of the idea of this special relativity theory and the completion of the corresponding publication, there elapsed five or six weeks. But [he added rather cryptically] it would be hardly correct to consider this as a birth date, because earlier the arguments and building blocks were being prepared over a period of years, although without bringing about the fundamental decision.[1]

Can we get some idea what may have happened during those years and what—or who—helped to bring about the "fundamental decision"? How large or how small was the effect of the work of earlier physicists? Is there some strong influence that has so far been overlooked?

The style of Einstein's great first paper of 1905 on relativity, *Zur Elektrodynamik bewegter Körper*,[2] was markedly different from what was then current, accepted practice; different, for example, from H. A. Lorentz's or Emil Cohn's, whose theories of electrodynamics of moving bodies were taken quite seriously at the time, for example, by Bucherer and Abraham. Nor are there really sufficient clues in the literature which biographers cite that help us to understand the structure of that paper. It starts with a curious question: why is there in Maxwell's theory

This article is slightly condensed from the version originally published in THE AMERICAN SCHOLAR, Volume 37, Winter, 1967–8, pp. 59–79.

one equation for finding the electromotive force generated in a moving conductor when it goes past a stationary magnet, and another equation when the conductor is stationary and the magnet is moving? After all, it is only the relative motion between conductor and magnet that counts. Then, without specifically mentioning by name any of the now-so-famous experiments, the introductory section of the paper ends by dismissing the conceptions of absolute motion and of the ether. There follows the crucial section, "Kinematischer Teil," which develops relativistic kinematics through a fundamental philosophical examination of the concepts of space and time. Only later on comes the treatment of Maxwell's equations, and finally, almost as an afterthought, some predictions about electron motion, ending with the equations "according to which the electron must move in conformity with the theory presented here." The only one who is given credit for being helpful in discussions leading to the writing of the paper is a friend, co-worker at the Patent Office, and former fellow-student of Einstein, an engineer named Michelange Besso.

This was a strange and unique way of writing a paper on electrodynamics in 1905. Max von Laue, one of the first and foremost partisans for and contributors to relativity theory, nevertheless confessed to Miss Margot Einstein in a letter of October 23, 1959, that he had felt fundamental difficulties for a long time. He wrote that after the publication of Einstein's paper in 1905, "slowly but steadily a new world opened before me. I had to spend a great deal of effort on it . . . . And particularly epistemological difficulties gave me much trouble. I believe that only since about 1950 have I mastered them." Leopold Infeld similarly writes:

> The title sounds modest, yet as we read it we notice almost immediately that it is different from other papers. There are no references; no authorities are quoted, and the few footnotes are of an explanatory character. The style is simple, and a great part of this article can be followed without advanced technical knowledge. But its full understanding requires a maturity of mind and taste that is more rare and precious than pedantic knowledge, for Einstein's paper deals with the most basic problems; it analyzes the meaning of concepts that might seem too simple to be scrutinized.[3]

So, in retrospect, it is not entirely surprising that it took also a long brooding period for Einstein himself before this remarkable work was hatched. But where, when, and from whom might Einstein have obtained some of his point of view, his questions, and his method?

Archival Sources

One way to begin to find out is, of course, to look at documents. Therefore, after these questions had raised themselves, I began to search for whatever documents might have survived that would be relevant. In the possession of Albert Einstein's Estate, and kept for the time being at the Institute for Advanced Study at Princeton, there are indeed documents—about twenty metal file drawers full, counting only the more scientific part of Einstein's manuscripts and the largely unpublished correspondence.

Einstein's devoted secretary, Miss Helen Dukas, has been putting the documents into systematic order; over the past few years, we have catalogued the material. Copies of correspondence are continually being received and added, but a preliminary archival calendar is now fairly well finished, and it has begun to be put to use in scholarly work. In this paper itself I shall be drawing on a number of hitherto unpublished documents, and thereby can hope to furnish one example of the many problems that can now be attacked with the aid of the archives.

In addition to the letters, there are eleven notebooks, starting from Einstein's student days; a few travel diaries; folders upon folders of published manuscripts, many in early draft; and several dozen unpublished manuscripts. All this survived more or less by good luck. On returning from a trip to the United States in the winter of 1932–33, Einstein found on reaching Europe that Hitler's supporters had taken over in Germany. Einstein never again set foot on German soil, and most of the correspondence was brought out by diplomatic pouch through the French Embassy.

One of the first things I looked for was, of course, any manuscript or draft of Einstein's 1905 paper on relativity theory. But particularly for the early papers, Einstein must have destroyed or discarded the manuscripts when they were returned from the printer, if they ever were. Einstein himself had occasion to regret this. During a War Bond drive in the United States during World War II, he was asked to donate the manuscript, and when he could not oblige in this way, he wrote the paper out all over again in longhand. Miss Dukas recalls that she dictated it to him from his published paper, and that he interrupted her, shaking his head and saying, "I could have said this more simply." The temptation to write something different, however, was overcome, and the manuscript was "auctioned" off to the person who promised to buy the largest amount of War Bonds (which turned out to be several million dollars), and who, in turn, deposited the manuscript in the Library of Congress in Washington.[4]

Thus there is no contemporaneous draft or manuscript of the 1905 paper from which one might learn something of its genesis. But there are two notebooks that Einstein kept while still a student at the Polytechnic Institute (E.T.H.) at Zurich during the period 1897 to 1900. Both are sets of lecture notes taken in the physics course given by Heinrich Friedrich Weber whose special field of work was alternating-current technology. One of them is on heat and thermodynamics, the other on technical problems such as liquefaction of gases (with detailed drawings), and electricity from Coulomb's Law to induction. But it does not even go to Maxwell's work! And on that hangs part of this tale. For what was left out in class was exactly what young Einstein was waiting for. The fact that he was thrown on his own devices had, as we shall see, some interesting consequences in the genesis of relativity theory.[5]

Reading at Home

As Besso wrote (in his notes of August 1946 for Strickelberg's article on Einstein in Switzerland), Einstein came to the Aarau Kanton-School in 1896 "with the [then much debated] questions concerning the palpability [*Greifbarkeit*] of ether and of atoms" in mind. When he went on to the Polytechnic Institute in Zurich, the lectures on physics made no great impression on Einstein, who found his teachers' discussions "self-explanatory." It was indeed his professor, Weber, who, Besso reports, said once to Einstein, "You are a clever fellow! But you have one fault: One can't tell you anything, one can't tell you anything." Clearly, Weber could not.[6]

This circumstance is corroborated by Einstein's classmate, Louis Kollros:

There was not very much theoretical physics done at the Poly, which was strong in mathematics. . . . [Weber's] lectures concerning classical physics were lively; but we waited in vain for an exposition of Maxwell's theory. We knew that the theory was founded in the identity of the transmission of electricity and light, and that the work of Hertz concerning electric waves had verified the theory. We would have gladly learned more about it. Above all, it was Einstein who was disappointed [for, as Einstein recalls in his *Autobiographical Notes*, it was "the most fascinating subject at the time"]. In order to fill this gap, he undertook to study on his own the works of Helmholtz, Maxwell, Hertz, Boltzmann, and Lorentz.[7]

Kollros's list of authors reminds us of the famous passage in Einstein's *Autobiographical Notes* which does seem relevant to the question of early influence shaping the thoughts expressed in the 1905 relativity paper. The passage concerns the period of 1897–1900:

I entered the Polytechnic Institute of Zürich as a student of mathematics and physics. There I had excellent teachers (for example, Hurwitz, Minkowski), so that I really could have gotten a sound mathematical education. However, I worked most of the time in the physical laboratory, fascinated by the direct contact with experience. The balance of the time I used in the main in order to study at home the works of Kirchhoff, Helmholtz, Hertz, etc.[8]

The really interesting part may well be the study of the last, the "etc." Who is hiding behind the phrase "et cetera"? Could it be somebody who prepared Einstein's way in presenting his relativity theory? We must, of course, not dismiss Kirchhoff, Helmholtz, and Hertz, or for that matter Boltzmann, Mach, Poincaré, and Lorentz. But these do not suffice to explain the form of Einstein's 1905 paper. If someone else exists, we should be able to find him.

Maxwell, Direct and Indirect

First, a look at the others. From June 1902 until October 1909, Einstein was at the Patent Office in Bern. According to a manuscript note from Besso in the Einstein Archives, the applicant for the position was expected to have an "intimate acquaintance with Maxwell's theory."[9] Einstein qualified on this score by the time he submitted his relativity paper in June 1905—of that there is no doubt—and he must have known Maxwell's theory earlier. There are a number of corroborating statements other than Kollros's and Besso's, for example, in a letter to von Laue, sent by Einstein from Princeton on January 17, 1952:

Dear Laue: I now have received your book concerning special relativity theory and find that it is very good. . . . [But] when one looks over your collection of proofs of the special relativity theory, one becomes of the opinion that Maxwell's theory is unquestionable. But in 1905 I already knew certainly that Maxwell's theory leads to false fluctuations of radiation pressure and, with it, to an incorrect Brownian motion of a mirror in a Planckian radiation cavity. In my view, one could not get around ascribing to radiation an objective atomistic structure which, of course, does not fit into the frame of Maxwell's theory. . . . Unfortunately, the fifty years which have since passed have not brought us closer to an understanding of atomistic structure of radiation.

Granting that Einstein obviously knew Maxwell's theory by 1905, the question is left through which books he learned it. It may have been by direct study of Maxwell's work, although there is no documentary evidence for this. At any rate, direct study would not have been the only or even the most important way. Maxwell came to most German students of physics first through the works of Helmholtz, Boltzmann, and Hertz.[10] They are in many ways quite different, but they also have at least one element in common: that these presentations of Maxwell's theory are quite un-Maxwellian, that, in different degrees, their style is even further from that of Maxwell than from Einstein's paper. On this point, a brief word must suffice here. For example, to a contemporaneous physicist in England and France, Helmholtz's way of thinking must have looked quite terrifying. Fully half of his introductory volume of the LECTURES ON THEORETICAL PHYSICS is spent on the following topics: philosophy and science; physical science; critique of the old logic; concepts and their connotations; hypotheses as bases for the laws; the completeness of scientific experience and its practical significance; and so forth.[11]

Maxwell's work proper is presented in Volume 5 of Helmholtz's LECTURES ON THEORETICAL PHYSICS, issued in 1897. The terminology there is one Einstein used to some extent later. What catches our eye is that there is very little attention to experimentation. One cannot, for example, find a reference to the Michelson experiments which, after all, were first tried under the sympathetic eye of Helmholtz himself. Even the section entitled "The Necessary Properties of the Ether" has no reference to experiments. And in the only paper that Helmholtz wrote specifically on the subject of Maxwell's theory, called *Consequences of Maxwell's Theory Concerning the Motion of the Pure Ether,* there is not a single mention of actual experiments.

What Einstein might have obtained from studying Helmholtz's version of Maxwell's theory is first of all a reinforcement of a taste for a consciously epistemological approach, and a confirmation that in this area experiments do not count crucially.

Reading Hertz, whose collected works were available by 1895, and whose notation in electromagnetism he used to a large degree, Einstein will have seen Hertz's first thorough essay on *The Fundamental Equations of Maxwell's Electrodynamics,* published in 1884, and the article significantly entitled *Concerning the Fundamental Equations of Electrodynamics for Moving Bodies* of 1890. Even this greatest of experimenters in the field of electromagnetism makes no explicit mention of the "ether" experiments that have loomed so large in recent discussions of the origins of relativity theory. On the other hand, the main effect that a study of Hertz's work might have had upon a reader like Einstein is perhaps best characterized by Hertz's own remarks in the PRINCIPLES OF MECHANICS: "In general, I owe very much to the fine book concerning the development of mechanics by Mach." It was one of very many forces urging young Einstein toward Ernst Mach. As he said later in his *Autobiographical Notes,* Mach's SCIENCE OF MECHANICS "shook this dogmatic faith" in "mechanics as the final basis of all physical thinking. . . . This book exercised a profound influence upon me in this regard while I was a student. . . . Mach's epistemological position . . . influenced me very greatly. . . ."[12]

Ernst Mach

Indeed, it is an ironic circumstance that the state of contemporary research physics during the period when the young Einstein began to work on special relativity was really not characterized by such a degree of dogmatic rigidity as he thought. As Stephen Brush has recently pointed out, the mechanistic view of physical reality was then defended by only a "few lonely men such as Boltzmann. . . . The most 'advanced' and 'sophisticated' theories were those that took a purely phenomenological viewpoint: scientific theories should deal only with the relations of observable quantities, and should strive for economy of thought rather than trying to explain phenomena in terms of unobservable entities. . . ."[13] In short, around 1900 Mach's views were no longer those of an isolated fighter, the role in which he rather liked to see himself and in which he appeared in his books that young Einstein read.[14]

No matter how some of the younger physicists of the time wrestled with the problems of physics, the use of conceptions developed in nineteenth-century physics seemed to them merely to produce failure and despair. Something of this flavor comes through in letters in Einstein's correspondence, and a famous passage in Einstein's *Autobiographical Notes*.[15] It is not too much to say that the new physics they fashioned was first of all "*eine Physik der Verzweiflung*" [a physics of despair]. And here the role of Mach as iconoclast and critic of classical conceptions was particularly important; for whether or not Einstein's assessment of the contemporary scene was right, it is certain that Mach's critical force and courage made a strong impression on him, as on so many others.[16]

Poincaré and Lorentz

The influence on and response to Einstein's work on the part of both Poincaré and Lorentz have also been a fascinating problem for the historian of recent science. Although the old myths will not die quickly, they have been pretty well exploded.[17] To put it briefly, and without intending in the slightest to denigrate Poincaré's enormous accomplishments, we may say that Poincaré saw the "crisis" in physics as one primarily revolving about experimental difficulties, and therefore involving neither epistemological nor fundamentally different theoretical reorientation. This is, of course, directly antithetical to Einstein's view of the matter at about the same time: the new *experimental* findings, such as the Michelson-Morley experiment, neither provoked the crisis as Einstein saw it, nor were they guides to the new orientation needed. That Einstein's work in 1905 was independent of Poincaré's investigations on electromagnetism in 1904–05 has now been repeatedly and adequately established.

When it comes to the debt or independence of Einstein with respect to Lorentz's work, and the response of Lorentz to Einstein's early papers, the record is also quite clear. Einstein and others have repeatedly said that he did not know of Lorentz's 1904 paper on electromagnetic phenomena.[18] On this well-worked ground, perhaps one need only to point out anecdotally how difficult it would have been for an almost unknown Patent Office employee in a Swiss town such as Bern to have had direct access to the PROCEEDINGS OF THE AMSTERDAM ACADEMY in which Lorentz published the 1904 paper. In the Rijksarchief at The Hague, Holland, I found a letter from von Laue, writing to Lorentz on November 30, 1905, from Berlin, apparently for the first time, and in his capacity as *Assistent* at the Institute for Theoretical Physics (therefore, as Planck's assistant):

> Since the Proceedings of the Amsterdam Academy are here more difficult to obtain than other journals—they exist only in the Royal Library, and it lends out recent journals only for a day—I take the liberty of expressing to you the request to send me, if possible, a reprint of your publication, *Electromagnetic Phenomena in a System Moving with Any Velocity Smaller Than That of Light*. . . .[19]

If one had to summarize the difference between Lorentz's and Einstein's relativity physics in a sentence, one might say this: Lorentz's work can be seen somewhat as that of a valiant and extraordinary captain rescuing a patched ship that is being battered against the rocks of experimental fact, whereas Einstein's work, far from being a direct theoretical response to unexpected experimental results, is a creative act of disenchantment with the mode of transportation itself—an escape to a rather different vehicle altogether.

An Almost Forgotten Teacher

This brings us back for a final assault on the problem of the possible antecedents of Einstein's work. Neither the shape nor the content of Einstein's 1905 paper is adequately explained as a sequel to the chain Lorentz-Poincaré, or Maxwell-Helmholtz-Boltzmann, or even Kirchhoff-Mach-Hertz. It is, of course, possible that Einstein's 1905 paper was a Minerva-like creation with no direct preparatory antecedent. And not having found any models in the works of the major contributors of the time, we may be tempted to make this assumption, even if reluctantly.

But it turns out that we do not have to do this. Working with the documents in the Princeton Archives, I came across a clue that raises the possibility of entertaining a quite different and unconventional view of the influences on Einstein's thought processes leading to his 1905 paper. Once, and quite casually, there appeared the name of a now almost unknown physicist who has here not yet been mentioned. It is August Föppl—a name vaguely familiar to a number of older German scientists and engineers, but to almost nobody else.

He sounds very much out of place compared to "Kirchhoff, Helmholtz, Hertz, etc."—so much so that he might well have ended up among the "et ceteras" mentioned in the *Autobiographical Notes* of Einstein. And indeed, the search for the identity of August Föppl starts very badly: born in 1854, Föppl was, at the age of thirty-six, a technical high school teacher and administrator in Leipzig when he published his first book, a rather pedestrian little outline of elementary physics. From the first exercise of this *Leitfaden* (how rapidly must a disc spin to throw off a lightly adhering object?) to the last (explain parallel winding in a.c. machines), there is nothing to indicate that this man could ever enter our story.

Two years later, now a civil engineer in Leipzig, Föppl published his first real book, DAS FACHWERK IM RAUME. The book works up some previous essays that Föppl had used for his degree candidacy at the University of Leipzig in 1886, and, one supposes, in connection with this subsequent task of helping to design the Markthalle of Leipzig. Yet, the book is by no means intended as a mere practical manual. On the contrary, Föppl objects to the definition by which *Fachwerk* usually is regarded as a structure made of solid straight rods, to carry loads. "For me it is a purely ideal structure" (p. 3). And in defense of this point he plunges into an epistemological digression concerning the process and warrant of introducing concepts such as rigid bodies or ether, "which by no means in every respect coincide with their *'realen Urbildern.'*"

And then, in 1892, Föppl was called to the University of Leipzig to teach, of all things, agricultural machinery. As he later confessed cheerfully, he knew very little about this subject, so he spent the summer touring factories to find out. His versatile intelligence seems to have helped him to absorb in a short time enough to enable him to teach the course soon thereafter. But the subject was not what his mind reached out for. And so, perhaps largely out of boredom, he began to write a book in his spare time. It became a treatise entitled INTRODUCTION TO MAXWELL'S THEORY OF ELECTRICITY, published in 1894. The work was a success: upon its publication, it "aroused astonishment [*Aufsehen*] in the profession, for at that time the electrodynamic considerations of the great English physicist, Maxwell, had hardly gained any ground"—to cite the introductory essay of the editors of the *Festschrift* published in Föppl's honor on his seventieth birthday, January 25, 1924.[20] This was the ancestor of a series of revisions, known to students of electricity as Abraham-Föppl (1904, 1907, 1911), later Abraham-Becker (1930), later Becker-Sauter (1958, 1964)—although the original was very different from all these revised editions.

Föppl's book was widely bought, particularly because of the author's ability to put Maxwell's theory clearly to engineers. Perhaps as a result of this success, Föppl was called in 1894 to the Technical University at Munich—the very city in which Einstein was then living, still a boy of fifteen—and there Föppl stayed and wrote volubly, although as far as I can find out, he never taught from his book on Maxwell's theory.

Before we look at Föppl's MAXWELL, we must size up the particular style that characterized his thinking, and this is done most simply by considering Föppl's immensely successful next work, the VORLESUNGEN ÜBER TECHNISCHE MECHANIK, published from 1898 on in several parts. (The FESTSCHRIFT, page vi, notes that up to 1924 nearly 100,000 volumes of this work were sold all over the world.) Föppl himself sent seven editions through the press, and others after him continued his work.

The foreword of Föppl's MECHANIK, dated June 1898, tells us something rather revealing about his special talent as a teacher and writer. His students, he confesses, have sometimes complained that he "proceeds too slowly rather than too fast," but he places very special emphasis on laying the foundations carefully. It is almost as if he had a special eye for a reader who might not have the benefit of formal lectures on the subject, and who might even have bad holes in his formal background.

After the encouraging foreword, the reader comes up against the first two sentences of the text, typical in their mixture of straightforwardness and discursiveness: "Mechanics is a part of physics. Its teaching rests, as does that of all natural sciences, in the last analysis on experience." And with this, he turns to a discussion of the meaning of the term "experience" [*Erfahrung*]. By page 4, he confesses:

> It is now, of course, no longer a question of mechanics, but a philosophical and epistemological question. Its discussion can, however, not be circumvented in an introduction to mechanics, no matter how, on the basis of earlier unfavorable experiences, one may shy away from touching on philosophical questions in the exact sciences.

Föppl announces that his exposition of antimetaphysical and self-conscious empiricism is shared by leading scientists generally, and he specifically calls on three by name, in whose spirit he believes he is proceeding: Kirchhoff, Heinrich Hertz (once at Munich University), and Ernst Mach. Indeed the volume on dynamics starts with the section entitled "Relative Motion," and in the preface Föppl says again: "One will notice that the [early part of the volume] is strongly influenced by the work of Mach, which made a persistent impression on me."[21]

Föppl's Maxwell

We are beginning to see some evidences of the kind of approach to physics which would appeal to a young reader with the kind of background, or lack of background, of Einstein in the late 1890's. This impression is much reinforced when we now return to Föppl's INTRODUCTION TO MAXWELL'S THEORY OF ELECTRICITY of 1894. He writes in his foreword that now not only the professional physicist, the teacher, and the student in physics, but also

> the scientifically trained electro-technical engineer[22] is attempting to make himself acquainted with the foundations of this [Maxwell's] theory in which today one may see with great probability the permanent foundation of every physical research in this domain. . . . With this there is a recent demand for an exposition of Maxwell's theory that is as widely understandable as possible, but also scientifically correct.

Maxwell's original work, Föppl reports, is too difficult, and it has many mistakes or incompletenesses which in the meantime have been removed. Boltzmann, he says, has written an exposition, but although nothing better of its kind can be done, Föppl sees need for another, different attempt. What Föppl particularly wants to provide is a "clear understanding of the concepts and considerations of this theory in order to give the reader the ability for his own, unsupervised work" ["*selbstständigen Arbeiten*"]—in short, just the kind of book an interested student would want if he were deprived of Maxwell's theory in course lectures.[23]

In Föppl's book we find six main sections: the first is on vector calculus; the second on fundamental electricity (Gauss's Theorem, Coulomb's Law, magnetism, induction, et cetera); the third and fourth are the usual extensions (pondermotive forces, vector potential, energy relations in the electromagnetic field between stationary conductors). So far, it is all done competently and patiently, but as if it were merely a prelude to something else.

Then we come to the fifth main section, which turns out to be of particular interest to us. It is entitled *The Electrodynamics of Moving Conductors* [*Die Elektrodynamik bewegter Leiter* (pp. 307–356)], and the first chapter in it is entitled "Electromotive-Force Induction by Movement." The first paragraph in this first chapter is "Relative and Absolute Motion in Space," and starts in an unusual way:

> The discussions of kinematics, namely of the general theory of motion, usually rest on the axiom that in the relationship of bodies to one another only relative motion is of importance. There can be no recourse to an absolute motion in space since there is absent any means to find such a motion if there is no reference object at hand from which the motion can be observed and measured.... According to both Maxwell's theory and the theory of optics, empty space in actuality does not exist at all. Even the so-called vacuum is filled with a medium, the ether.
>
> ... the conception of space without this content [ether] is a contradiction, somewhat as if one tries to think of a forest without trees. The notion of completely empty space would be not at all subject to possible experience; or, in other words, we would first have to make a deep-going revision of that conception of space which has been impressed upon human thinking in its previous period of development. *The decision on this question forms perhaps the most important problem of science of our time.* (Italics supplied.)

Föppl is not ready to give up the ether or absolute motion, but he knows where the physically important problems lie. He continues a few lines later in this way:

> When in the following we make use of laws of kinematics for relative motion, we must proceed with caution. We must not consider it as a priori settled that it is, for example, all the same whether a magnet [moves] in the vicinity of a resting electric circuit or whether it is the latter that moves while the magnet is at rest.

This, we recall, describes precisely the experimental situation with which Einstein's paper starts—and Föppl adds immediately a rather familiar kind of *Gedanken*-experiment: "To decide this question, we can consider a third case." He proposes to think of both magnet and conductor moving together, with no relative motion between them. Experience shows, he says, that in this case the "absolute motion" in itself causes no electric or magnetic force in either body. This thought experiment is then quickly developed to show that in the previous two cases what counts is only relative motion.

Later, Föppl goes on to discuss the interaction of moving magnets and resting conductors (pp. 314–320), and resting magnets and moving conductors (pp. 321–324). The rest of this section, too, may be directed first of all to engineers (unipolar induction, emf for a moving conductor, magnetomotive force, motion of a wire loop in a magnetic field, et cetera).

There is a rather brief last (sixth) part, a summary of the other aspects of Maxwell's work, including electromagnetic waves—again, virtually without a reference to the actual ether experiments. But our attention stays fixed on the fifth section of Föppl's book; there, and in portions of the rest of the book, is the kind of thinking that would indeed have appealed to Einstein, and that in fact is far closer to the sequence and style of argument of Einstein's 1905 paper than the work of any of the others who are more familiar—far more so than the books on electromagnetism by Maxwell, Helmholtz, Boltzmann, Hertz, or Runge, for example.

Other References to Föppl

But before the parallelism with Föppl carries the day, we must ask for more evidence. After all, earlier we dismissed the suggestion that Einstein built on Poincaré's and Lorentz's work of 1904, even though there are certain parallels.

We therefore must ask, why did not anyone else who knew Einstein intimately vouch in print for the fact that Einstein had read Föppl's book? Thus I asked my former teacher and colleague, Einstein's biographer, Philipp Frank, why he had made no mention of Föppl in his book.[24] Frank replied that he thought he had mentioned Föppl, and I showed him my copy of the biography in which it was plain that he had not. This was a considerable surprise to Frank, but after some thought he referred me to the German edition of his book.[25] In the foreword, Frank explained that this, the German edition, is the first complete edition of his manuscript as written in 1939–1941 (whereas the English-language publisher had made many cuts). And there, on page 38, Frank writes that during his years as a student at the Polytechnicum in Zurich,

> Einstein threw himself into the work of these classics of theoretical physics [of the late nineteenth century], the lectures of Helmholtz, Kirchhoff, Boltzmann, the electricity theory of J. C. Maxwell and H. Hertz, and their exposition in the textbook of Abraham-Föppl. Einstein buried himself with a certain fanaticism day and night in these books from which he learned how one builds up the mathematical framework and then with its help constructs the edifice of physics.[26]

And there is one other guide that leads us to Föppl. There exist, of course, dozens of biographies of Einstein—most of them written at second or third hand. Here, the Archives at Princeton held a surprise for me. I knew that a certain Anton Reiser had published an Einstein biography in English in 1930,[27] when Einstein was still in Berlin. Despite a pleasant foreword by Einstein ("the author of this book knows me rather intimately . . . I found the facts of the book duly accurate. . . ."), at first glance it can hardly be considered a reliable book: for quite apart from the suspicious circumstance that no German edition was ever brought out, there are also no credentials for the author of the book. No other publications by Reiser exist, and a search in the usual sources leads one to suspect that he simply does not exist. Now the material in the Archives shows that the name Anton Reiser was a pseudonym for Rudolf Kayser; and Rudolf Kayser was Einstein's own son-in-law who had proceeded with the biography with Einstein's acquiescence. We return therefore to Reiser's obscure and difficult-to-find book with new respect—and sure enough, there we find Föppl again: ". . . the scientific courses offered to him in Zurich soon seemed insufficient and inadequate, so that he habitually cut his classes. His development as a scientist did not suffer thereby. With veritable mania for reading, day and night, he went through the works of the great physicists—Kirchhoff, Hertz, Helmholtz, Föppel" [*sic*].

So we may perhaps feel that the missing signal has been recovered from the noise level of the "et cetera." But in a real sense, genius does not have predecessors. It would of course be absurd to claim that Föppl in any way "explains" Einstein, or even that there is a simple chain connecting Föppl's book and Einstein's relativity theory. No, the proper attitude here is exactly Philipp Frank's in the passage cited: from books such as these, Einstein learned how one builds up the framework, and then with its help constructs the edifice of physics.

In balance we may say the role of Föppl was that he, with Helmholtz on the one hand and Mach on the other, reinforced the unique aspects that made Einstein's 1905 paper so important (and, for his contemporaries, so difficult—unlikely though this now seems to a generation of physicists brought up on Dirac and Feynman). As the various contemporaneous treatments of electrodynamics show, in principle a great diversity of possible roads were open to Einstein. What Föppl was capable of providing in helping to shape Einstein's thought processes as he was fashioning the relativity theory was, first of all, encouragement to go ahead in a manner so very different from that taught to him at school and presented in all the respectable books by the foremost physicists. It helps us to understand better what to this day remains as the most startling part of Einstein's relativity paper: a mixture that contains a good share of youthful philosophizing as a part of doing physics; the recognition that the fundamental problem to be cracked is how to achieve a new point of view on the conceptions of time and space; the attention to *Gedanken*-experiments, and, conversely, a quite low interest in the actual detailed experiments which so many of our texts make appear to be the point of departure of relativity theory.[28]

And there is also, I believe, some poignancy in the discovery of how Föppl may have reached across to Einstein—the book of an "outsider" who did not have students to whom to teach its contents in lectures, falling into the hands of a student who, regarded as an "outsider" by his teachers, was looking to this book for the material and the stimulation that he could not get in their lectures.

In this study, we have looked at some documents that Einstein surely did not initially mean to be used for historic research. We can nevertheless be sure that Einstein would have understood and not objected to this purpose. For as he wrote to Besso (November 30, 1949): "When I write you something, you can show it to anyone you like. I have long been above making secrets." And in another unpublished manuscript (No. 17, undated, not before 1931): "Science as an existing, finished [corpus of knowledge] is the most objective, most unpersonal [thing] human beings know, [but] science as something coming into being, as aim, is just as subjective and psychologically conditioned as any other of man's efforts. . . ."And that aspect, he went on to say, one should certainly "permit oneself also." Happily, he and his friends and colleagues have done just that. They have left us the record of "science coming into being," and thereby they have enriched our understanding of scientific work in the larger sense.

I wish to acknowledge with gratitude the permission to quote excerpts, and other help received from the Trustees of the Albert Einstein Estate, and particularly from Miss Helen Dukas. My thanks go to the Rockefeller Foundation for partial financial support for cataloguing the documents; the director and staff of the Institute for Advanced Study at Princeton have been most hospitable throughout this work.

NOTES

1. The literary rights to quotations from the writings of Albert Einstein belong to the Estate of Albert Einstein; permission to use quotations should be secured from the Executor.

2. Albert Einstein, *Zur Elektrodynamik bewegter Körper*, ANNALEN DER PHYSIK, 17:891–921, 1905.

3. Leopold Infeld, ALBERT EINSTEIN, HIS WORK AND ITS INFLUENCE ON OUR TIMES (New York: Charles Scribner's Sons, 1950), p. 23.

4. And so it is both true and false that "Einstein's manuscript is now in the Library of Congress, Washington," as Ernst Weil tells us in ALBERT EINSTEIN, 14TH MARCH 1879 (ULM)—18TH APRIL 1955 (PRINCETON, N.J.): A BIBLIOGRAPHY OF HIS SCIENTIFIC PAPERS, 1901–1954 (London: Robert Stockwell, 1960), p. 6.

5. But lest we regard the Swiss school uncommonly behind the times, we should recall the situation was not greatly different at Cambridge University, among many others. Ebenezer Cunningham wrote recently (Private communication to Mr. Stanley Goldberg): "You ask about books in use in my early days. The Tripos which I took in 1902 hardly dealt at all with Electrodynamics. Maxwell's work was too recent and had not reached text-book stage. Abraham's book was unknown, though it became the background of my own teaching on the subject on my return to Cambridge in 1911. I am tempted to think that I really introduced systematic teaching on the subject to Cambridge."

6. Weber's successor, however, was the well-known physicist Pierre Weiss, who brought Einstein back to Zurich from Prague in 1912.

7. Louis Kollros, *Erinnerungen eines Kommilitonen*, in Carl Seelig, HELLE ZEIT—DUNKLE ZEIT (Zurich: Europa Verlag, 1956), pp. 17–31.

8. Albert Einstein, *Autobiographical Notes*, in P. A. Schilpp, ed., ALBERT EINSTEIN: PHILOSOPHER-SCIENTIST (Evanston, Ill.: Library of Living Philosophers, 1949), p. 15.

9. A pamphlet entitled *Erinnerungen an Albert Einstein*, covering the period of his employment at the Patent Office in Bern, and published by that office in 1965, contains a note by Einstein's colleague, Joseph Sauter, written in 1955. Sauter recalled that Einstein was admitted to the post of Technical Expert (Third Class) "without possessing the Diploma of Engineering, but as a physicist familiar with the theory of Maxwell. That theory was not yet on the official syllabus of the Polytechnic Institute of Zurich."

10. It is technically possible that Einstein might have seen Henri Poincaré's ÉLECTRICITÉ ET OPTIQUE, Paris: Georges Caré, 1890. Volume I, on Maxwell's

work, was the set of lectures of 1888–1889, published in 1890; and Volume II, on the work of Helmholtz and Hertz, was the lectures of 1889–1890, published in 1891. Both works were translated into German and appeared in 1891. Articles by Poincaré were reviewed in the literature by Minkowski in 1890 and Hurwitz in 1896. But I have seen no evidence that Einstein read Poincaré's original or translated lectures.

11. Although the great experimentalists Helmholtz and Hertz are least guilty in this respect, their books on electromagnetism do remind one of the passage by J. T. Merz, in which he compared the scientific spirit in British and Continental science: "Continental thinkers, whose lives are devoted to the realisation of some great ideal, complain of the want of method, of the erratic absence of discipline, which is peculiar to English genius. The fascination which practical interests exert in this country appears to them an absence of full devotedness to purely ideal pursuits. The English man of science would reply that it is unsafe to trust exclusively to the guidance of a pure idea, that the ideality of German research has frequently been identical with unreality, and that in no country has so much time and power been frittered away in following phantoms, and in systematising empty notions, as in the Land of the Idea; but he would as readily admit that his own country is greatly deficient in such organisations for combined scientific labour as exist abroad, and that England possesses no well-trained army of intellectual workers." A History of European Thought in the Nineteenth Century (London: William Blackwood & Sons, 1904), Volume I, pp. 251–252.

12. Einstein, *Autobiographical Notes*, p. 21. Besso, writing in late 1947 to Einstein, reminds him that during the year of 1897 or 1898, Besso had drawn Einstein's attention to Mach. The correspondence of Einstein abounds with references to the influence of Mach in the formative years of relativity theory. For example, in a letter of August 8, 1942, to A. S. Nash, Einstein wrote: "In the case of Mach, the influence was not only through his philosophy, but also through his critique concerning the fundamentals of physics."

13. Stephen Brush, *Thermodynamics and History*, The Graduate Journal, 7:477–566, 1967.

14. On the contrary, it was the great H. A. Lorentz and Henri Poincaré whose styles were coming to be out of step with the daring new, nonclassical physics exemplified by the Curies, Rutherford, Einstein himself, and at least at one point, even by Planck. Thus in an unpublished letter from Berlin (1931) from Max Planck to R. W. Wood, kindly made available by Professor Wood's son to the American Institute of Physics Center for the History and Philosophy of Physics, and on deposit at their Archives in New York: "Dear Colleague: You expressed recently, at our nice dinner at Trinity Hall, the wish that I should

describe to you more concerning the psychological side of the considerations which led me at the time to postulate the hypothesis of energy quanta. Here I want to accommodate your wish. Briefly put, I can describe the whole effort as an act of desperation, for by nature I am peaceful and against dubious adventures. But I had been fighting already for six years, from 1894 on, with the problem of equilibrium between radiation and matter without having any success; I knew that this problem is of fundamental significance for physics; I knew the formula which provides the energy distribution in the normal spectrum; a theoretical explanation, therefore, *had* to be found at all cost, whatever the price. Classical physics was not sufficient, that was clear to me. . . . [Except for the two laws of thermodynamics] I was ready for any sacrifice of my established physical convictions. Now Boltzmann had explained that thermodynamic equilibrium comes about through statistical equilibrium, and when one applied these considerations to the equilibrium between matter and radiation, one finds that one can prevent the deterioration of energy in radiation by means of the supposition that energy is from the beginning forced to remain in certain quanta. This was a purely formal assumption, and I did not really think much about it except just this: No matter what the circumstances, may it cost what it will, I had to bring about a positive result." (Translation by G. H.)

15. Einstein, *Autobiographical Notes*, pp. 51–53: "Reflections of this type made it clear to me as long ago as shortly after 1900, i.e., shortly after Planck's trail-blazing work, that neither mechanics nor thermodynamics could (except in limiting cases) claim exact validity. By and by I despaired of the possibility of discovering the true laws by means of constructive efforts based on known facts. The longer and the more despairingly I tried, the more I came to the conviction that only the discovery of a universal formal principle could lead us to assured results."

16. We cannot go here into the vast and important topic of the relation between Einstein and Mach. Suffice it to say that the archival materials help to illuminate each of the stages in the drama.

17. For example, Gerald Holton, *On the Origins of the Special Theory of Relativity*, AMERICAN JOURNAL OF PHYSICS, 28:627–636, 1960, and reprinted here as Essay 5; and Gerald Holton, *On the Thematic Analysis of Science: The Case of Poincaré and Relativity*, in MÉLANGES ALEXANDRE KOYRÉ (Paris: Hermann, 1964), Volume II, pp. 257–268, and reprinted here as Essay 6.

18. For the evidences, see, for example, Holton, *On the Origins of the Special Theory of Relativity*. In his lectures and publications Lorentz repeatedly gave generous credit to the novelty and independence of Einstein's work. See also, for example, Lorentz's remarks quoted in Ludwik Silberstein, THE THEORY OF RELATIVITY (London: Macmillan & Co., 1924), pp. 115–116, and

Lorentz's footnote addendum in the 1912 edition of his 1904 essay, in H. A. Lorentz *et al.*, DAS RELATIVITÄTSPRINZIP (Stuttgart: B. G. Teubner, 1912), p. 10.

19. H. A. Lorentz, *Electromagnetic Phenomena in a System Moving with Any Velocity Smaller Than That of Light*, KON. AKADEMIE VAN WETENSCHAPPEN AMSTERDAM, PROCEEDINGS OF THE SECTION OF SCIENCES, 6:809–831, 1904 (English-language edition), originally published June, 1904 in the Dutch-language edition. At that time, incidentally, Einstein probably knew no Dutch and little, if any, English. In a letter to Besso, dating no earlier than 1913, Einstein writes: "Ich lerne englisch (bei Wohlwand), langsam aber gründlich."

20. BEITRÄGE ZUR TECHNISCHEN MECHANIK UND TECHNISCHEN PHYSIK (Berlin: Julius Springer, 1924), with essays by some of Föppl's students, including Theodor von Karman, *Die mittragende Breite*, pp. 114–127; L. Prandtl, *Elastisch bestimmte und elastisch unbestimmte Systeme*, pp. 52–61; T. Thoma, *Neuzeitliche Hydrodynamik und praktische Technik*, pp. 128–129; S.Timoschenko, *Über die Biegung von Stäben, die eine kleine anfängliche Krümmung haben*, pp. 74–81; and Föppl's two sons, Ludwig Föppl, *Bestimmung der Knicklast eines Stabes aus Schwingungsversuchen*, pp. 74–78; and Otto Föppl, *Drehschwingungfestigkeit und Schwingungdämpfungsfestigkeit von Baustoffen*, pp. 10–16. As this list alone shows, his influence was large, although predominantly in technical mechanics. In 1904 Föppl made a gyroscopic experiment to measure the rate of rotation of the earth, a work "which made him familiar with questions of absolute and relative motion." And in 1914, he wrote an essay *Über absolute und relative Bewegung,* a field "in which A. Föppl already, before Einstein, occupied himself with the relativity theory, though not with such remarkable success"—according to the editor's introduction in the *Festschrift.*

21. Föppl had the satisfaction later that Mach, in the revised editions of THE SCIENCE OF MECHANICS, referred to Föppl's contributions; e.g., in the preface to the fourth edition, dated January, 1901, Mach wrote, quite typically: ". . . the partial considerations which my expositions have received in the writings of Boltzmann, Föppl, Hertz, Love, Maggi, Pearson, and Slate, have awakened in me the hope that my work shall not have been in vain."

22. It is, incidentally, worth noting that Einstein came to the Zurich E. T. H. initially planning to study engineering, and that both Einstein's father and closest uncle were in electrical engineering and manufacturing.

23. One idiosyncracy of the book that interests us is explained in the following manner: "In this book I have left out citation of sources as a matter of principle. . . . I wanted to write not a Handbook but a *Lehrbuch* which should as far as possible be cast in one piece. Therefore, I avoided as far as at all

possible during the writing going back to publications which I had read earlier in order that I might not be directly influenced by them. I wanted to be led by the developments and results of other authors only insofar as these matters had firmly penetrated into my memory and had become an intimate part of my own views. In this manner I hoped to attain a more unified and coherent exposition of the whole system than would have been possible by going another way."

As a consequence, there is a remarkable paucity of references to actual experimental situations (of course, none is made to the Michelson or other ether-drift experiments; but almost all references to any others are also missing).

24. Philipp Frank, EINSTEIN, HIS LIFE AND TIMES, trans. George Rosen, ed. and rev. Suichi Kusaka, New York: Alfred A. Knopf, 1947.

25. Philipp Frank, EINSTEIN, SEIN LEBEN UND SEINE ZEIT (Munich: Paul List Verlag, 1949), p. 38.

26. Since Max Abraham's version of August Föppl's book did not get done until 1904, it would be Föppl's original work that must be meant here; but this slip does remind us that a substantial fraction of several successive generations of physicists were brought up on—and then taught from—Abraham–Föppl and later Abraham–Becker.

27. Anton Reiser, ALBERT EINSTEIN, A BIOGRAPHICAL PORTRAIT, New York: A. & C. Boni, 1930.

28. [The last point is developed in Essay 9.]

8 MACH, EINSTEIN, AND THE SEARCH FOR REALITY

In the history of ideas of our century, there is a chapter that might be entitled "The Philosophical Pilgrimage of Albert Einstein," a pilgrimage from a philosophy of science in which sensationism and empiricism were at the center, to one in which the basis was a rational realism. This essay, a portion of a more extensive study,[1] is concerned with Einstein's gradual philosophical reorientation, particularly as it has become discernible during the work on his largely unpublished scientific correspondence.[2]

The earliest known letter by Einstein takes us right into the middle of the case. It is dated 19 March 1901 and addressed to Wilhelm Ostwald.[3] The immediate cause for Einstein's letter was his failure to receive an assistantship at the school where he had recently finished his formal studies, the Polytechnic Institute in Zürich; he now turned to Ostwald to ask for a position at his laboratory, partly in the hope of receiving "the opportunity for further education." Einstein included a copy of his first publication, *Folgerungen aus den Capillaritätserscheinungen*[4] (1901), which he said had been inspired (*angeregt*) by Ostwald's work; indeed, Ostwald's ALLGEMEINE CHEMIE is the first book mentioned in all of Einstein's published work.

This article was originally published in DAEDALUS, Spring, 1968, pp. 636–673.

Not having received an answer, Einstein wrote again to Ostwald on 3 April 1901. On 13 April 1901 his father, Hermann Einstein, sent Ostwald a moving appeal, evidently without his son's knowledge. Hermann Einstein reported that his son esteems Ostwald "most highly among all scholars currently active in physics."[5]

The choice of Ostwald was significant. He was, of course, not only one of the foremost chemists, but also an active "philosopher-scientist" during the 1890's and 1900's, a time of turmoil in the physical sciences as well as in the philosophy of science. The opponents of kinetic, mechanical, or materialistic views of natural phenomena were vociferous. They objected to atomic theory and gained great strength from the victories of thermodynamics, a field in which no knowledge or assumption was needed concerning the detailed nature of material substances (for example, for an understanding of heat engines).

Ostwald was a major critic of the mechanical interpretation of physical phenomena, as were Helm, Stallo, and Mach. Their form of positivism—as against the sophisticated logical positivism developed later in Carnap and Ayer's work—provided an epistemology for the new phenomenologically based science of correlated observations, linking energetics and sensationism. In the second (1893) edition of his influential textbook on chemistry, Ostwald had given up the mechanical treatment of his first edition for Helm's "energetic" one. "Hypothetical" quantities such as atomic entities were to be omitted; instead, these authors claimed they were satisfied, as Merz wrote around 1904, with "measuring such quantities as are presented in observation, such as energy, mass, pressure, volume, temperature, heat, electrical potential, etc., without reducing them to imaginary mechanisms or kinetic quantities." They condemned such conceptions as the ether, with properties not accessible to direct observation, and they issued a call "to consider anew the ultimate principles of all physical reasoning, notably the scope and validity of the Newtonian laws of motion and of the conceptions of force and action, of absolute and relative motion."[6]

All these iconoclastic demands—except anti-atomism—must have been congenial to the young Einstein who, according to his colleague Joseph Sauter, was fond of calling himself "a heretic."[7] Thus, we may well suspect that Einstein felt sympathetic to Ostwald who denied in the ALLGEMEINE CHEMIE that "the assumption of that medium, the ether, is unavoidable. To me it does not seem to be so. . . . There is no need to inquire for a carrier of it when we find it anywhere. This enables us to look upon radiant energy as independently existing in space."[8] It is a position quite consistent with that shown later in Einstein's papers of 1905 on photon theory and relativity theory.

In addition, it is worth noting that Einstein, in applying to Ostwald's laboratory, seemed to conceive of himself as an experimentalist. We know from many sources that in his student years in Zürich, Einstein's earlier childhood interest in mathematics had slackened considerably. In the *Autobiographical Notes*, Einstein reported: "I really could have gotten a sound mathematical education. However, I worked most of the time in the physical laboratory, fascinated by the direct contact with experience."[9] To this, one of his few reliable biographers adds: "No one could stir him to visit the mathematical seminars. . . . He did not yet see the possibility of seizing that formative power resident in mathematics, which later became the guide of his work. . . . He wanted to proceed quite empirically, to suit his scientific feeling of the time. . . . As a natural scientist, he was a pure empiricist."[10]

Ostwald's main philosophical ally was the prolific and versatile Austrian physicist and philosopher Ernst Mach (1838–1916), whose main work Einstein had read avidly in his student years and with whom he was destined to have later the encounters that form a main concern of this paper. Mach's major book, THE SCIENCE OF MECHANICS,[11] first published in 1883, is perhaps most widely known for its discussion of Newton's PRINCIPIA, in particular for its devastating critique of what Mach called the "conceptual monstrosity of absolute space"—a conceptual monstrosity because it is "purely a thought-thing which cannot be pointed to in experience."[12] Starting from his analysis of Newtonian presuppositions, Mach proceeded in his announced program of eliminating all metaphysical ideas from science. As Mach said quite bluntly in the preface to the first edition of THE SCIENCE OF MECHANICS: "This work is not a text to drill theorems of mechanics. Rather, its intention is an enlightening one—or to put it still more plainly, an anti-metaphysical one."

It will be useful to review briefly the essential points of Mach's philosophy. Here we can benefit from a good, although virtually unknown, summary presented by his sympathetic follower, Moritz Schlick, in the essay *Ernst Mach, der Philosoph.*

Mach was a physicist, physiologist, and also psychologist, and his philosophy . . . arose from the wish to find a principal point of view to which he could hew in any research, one which he would not have to change when going from the field of physics to that of physiology or psychology. Such a firm point of view he reached by going back to that which is given before all scientific research: namely, the world of sensations. . . . Since all our testimony concerning the so-called external world relies only on sensations, Mach held that we can and must take these sensations and complexes of sensations to be the sole contents [*Gegenstände*] of those testimonies, and, therefore, that there is no need to assume in addition an unknown reality hidden behind the sensations. With that, the existence *der Dinge an sich* is removed as an unjustified and unnecessary assumption. A body, a physical object, is nothing else than a complex, a more or less firm [we would say, invariant] pattern of sensations, i.e., of colors, sounds, sensations of heat and pressure, etc.

There exists in this world nothing whatever other than sensations and their connections. In place of the word "sensations," Mach liked to use rather the more neutral word "elements." . . . [As is particularly clear in Mach's book ERKENNTNIS UND IRRTUM] scientific knowledge of the world consists, according to Mach, in nothing else than the simplest possible description of the connections between the elements, and it has as its only aim the intellectual mastery of those facts by means of the least possible effort of thought. This aim is reached by means of a more and more complete "accommodation of the thoughts to one another." This is the formulation by Mach of his famous "principle of the economy of thought."[13]

The influence of Mach's point of view, particularly in the German-speaking countries, was enormous—on physics, on physiology, on psychology, and on the fields of the history and the philosophy of science[14] (not to mention Mach's profound effect on the young Lenin, Hofmannsthal, Musil, among many others outside the sciences). Strangely neglected by recent scholarship—there is not even a major biography—Mach has in the last two or three years again become the subject of a number of promising studies. To be sure, Mach himself always liked to insist that he was beleaguered and neglected, and that he did not have, or wish to have, a philosophical system; yet his philosophical ideas and attitudes had become so widely a part of the intellectual equipment of the period from the 1880's on that Einstein was quite right in saying later that "even those who think of themselves as Mach's opponents hardly know how much of Mach's views they have, as it were, imbibed with their mother's milk."[15]

The problems of physics themselves at that time helped to reinforce the appeal of the new philosophical attitude urged by Mach. The great program of nineteenth-century physics, the reconciliation of the notions of ether, matter, and electricity by means of mechanistic pictures and hypotheses, had led to enormities—for example, Larmor's proposal that the electron is a permanent but movable state of twist or strain in the ether, forming discontinuous particles of electricity and possibly of all ponderable matter. To many of the younger physicists of the time, attacking the problems of physics with conceptions inherited from classical nineteenth-century physics did not seem to lead anywhere. And here Mach's iconoclasm and incisive critical courage, if not the details of his philosophy, made a strong impression on his readers.

Mach's Early Influence on Einstein

As the correspondence at the Einstein Archives at Princeton reveals, one of the young scientists deeply caught up in Mach's point of view was Michelange (Michele) Besso—Einstein's oldest and closest friend, fellow student, and colleague at the Patent Office in Bern, the only person to whom Einstein gave public credit for help (*manche wertvolle Anregung*) when he published his basic paper on relativity in 1905. It was Besso who introduced Einstein to Mach's work. In a letter of 8 April 1952 to Carl Seelig, Einstein wrote: "My attention was drawn to Ernst Mach's SCIENCE OF MECHANICS by my friend Besso while a student, around the year 1897. The book exerted a deep and persisting impression upon me . . . , owing to its physical orientation toward fundamental concepts and fundamental laws." As Einstein noted in his *Autobiographical Notes* written in 1946, Ernst Mach's THE SCIENCE OF MECHANICS "shook this dogmatic faith" in "mechanics as the final basis of all physical thinking. . . . This book exercised a profound influence upon me in this regard while I was a student. I see Mach's greatness in his incorruptible skepticism and independence; in my younger years, however, Mach's epistemological position also influenced me very greatly."[16]

As the long correspondence between those old friends shows, Besso remained a loyal Machist to the end. Thus, writing to Einstein on 8 December 1947, he still said: "As far as the history of science is concerned, it appears to me that Mach stands at the center of the development of the last 50 or 70 years." Is it not true, Besso also asked, "that this introduction [to Mach] fell into a phase of development of the young physicist [Einstein] when the Machist style of thinking pointed decisively at observables—perhaps even, indirectly, to clocks and meter sticks?"

Turning now to Einstein's crucial first paper on relativity in 1905, we can discern in it influences of many, partly contradictory, points of view—not surprising in a work of such originality by a young contributor. Elsewhere I have examined the effect—or lack of effect—on that paper of three contemporary physicists: H. A. Lorentz,[17] Henri Poincaré,[18] and August Föppl. Here we may ask in what sense and to what extent Einstein's initial relativity paper of 1905 was imbued with the style of thinking associated with Ernst Mach and his followers—apart from the characteristics of clarity and independence, the two traits in Mach which Einstein always praised most.

In brief, the answer is that the Machist component—a strong component, even if not the whole story—shows up prominently in two related respects: first, by Einstein's insistence from the beginning of his relativity paper that the fundamental problems of physics cannot be understood until an epistemological analysis is carried out, particularly so with respect to the meaning of the conceptions of space and time;[19] and second, by Einstein's identification of reality with what is given by sensations, the "events," rather than putting reality on a plane beyond or behind sense experience.

From the outset, the instrumentalist, and hence sensationist, views of measurement and of the concepts of space and time are strikingly evident. The key concept in the early part of the 1905 paper is introduced at the top of the third page in a straightforward way. Indeed, Leopold Infeld in his biography of Einstein called them "the simplest sentence[s] I have ever encountered in a scientific paper." Einstein wrote: "We have to take into account that all our judgments in which time plays a part are always judgments of *simultaneous events*. If for instance I say, 'that train arrived here at seven o'clock,' I mean something like this: 'The pointing of the small hand of my watch to seven and the arrival of the train are simultaneous events.' "[20]

The basic concept introduced here, one that overlaps almost entirely Mach's basic "elements," is Einstein's concept of *events* [*Ereignisse*]—a word that recurs in Einstein's paper about a dozen times immediately following this citation. Transposed into Minkowski's later formulation of relativity, Einstein's "events" are the intersections of particular "world lines," say that of the train and that of the clock. The time (*t* coordinate) of an event by itself has no operational meaning. As Einstein says: "The 'time' of an event is that which is given simultaneously with the event by a stationary clock located at the place of the event."[21] We can say that just as the *time* of an event assumes meaning only when it connects with our consciousness through sense experience (that is, when it is subjected to measurement-in-principle by means of a clock present at the same place), so also is the *place*, or space coordinate, of an event meaningful only if it enters our sensory experience while being subjected to measurement-in-principle (that is, by means of meter sticks present on that occasion at the same time).[22]

This was the kind of operationalist message which, for most of his readers, overshadowed all other philosophical aspects in Einstein's paper. His work was enthusiastically embraced by the groups who saw themselves as philosophical heirs of Mach, the Vienna Circle of neopositivists and its predecessors and related followers,[23] providing a tremendous boost for the philosophy that had initially helped to nurture it. A typical response welcoming the relativity theory as "the victory over the metaphysics of absolutes in the conceptions of space and time . . . a mighty impulse for the development of the philosophical point of view of our time," was extended by Joseph Petzoldt in the inaugural session of the *Gesellschaft für positivistische Philosophie* in Berlin, 11 November 1912.[24] Michele Besso, who had heard the message from Einstein before anyone else, had exclaimed: "In the setting of Minkowski's space-time framework, it was now first possible to carry through the thought which the great mathematician, Bernhard Riemann, had grasped: 'The space-time framework itself is formed by the events in it.' "[25]

To be sure, re-reading Einstein's paper with the wisdom of hindsight, as we shall do presently, we can find in it also very different trends, warning of the possibility that "reality" in the end is not going to be left identical with "events." There are premonitions that sensory experiences, in Einstein's later work, will not be regarded as the chief building blocks of the "world," that the laws of physics themselves will be seen to be built into the event-world as the undergirding structure "governing" the pattern of events.

Such precursors appear even earlier, in one of Einstein's first letters in the Archives. Addressed to his friend Marcel Grossmann, it is dated

14 April 1901, when Einstein believed he had found a connection between Newtonian forces and the forces of attraction between molecules: "It is a wonderful feeling to recognize the unity of a complex of appearances which, to direct sense experience, seem to be separate things." Already there is a hint here of the high value that will be placed on intuited unity and the limited role seen for evident sense experience.

But all this was not yet ready to come into full view, even to the author. Taking the early papers as a whole, and in the context of the physics of the day, we find that Einstein's philosophical pilgrimage did start on the historic ground of positivism. Moreover, Einstein thought so himself, and confessed as much in letters to Ernst Mach.

The Einstein-Mach Letters

In the history of recent science, the relation between Einstein and Mach is an important topic that has begun to interest a number of scholars. Indeed, it is a drama of which we can sketch here four stages: Einstein's early acceptance of the main features of Mach's doctrine; the Einstein-Mach correspondence and meeting; the revelation in 1921 of Mach's unexpected and vigorous attack on Einstein's relativity theory; and Einstein's own further development of a philosophy of knowledge in which he rejected many, if not all, of his earlier Machist beliefs.

Happily, the correspondence is preserved at least in part. A few letters have been found, all from Einstein to Mach. Those of concern here are part of an exchange between 1909 and 1913, and they testify to Einstein's deeply felt attraction to Mach's viewpoint, just at a time when the mighty Mach himself—forty years senior to the young Einstein whose work was just becoming widely known—had for his part embraced the relativity theory publicly by writing in the second (1909) edition of CONSERVATION OF ENERGY: "I subscribe, then, to the principle of relativity, which is also firmly upheld in my MECHANICS and WÄRMELEHRE.[26] In the first letter, Einstein writes from Bern on 9 August 1909. Having thanked Mach for sending him the book on the Law of Conservation of Energy, he adds: "I know, of course, your main publications very well, of which I most admire your book on Mechanics. You have had such a strong influence upon the epistemological conceptions of the younger generation of physicists that even your opponents today, such as Planck, undoubtedly would have been called Mach followers by physicists of the kind that was typical a few decades ago."

It will be important for our analysis to remember that Planck was Einstein's earliest patron in scientific circles. It was Planck who, in 1905, as editor of the ANNALEN DER PHYSIK, received Einstein's first relativity paper and thereupon held a review seminar on the paper in Berlin. Planck defended Einstein's work on relativity in public meetings from the beginning, and by 1913 had succeeded in persuading his German colleagues to invite Einstein to the Kaiser-Wilhelm-Gesellschaft in Berlin. With a polemical essay *Against the New Energetics* in 1896, he had made clear his position, and by 1909 Planck was one of the few opponents of Mach, and scientifically the most prominent one. He had just written a famous attack, *Die Einheit des physikalischen Weltbildes.* Far from accepting Mach's view that, as he put it, "Nothing is real except the perceptions, and all natural science is ultimately an economic adaptation of our ideas to our perceptions," Planck held to the entirely antithetical position that a basic aim of science is "the finding of a *fixed* world picture independent of the variation of time and people," or, more generally, "the complete liberation of the physical picture from the individuality of the separate intellects."[27]

At least by implication, Einstein's remarks to Mach show that he dissociated himself from Planck's view. It may also not be irrelevant that just at that time Einstein, who since 1906 had been objecting to inconsistencies in Planck's quantum theory, was preparing his first major invited paper before a scientific congress, the eighty-first meeting of the Naturforscherversammlung, announced for September, 1909, in Salzburg. Einstein's paper called for a revision of Maxwell's theory to accommodate the probabilistic character of the emission of photons—none of which Planck could accept—and concluded: "To accept Planck's theory means, in my view, to throw out the bases of our [1905] theory of radiation."

Mach's reply to Einstein's first letter is now lost, but it must have come quickly, because eight days later Einstein sends an acknowledgment:

> Bern, 17 August 1909. Your friendly letter gave me enormous pleasure. . . . I am very glad that you are pleased with the relativity theory. . . . Thanking you again for your friendly letter, I remain, your student [indeed: *Ihr Sie verehrender Schüler*], A. Einstein.

Einstein's next letter was written as physics professor in Prague, where Mach before him had been for twenty-eight years. The post had been offered to Einstein on the basis of recommendations of a faction (Lampa, Pick) who regarded themselves as faithful disciples of Mach. The letter was sent out about New Year's 1911–12, perhaps just before or after Einstein's sole (and, according to Philipp Frank's account in EINSTEIN, HIS LIFE AND TIMES, not very successful) visit to Mach, and after the first progress toward the general relativity theory:

> . . . I can't quite understand how Planck has so little understanding for your efforts. His stand to my [general relativity] theory is also one of refusal. But I can't take it amiss; so far, that one single epistemological argument is the only thing which I can bring forward in favor of my theory.[28]

Here, Einstein is referring delicately to the Mach Principle, which he had been putting at the center of the developing theory.[29] Mach responded by sending Einstein a copy of one of his books, probably the ANALYSIS OF SENSATIONS.

In the last of these letters to Mach (who was now seventy-five years old, and for some years had been paralyzed), Einstein writes from Zürich on 25 June 1913:

> Recently you have probably received my new publication on relativity and gravitation which I have at last finished after unending labor and painful doubt. [This must have been the *Entwurf einer verallgemeinerten Relativitätstheorie und einer Theorie der Gravitation*, written with Marcel Grossmann.[30]] Next year at the solar eclipse it will turn out whether the light rays are bent by the sun, in other words whether the basic and fundamental assumption of the equivalence of the acceleration of the reference frame and of the gravitational field really holds. If so, then your inspired investigations into the foundations of mechanics—despite Planck's unjust criticism—will receive a splendid confirmation. For it is a necessary consequence that inertia has its origin in a kind of mutual interaction of bodies, fully in the sense of your critique of Newton's bucket experiment.[31]

The Paths Diverge

The significant correspondence stops here, but Einstein's public and private avowals of his adherence to Mach's ideas continue for several years more. For example, there is his well-known, moving eulogy of Mach, published in 1916.[15] In August, 1918, Einstein writes to Besso quite sternly about an apparent—and quite temporary—lapse in Besso's positivistic epistemology; it is an interesting letter, worth citing in full:

28 August 1918.
Dear Michele:
In your last letter I find, on re-reading, something which makes me angry: That speculation has proved itself to be superior to empiricism. You are thinking here about the development of relativity theory. However, I find that this development teaches something else, that it is practically the opposite, namely that a theory which wishes to deserve trust must be built upon generalizable facts.

Old examples: Chief postulates of thermodynamics [based] on impossibility of perpetuum mobile. Mechanics [based] on a grasped [*ertasteten*] law of inertia. Kinetic gas theory [based] on equivalence of heat and mechanical energy (also historically). Special Relativity on the constancy of light velocity and Maxwell's equation for the vacuum, which in turn rest on empirical foundations. Relativity with respect to uniform [?] translation is a *fact of experience.*

General Reality: *Equivalence of inertial and gravitational mass.* Never has a truly useful and deep-going theory really been found purely speculatively. The nearest case is Maxwell's hypothesis concerning displacement current; there the problem was to do justice to the fact of light propagation. . . . With cordial greetings, your Albert. [Emphasis in the original]

Careful reading of this letter shows us that already here there is evidence of divergence between the conception of "fact" as understood by Einstein and "fact" as understood by a true Machist. The impossibility of the perpetuum mobile, the first law of Newton, the constancy of light velocity, the validity of Maxwell's equations, the equivalence of inertial and gravitational mass—none of these would have been called "facts of experience" by Mach. Indeed, Mach might have insisted that—to use one of his favorite battle words—it is evidence of "dogmatism" not to regard all these conceptual constructs as continually in need of probing re-examinations; thus, Mach had written:

. . . for me, matter, time and space are still *problems,* to which, incidentally, the physicists (Lorentz, Einstein, Minkowski) are also slowly approaching.[32]

Similar evidence of Einstein's gradual apostasy appears in a letter of 4 December 1919 to Paul Ehrenfest. Einstein writes:

I understand your difficulties with the development of relativity theory. They arise simply because you want to base the innovations of 1905 on epistemological grounds (nonexistence of the stagnant ether) instead of empirical grounds (equivalence of all inertial systems with respect to light).

Mach would have applauded Einstein's life-long suspicion of formal epistemological systems, but how strange would he have found this use of the word *empirical* to characterize the hypothesis of the equivalence of all inertial systems with respect to light! What we see forming slowly here is Einstein's view that the fundamental role played by experience in the construction of fundamental physical theory is, after all, not through the "atom" of experience, not through the individual sensation or the protocol sentence, but through some creative digest or synthesis of "*die gesammten Erfahrungstatsachen,*" the *totality* of physical experience.[33] But all this was still hidden. Until Mach's death, and for several years after, Einstein considered and declared himself a disciple of Mach.

In the meantime, however, unknown to Einstein and everyone else, a time bomb had been ticking away. Set in 1913, it went off in 1921, five years after Mach's death, when Mach's THE PRINCIPLES OF PHYSICAL OPTICS was published at last. Mach's preface was dated July, 1913—perhaps a few days or, at most, a few weeks after Mach had received Einstein's last, enthusiastic letter and the article on general relativity theory. In a well-known passage in the preface (but one usually found in an inaccurate translation), Mach had written:

I am compelled, in what may be my last opportunity, to cancel my views[*Anschauungen*] of the relativity theory.

I gather from the publications which have reached me, and especially from my correspondence, that I am gradually becoming regarded as the forerunner of relativity. I am able even now to picture approximately what new expositions and interpretations many of the ideas expressed in my book on Mechanics will receive in the future from this point of view. It was to be expected that philosophers and physicists should carry on a crusade against me, for, as I have repeatedly observed, I was merely an unprejudiced rambler endowed with original ideas, in varied fields of knowledge. I must, however, as assuredly disclaim to be a forerunner of the relativists as I personally reject the atomistic doctrine of the present-day school, or church. The reason why, and the extent to which, I reject [*ablehne*] the present-day relativity theory, which I find to be growing more and more dogmatical, together with the particular reasons which have led me to such a view—considerations based on the physiology of the senses, epistemological doubts, and above all the insight resulting from my experiments—must remain to be treated in the sequel [a sequel which was never published].[34]

Certainly, Einstein was deeply disappointed by this belated disclosure of Mach's sudden dismissal of the relativity theory. Some months later, during a lecture on 6 April 1922 in Paris, in a discussion with the anti-Machist philosopher Emile Meyerson, Einstein allowed in a widely reported remark that Mach was "*un bon mécanicien,*" but "*déplorable philosophe.*"[35]

We can well understand that Mach's rejection was at heart very painful, the more so as it was somehow Einstein's tragic fate to have the contribution he most cared about rejected by the very men whose approval and understanding he would have most gladly had—a situation not unknown in the history of science. In addition to Mach, the list includes these four: *Henri Poincaré,* who, to his death in 1912, only once deigned to mention Einstein's name in print, and then only to register an objection; *H. A. Lorentz,* who gave Einstein personally every possible encouragement—short of fully accepting the theory of relativity for himself; *Max Planck,* whose support of the special theory of relativity was unstinting, but who resisted Einstein's ideas on general relativity and the early quantum theory of radiation; and *A. A. Michelson,* who to the end of his days did not believe in relativity theory, and once said to Einstein that he was sorry that his own work may have helped to start this "monster."[36]

Soon Einstein's generosity again took the upper hand and resulted, from then to the end of his life, in many further personal testimonies to Mach's earlier influence.[37] A detailed analysis was provided in Einstein's letter of 8 January 1948 to Besso:

As far as Mach is concerned, I wish to differentiate between Mach's influence in general and his influence on me. . . . Particularly in the MECHANICS and the WÄRMELEHRE he tried to show how concepts arose out of experience. He took convincingly the position that these conceptions, even the most fundamental ones, obtained their warrant only out of empirical knowledge, that they are in no way logically necessary. . . .

I see his weakness in this, that he more or less believed science to consist in a mere ordering of empirical material; that is to say, he did not recognized the freely constructive element in formation of concepts. In a way he thought that theories arise through *discoveries* and not through inventions. He even went so far that he regarded "sensations" not only as material which has to be investigated, but, as it were, as the building blocks of the real world; thereby, he believed, he could overcome the difference between psychology and physics. If he had drawn the full consequences, he would have had to reject not only atomism but also the idea of a physical reality.

Now, as far as Mach's influence on my own development is concerned, it certainly was great. I remember very well that you drew my attention to his MECHANICS and WÄRMELEHRE during my first years of study, and that both books made a great impression on me. The extent to which they influenced my own work is, to say the truth, not clear to me. As far as I am conscious of it, the immediate influence of Hume on me was greater. . . . But, as I said, I am not able to analyze that which lies anchored in unconscious thought. It is interesting, by the way, that Mach rejected the special relativity theory passionately (he did not live to see the general relativity theory [in the developed form]). The theory was, for him, inadmissibly speculative. He did not know that this speculative character belongs also to Newton's mechanics, and to every theory which thought is capable of. There exists only a gradual difference between theories, insofar as the chains of thought from fundamental concepts to empirically verifiable conclusions are of different lengths and complications.[38]

Antipositivistic Component of Einstein's Work

Ernst Mach's harsh words in his 1913 preface leave a tantalizing mystery. Ludwig Mach's destruction of his father's papers has so far made it impossible to find out more about the "experiments" (possibly on the constancy of the velocity of light) at which Ernst Mach hinted. Since 1921, many speculations have been offered to explain Mach's remarks.[39] They all leave something to be desired. Yet, I believe, it is not so difficult to reconstruct the main reasons why Mach ended up rejecting the relativity theory. To put it very simply, Mach had recognized more and more clearly, years before Einstein did so himself, that Einstein had indeed fallen away from the faith, had left behind him the confines of Machist empiriocriticism.

The list of evidences is long. Here only a few examples can be given, the first from the 1905 relativity paper itself: what had made it really work was that it contained and combined elements based on two entirely different philosophies of science—not merely the empiricist-operationist component, but the courageous initial postulation, in the second paragraph, of two thematic hypotheses (one on the constancy of light velocity and the other on the extension of the principle of relativity to all branches of physics), two postulates for which there was and can be no direct empirical confirmation.

For a long time, Einstein did not draw attention to this feature. In a lecture at King's College, London, in 1921, just before the posthumous publication of Mach's attack, Einstein still was protesting that the origin of relativity theory lay in the facts of direct experience:

... I am anxious to draw attention to the fact that this theory is not speculative in origin; it owes its invention entirely to the desire to make physical theory fit observed fact as well as possible. We have here no revolutionary act, but the natural continuation of a line that can be traced through centuries. The abandonment of certain notions connected with space, time, and motion, hitherto treated as fundamentals, must not be regarded as arbitrary, but only as conditioned by observed facts.[40]

By June, 1933, however, when Einstein returned to England to give the Herbert Spencer Lecture at Oxford entitled *On the Method of Theoretical Physics*, the more complex epistemology that was in fact inherent in his work from the beginning had begun to be expressed. He opened this lecture with the significant sentence: "If you want to find out anything from the theoretical physicists about the methods they use, I advise you to stick closely to one principle: Don't listen to their words, fix your attention on their deeds." He went on to divide the tasks of experience and reason in a very different way from that advocated in his earlier visit to England:

We are concerned with the eternal antithesis between the two inseparable components of our knowledge, the empirical and the rational. . . . The structure of the system is the work of reason; the empirical contents and their mutual relations must find their representation in the conclusions of the theory. In the possibility of such a representation lie the sole value and justification of the whole system, and especially the concepts and fundamental principles which underlie it. Apart from that, these latter are free inventions of the human intellect, which cannot be justified either by the nature of that intellect or in any other fashion *a priori*.

In the summary of this section, he draws attention to the "purely fictitious character of the fundamentals of scientific theory." It is this penetrating insight which Mach must have smelled out much earlier and dismissed as "dogmatism."

Indeed, Einstein, in his 1933 Spencer Lecture—widely read, as were and still are so many of his essays—castigates the old view that "the fundamental concepts and postulates of physics were not in the logical sense inventions of the human mind but could be deduced from experience by 'abstraction'—that is to say, by logical means. A clear recognition of the erroneousness of this notion really only came with the general theory of relativity."

Einstein ends this discussion with the enunciation of his current credo, so far from that he had expressed earlier:

> Nature is the realization of the simplest conceivable mathematical ideas. I am convinced that we can discover, by means of purely mathematical constructions, those concepts and those lawful connections between them which furnish the key to the understanding of natural phenomena. Experience may suggest the appropriate mathematical concepts, but they most certainly cannot be deduced from it. Experience remains, of course, the sole criterion of physical utility of a mathematical construction. But the creative principle resides in mathematics. In a certain sense, therefore, I hold it true that pure thought can grasp reality, as the ancients dreamed.[41]

Technically, Einstein was now at—or rather just past—the midstage of his pilgrimage. He had long ago abandoned his youthful allegiance to a primitive phenomenalism that Mach would have commended. In the first of the two passages just cited and others like it, he had gone on to a more refined form of phenomenalism which many of the logical positivists could still accept. He has, however, gone beyond it in the second passage, turning toward interests that we shall see later to have matured into clearly metaphysical conceptions.

Later, Einstein himself stressed the key role of what we have called thematic rather than phenomenic elements[42]—and thereby he fixed the early date at which, in retrospect, he found this need to have arisen in his earliest work. Thus he wrote in his *Autobiographical Notes* of 1946 that "shortly after 1900 . . . I despaired of the possibility of discovering the true laws by means of constructive efforts based on known facts. The longer and the more despairingly I tried, the more I came to the conviction that only the discovery of a *universal formal principle* could lead us to assured results."[43]

Another example of evidence of the undercurrent of disengagement from a Machist position is an early one: it comes from Einstein's article on relativity in the 1907 Jahrbuch der Radioaktivität und Elektronik,[44] where Einstein responds, after a year's silence, to Walter Kaufmann's 1906 paper in the Annalen der Physik.[45] That paper had been the first publication in the Annalen to mention Einstein's work on the relativity theory, published there the previous year. Coming from the eminent experimental physicist Kaufmann, it had been most significant that this very first discussion was announced as a categorical, experimental disproof of Einstein's theory. Kaufmann had begun his attack with the devastating summary:

I anticipate right here the general result of the measurements to be described in the following: *the measurement results are not compatible with the Lorentz-Einsteinian fundamental assumption.*[46]

Einstein could not have known that Kaufmann's equipment was inadequate. Indeed, it took ten years for this to be fully realized, through the work of Guye and Lavanchy in 1916. So in his discussion of 1907, Einstein had to acknowledge that there seemed to be small but significant differences between Kaufmann's results and Einstein's predictions. He agreed that Kaufmann's calculations seemed to be free of error, but "whether there is an unsuspected systematic error or whether the foundations of relativity theory do not correspond with the facts one will be able to decide with certainty only if a great variety of observational material is at hand."[47]

Despite this prophetic remark, Einstein does not rest his case on it. On the contrary, he has a very different, and what for his time and situation must have been a very daring, point to make: he acknowledges that the theories of electron motion given earlier by Abraham and by Bucherer do give predictions considerably closer to the experimental results of Kaufmann. But Einstein refuses to let the "facts" decide the matter: "In my opinion both [their] theories have a rather small probability, because their fundamental assumptions concerning the mass of moving electrons are not explainable in terms of theoretical systems which embrace a greater complex of phenomena."[48]

This is the characteristic position—the crucial difference between Einstein and those who make the correspondence with experimental fact the chief deciding factor for or against a theory: even though the "experimental facts" at that time very clearly seemed to favor the theory of his opponents rather than his own, he finds the ad hoc character of their theories more significant and objectionable than an apparent disagreement between his theory and their "facts."[49]

So already in this 1907 article—which, incidentally, Einstein mentions in his postcard of 17 August to Ernst Mach, with a remark regretting that he has no more reprints for distribution—we have explicit evidence of a hardening of Einstein against the epistemological priority of experiment, not to speak of sensory experience. In the years that followed, Einstein more and more openly put the consistency of a simple and convincing theory or of a thematic conception higher in importance than the latest news from the laboratory—and again and again he turned out to be right.

Thus, only a few months after Einstein had written in his fourth letter to Mach that the solar eclipse experiment will decide "whether the basic and fundamental assumption of the equivalence of the acceleration of the reference frame and of the gravitational field really holds," Einstein writes to Besso in a very different vein (in March 1914), before the first, ill-fated eclipse expedition was scheduled to test the conclusions of the preliminary version of the general relativity theory: "Now I am fully satisfied, and I do not doubt any more the correctness of the whole system, may the observation of the eclipse succeed or not. The sense of the thing [*die Vernunft der Sache*] is too evident." And later, commenting on the fact that there remains up to 10 per cent discrepancy between the measured deviation of light owing to the sun's field and the calculated effect based on the general relativity theory: "For the expert, this thing is not particularly important, because the main significance of the theory does not lie in the verification of little effects, but rather in the great simplification of the theoretical basis of physics as a whole."[50] Or again, in Einstein's *Notes on the Origin of the General Theory of Relativity*,[51] he reports that he "was in the highest degree amazed" by the existence of the equivalence between inertial and gravitational mass, but that he "had no serious doubts about its strict validity, even without knowing the results of the admirable experiment of Eötvös."

The same point is made again in a revealing account given by Einstein's student, Ilse Rosenthal-Schneider. In a manuscript, *Reminiscences of Conversation with Einstein*, dated 23 July 1957, she reports:

> Once when I was with Einstein in order to read with him a work that contained many objections against his theory . . . he suddenly interrupted the discussion of the book, reached for a telegram that was lying on the windowsill, and handed it to me with the words, "Here, this will perhaps interest you." It was Eddington's cable with the results of measurement of the eclipse expedition [1919]. When I was giving expression to my joy that the results coincided with

his calculations, he said quite unmoved, "But I knew that the theory is correct"; and when I asked, what if there had been no confirmation of his prediction, he countered: "Then I would have been sorry for the dear Lord—the theory *is* correct."[52]

Minkowski's "World" and the World of Sensations

The third major point at which Mach, if not Einstein himself, must have seen that their paths were diverging is the development of relativity theory into the geometry of the four-dimensional space-time continuum, begun in 1907 by the mathematician Hermann Minkowski (who, incidentally, had had Einstein as a student in Zürich). Indeed, it was through Minkowski's semipopular lecture, "Space and Time," on 21 September 1908 at the eightieth meeting of the Naturforscherversammlung,[53] that a number of scientists first became intrigued with relativity theory. We have several indications that Mach, too, was both interested in and concerned about the introduction of four-dimensional geometry into physics (in Mach's correspondence around 1910, for example, with August Föppl); according to Friedrich Herneck,[54] Ernst Mach specially invited the young Viennese physicist Philipp Frank to visit him "in order to find out more about the relativity theory, above all about the use of four-dimensional geometry." As a result, Frank, who had recently finished his studies under Ludwig Boltzmann and had begun to publish noteworthy contributions to relativity, published the "presentation of Einstein's theory to which Mach gave his assent" under the title *Das Relativitätsprinzip und die Darstellung der physikalischen Erscheinungen im vierdimensionalen Raum.*[55] It is an attempt, addressed to readers "who do not master modern mathematical methods," to show that Minkowski's work brings out the "empirical facts far more clearly by the use of four-dimensional world lines." The essay ends with the reassuring conclusion: "In this four-dimensional world the facts of experience can be presented more adequately than in three-dimensional space, where always only an arbitrary and one-sided projection is pictured."

Following Minkowski's own papers on the whole, Frank's treatment can make it nevertheless still appear that in most respects the time dimension is equivalent to the space dimensions. Thereby one could think that Minkowski's treatment based itself not only on a functional and operational interconnection of space and time, but also—fully in accord with Mach's own views—on the primacy of ordinary, "experienced" space and time in the relativistic description of phenomena.

Perhaps as a result of this presentation, Mach invoked the names of Lorentz, Einstein, and Minkowski in his reply of 1910 to Planck's first attack, citing them as physicists "who are moving closer to the problems of matter, space, and time." Already a year earlier, Mach seems to have been hospitable to Minkowski's presentation, although not without reservations. Mach wrote in the 1909 edition of CONSERVATION OF ENERGY:[26] "Space and time are here conceived not as independent entities, but as forms of the dependence of the phenomena on one another"; he also added a reference to Minkowski's lecture of 1908.[53] But a few lines earlier, Mach had written: "Spaces of many dimensions seem to me not so essential for physics. I would only uphold them if things of thought [*Gedankendinge*] like atoms are maintained to be indispensable, and if, then, also the freedom of working hypotheses is upheld."

It was correctly pointed out by C. B. Weinberg[56] that Mach may eventually have had two sources of suspicion against the Minkowskian form of relativity theory. As was noted above, Mach regarded the fundamental notions of mechanics as problems to be continually discussed with maximum openness within the frame of empiricism, rather than as questions that can be solved and settled—as the relativists, seemingly dogmatic and sure of themselves, were in his opinion more and more inclined to do. In addition, Mach held that the questions of physics were to be studied in a broader setting, encompassing biology and psychophysiology. Thus Mach wrote: "Physics is not the entire world; biology is there too, and belongs essentially to the world picture.[57]

But I see also a third reason for Mach's eventual antagonism against such conceptions as Minkowski's (unless one restricted their application to "mere things of thought like atoms and molecules, which by their very nature can never be made the objects of sensuous contemplations"[58]). If one takes Minkowski's essay seriously—for example, the abandonment of space and time separately, with identity granted only to "a kind of union of the two"—one must recognize that it entails the abandonment of the conceptions of experiential space and experiential time; and that is an attack on the very roots of sensations-physics, on the meaning of actual measurements. If identity, meaning, or "reality" lies in the four-dimensional space-time interval *ds*, one is dealing with a quantity which is hardly *denkökonomisch,* nor one that preserves the primacy of measurements in "real" space and time. Mach may well have seen the warning flag; and worse was soon to come, as we shall see at once.

In his exuberant lecture of 1908,[58] Minkowski had announced that "three-dimensional geometry becomes a chapter in four-dimensional physics. . . . Space and time are to fade away into the shadows, and only *eine Welt an sich* will subsist." In this "world" the crucial innovation is the conception of the *"zeitartige Vektorelement," ds,* defined as $(1/c)\sqrt{c^2dt^2\text{-}dx^2\text{-}dy^2\text{-}dz^2}$ with imaginary components. To Mach, the word *Element* had a pivotal and very different meaning. As we saw in Schlick's summary, elements were nothing less than the sensations and complexes of sensations of which the world consists and which completely define the world. Minkowski's rendition of relativity theory was now revealing the need to move the ground of basic, elemental truths from the plane of direct experience in ordinary space and time to a mathematicized, formalistic model of the world in a union of space and time that is not directly accessible to sensation—and, in this respect, is reminiscent of absolute space and time concepts that Mach had called "metaphysical monsters."[59]

Here, then, is an issue which, more and more, had separated Einstein and Mach even before they realized it. To the latter, the fundamental task of science was economic and descriptive; to the former, it was speculative-constructive and intuitive. Mach had once written: "If all the individual facts—all the individual phenomena, knowledge of which we desire—were immediately accessible to us, science would never have risen."[60] To this, with the forthrightness caused perhaps by his recent discovery of Mach's opposition, Einstein countered during his lecture in Paris of 6 April 1922: "Mach's system studies the existing relations between data of experience: for Mach, science is the totality of these relations. That point of view is wrong, and in fact what Mach has done is to make a catalog, not a system."[61]

We are witnessing here an old conflict, one that has continued throughout the development of the sciences. Mach's phenomenalism brandished an undeniable and irresistible weapon for the critical reevaluation of classical physics, and in this it seems to hark back to an ancient position that looked upon sensuous appearances as the beginning and end of scientific achievement. One can read Galileo in this light, when he urges the primary need of *description* for the fall of bodies, leaving "the causes" to be found out later. So one can understand (or rather, misunderstand) Newton, with his too-well-remembered remark: "I feign no hypotheses."[62] Kirchhoff is in this tradition. Boltzmann wrote of him in 1888:

> The aim is not to produce bold hypotheses as to the essence of matter, or to explain the movement of a body from that of molecules, but to present equations which, free from hypotheses, are as far as possible true and quantitatively correct correspondents of the phenomenal world, careless of the essence of things and forces. In his book on mechanics, Kirchhoff will ban all metaphysical concepts, such as forces, the cause of a motion; he seeks only the equations which correspond so far as possible to observed motions.[63]

And so could, and did, Einstein himself understand the Machist component of his own early work.

Phenomenalistic positivism in science has always been victorious, but only up to a very definite limit. It is the necessary sword for destroying old error, but it makes an inadequate plowshare for cultivating a new harvest. I find it exceedingly significant that Einstein saw this during the transition phase of partial disengagement from the Machist philosophy. In the spring of 1917 Einstein wrote to Besso and mentioned a manuscript which Friedrich Adler had sent him. Einstein commented: "He rides Mach's poor horse to exhaustion." To this, Besso—the loyal Machist—responds on 5 May 1917: "As to Mach's little horse, we should not insult it; did it not make possible the infernal journey through the relativities? And who knows—in the case of the nasty quanta, it may also carry Don Quixote de la Einsta through it all!"

Einstein's answer of 13 May 1917 is revealing: "I do not inveigh against Mach's little horse; but you know what I think about it. It cannot give birth to anything living, it can only exterminate harmful vermin."

Toward a Rationalistic Realism

The rest of the pilgrimage is easy to reconstruct, as Einstein more and more openly and consciously turned Mach's doctrine upside down—minimizing rather than maximizing the role of actual details of experience, both at the beginning and at the end of scientific theory, and opting for a rationalism that almost inevitably would lead him to the conception of an objective, "real" world behind the phenomena to which our senses are exposed.

In the essay, *Maxwell's Influence on the Evolution of the Idea of Physical Reality* (1931), Einstein began with a sentence that could have been taken almost verbatim from Max Planck's attack on Mach in 1909, cited above: "The belief in an external world independent of the perceiving subject is the basis of all natural science." Again and again, in the period beginning with his work on the general relativity theory, Einstein insisted that between experience and reason, as well as between the world of sensory perception and the objective world, there are logically unbridgeable chasms. He characterized the efficacy of reason to grasp reality by the word *miraculous*; the very terminology in these statements would have been an anathema to Mach.

We may well ask when and under what circumstances Einstein himself became aware of his change. Here again, we may turn for illumination to one of the hitherto unpublished letters, one written to his old friend, Cornelius Lanczos, on 24 January 1938:

> Coming from sceptical empiricism of somewhat the kind of Mach's, I was made, by the problem of gravitation, into a believing rationalist, that is, one who seeks the only trustworthy source of truth in mathematical simplicity. The logically simple does not, of course, have to be physically true; but the physically true is logically simple, that is, it has unity at the foundation.

Indeed, all the evidence points to the conclusion that Einstein's work on general relativity theory was crucial in his epistemological development. As he wrote later in *Physics and Reality* (1936): "the first aim of the general theory of relativity was the preliminary version which, while not meeting the requirements for constituting a closed system, could be connected in as simple a manner as possible with 'directly observed facts.' "[64] But the aim, still apparent during the first years of correspondence with Mach, could not be achieved. In *Notes on the Origin of the General Theory of Relativity*, Einstein reported:

> I soon saw that the inclusion of non-linear transformation, as the principle of equivalence demanded, was inevitably fatal to the simple physical interpretation of the coordinate—i.e., that it could no longer be required that coordinate differences [*ds*] should signify direct results of measurement with ideal scales or clocks. I was much bothered by this piece of knowledge . . . [just as Mach must have been].

The solution of the above mentioned dilemma [from 1912 on] was therefore as follows: A physical significance attaches not to the differentials of the coordinates, but only to the Riemannian metric corresponding to them.[65]

And this is precisely a chief result of the 1913 essay of Einstein and Grossmann,[66] the same paper which Einstein sent to Mach and discussed in his fourth letter. This result was the final consequence of the Minkowskian four-space representation—the sacrifice of the primacy of direct sense perception in constructing a physically significant system. It was the choice that Einstein had to make—against fidelity to a catalogue of individual operational experiences, and in favor of fidelity to the ancient hope for a unity at the base of physical theory.[67]

Enough has been written in other places to show the connections that existed between Einstein's scientific rationalism and his religious beliefs. Max Born summarized it in one sentence: "He believed in the power of reason to guess the laws according to which God has built the world."[68] Perhaps the best expression of this position by Einstein himself is to be found in his essay, *Über den gegenwärtigen Stand der Feld-Theorie*, in the FESTSCHRIFT of 1929 for Aurel Stodola:

Physical Theory has two ardent desires, to gather up as far as possible all pertinent phenomena and their connections, and to help us not only to know *how* Nature is and *how* her transactions are carried through, but also to reach as far as possible the perhaps utopian and seemingly arrogant aim of knowing why Nature is *thus and not otherwise*. Here lies the highest satisfaction of a scientific person. . . . [On making deductions from a "fundamental hypothesis" such as that of the kinetic-molecular theory,] one experiences, so to speak, that God Himself could not have arranged those connections [between, for example, pressure, volume, and temperature] in any other way than that which factually exists, any more than it would be in His power to make the number 4 into a prime number. This is the promethean element of the scientific experience. . . . Here has always been for me the particular magic of scientific considerations; that is, as it were, the religious basis of scientific effort.[69]

This fervor is indeed far from the kind of analysis which Einstein had made only a few years earlier. It is doubly far from the asceticism of his first philosophic mentor, Mach, who had written in his day book: "Colors, space, tones, etc. These are the only realities. Others do not exist."[70] It is, on the contrary, far closer to the rational realism of his first scientific mentor, Planck, who had written: "The disjointed data of experience can never furnish a veritable science without the intelligent interference of a spirit actuated by faith. . . . We have a right to feel secure in surrendering to our belief in a philosophy of the world based upon a faith in the rational ordering of this world."[71] Indeed, we note the philosophical kinship of Einstein's position with seventeenth-century natural philosophers—for example, with Johannes Kepler who, in the preface of the MYSTERIUM COSMOGRAPHICUM, announced that he wanted to find out concerning the number, positions, and motions of the planets, "why they are as they are, and not otherwise," and who wrote to Herwart in April, 1599, that, with regard to numbers and quantity, "our knowledge is of the same kind as God's, at least insofar as we can understand something of it in this mortal life."

Not unexpectedly, we find that during this period (around 1930) Einstein's non-scientific writings began to refer to religious questions much more frequently than before. There is a close relation between his epistemology, in which reality does not need to be validated by the individual's sensorium, and what he called "Cosmic Religion," defined as follows: "The individual feels the vanity of human desires and aims, and the nobility and marvelous order which are revealed in nature and in the world of thought. He feels the individual destiny as an imprisonment and seeks to experience the totality of existence as a unity full of significance."[72]

Needless to say, Einstein's friends from earlier days sometimes had to be informed of his change of outlook in a blunt way. For example, Einstein wrote to Moritz Schlick on 28 November 1930:

> In general your presentation fails to correspond to my conceptual style insofar as I find your whole orientation so to speak too positivistic. . . .I tell you straight out: Physics is the attempt at the conceptual construction of a model of the *real world* and of its lawful structure. To be sure, it [physics] must present exactly the empirical relations between those sense experiences to which we are open; but only *in this way* is it chained to them. . . . In short, I suffer under the (unsharp) separation of Reality of Experience and Reality of Being. . . .
>
> You will be astonished about the "metaphysicist" Einstein. But every four- and two-legged animal is de facto in this sense metaphysicist. [Emphasis in the original.]

Similarly, Philipp Frank, Einstein's early associate and later his biographer, reports that the realization of Einstein's true state of thought reached Frank in a most embarrassing way, at the Congress of German physicists in Prague in 1929, just as Frank was delivering "an address in which I attacked the metaphysical position of the German physicists and defended the positivistic ideas of Mach." The very next speaker disagreed and showed Frank that he had been mistaken still to associate Einstein's views with that of Mach and himself. "He added that Einstein was entirely in accord with Planck's view that physical laws describe a reality in space and time that is independent of ourselves. At that time," Frank comments, "this presentation of Einstein's views took me very much by surprise."[73]

In retrospect it is, of course, much easier to see the evidences that this change was being prepared. Einstein himself realized more and more clearly how closely he had moved to Planck, from whom he earlier dissociated himself in three of the four letters to Mach. At the celebration of Planck's sixtieth birthday, two years after Mach's death, Einstein made a moving speech in which, perhaps for the first time, he referred publicly to the Planck-Mach dispute and affirmed his belief that "there is no logical way to the discovery of these elementary laws. There is only the way of intuition" based on *Einfühlung* in experience.[74] The scientific dispute concerning the theory of radiation between Einstein and Planck, too, had been settled (in Einstein's favor) by a sequence of developments after 1911—for example, by Bohr's theory of radiation from gas atoms. As colleagues, Planck and Einstein saw each other regularly from 1913 on. Among evidences of the coincidence of these outlooks there is in the Einstein Archives a handwritten draft, written on or just before 17 April 1931 and intended as Einstein's introduction to Planck's hard-hitting article *Positivism and the Real External World*.[75] In lauding Planck's article, Einstein concludes: "I presume I may add that both Planck's conception of the logical state of affairs as well as his subjective expectation concerning the later development of our science corresponds entirely with my own understanding."[76]

This essay gave a clear exposition of Planck's (and one may assume, Einstein's) views, both in physics and in philosophy more generally. Thus Planck wrote there:

The essential point of the positivist theory is that there is no other source of knowledge except the straight and short way of perception through the senses. Positivism always holds strictly to that. Now, the two sentences: (1) *there is a real outer world which exists independently of our act of knowing* and (2) *the real outer world is not directly knowable* form together the cardinal hinge on which the whole structure of physical science turns. And yet there is a certain degree of contradiction between those two sentences. This fact discloses the presence of the irrational, or mystic, element which adheres to physical science as to every other branch of human knowledge. The effect of this is that a science is never in a position completely and exhaustively to solve the problem it has to face. We must accept that as a hard and fast, irrefutable fact, and this fact cannot be removed by a theory which restricts the scope of science at its very start. Therefore, we see the task of science arising before us as an incessant struggle toward a goal which will never be reached, because by its very nature it is unreachable. It is of a metaphysical character, and, as such, is always again and again beyond our achievement.[77]

From then on, Einstein's and Planck's writings on these matters are often almost indistinguishable from each other. Thus, in an essay in honor of Bertrand Russell, Einstein warns that the "fateful 'fear of metaphysics' . . . has come to be a malady of contemporary empiricistic philosophizing."[78] On the other hand, in the numerous letters between the two old friends, Einstein and Besso, each to the very end touchingly and patiently tries to explain his position, and perhaps to change the other's. Thus, on 28 February 1952, Besso once more presents a way of making Mach's views again acceptable to Einstein. The latter, in answering on 20 March 1952, once more responds that the facts cannot lead to a deductive theory and, at most, can set the stage "for intuiting a general principle" as the basis of a deductive theory. A little later, Besso is gently scolded (in Einstein's letter of 13 July 1952) : "It appears that you do not take the four-dimensionality of reality seriously, but that instead you take the present to be the only reality. What you call 'world' is in physical terminology 'spacelike sections' for which the relativity theory—already the special theory—denies objective reality."

In the end, Einstein came to embrace the view which many, and perhaps he himself, thought earlier he had eliminated from physics in his basic 1905 paper on relativity theory: that there exists an external, objective, physical reality which we may hope to grasp—not directly, empirically, or logically, or with fullest certainty, but at least by an intuitive leap, one that is only guided by experience of the totality of sensible "facts." Events take place in a "real world," of which the space-time world of sensory experience, and even the world of multidimensional continua, are useful conceptions, but no more than that.

For a scientist to change his philosophical beliefs so fundamentally is rare, but not unprecedented. Mach himself underwent a dramatic transformation quite early (from Kantian idealism, at about age seventeen or eighteen, according to Mach's autobiographical notes). We have noted that Ostwald changed twice, once to anti-atomism and then back to atomism. And strangely Planck himself confessed in his 1910 attack on Mach[27] that some twenty years earlier, near the beginning of his own career when Planck was in his late twenties (and Mach was in his late forties), he, too, had been counted "one of the decided followers of the Machist philosophy," as indeed is evident in Planck's early essay on the conservation of energy (1887).

In an unpublished fragment apparently intended as an additional critical reply to one of the essays in the collection ALBERT EINSTEIN: PHILOSOPHER-SCIENTIST (1949), Einstein returned once more to deal —quite scathingly—with the opposition. The very words he used showed how complete was the change in his epistemology. Perhaps even without consciously remembering Planck's words in the attack on Mach of 1909 cited earlier—that a basic aim of science is "the complete liberation of the physical world picture from the individuality of the separate intellects"[32]—Einstein refers to a "basic axiom" in his own thinking:

> It is the postulation of a "real world" which so-to-speak liberates the "world" from the thinking and experiencing subject. The extreme positivists think that they can do without it; this seems to me to be an illusion, if they are not willing to renounce thought itself.

Einstein's final epistemological message was that the world of mere experience must be subjugated by and based in fundamental thought so general that it may be called cosmological in character. To be sure, modern philosophy did not gain thereby a major, novel, and finished corpus. Physicists the world over generally feel that today one must steer more or less a middle course in the area between, on the one hand, the Machist attachment to empirical data or heuristic proposals as the sole source of theory and, on the other, the aesthetic-mathematical attachment to persuasive internal harmony as the warrant of truth. Moreover, the old dichotomy between rationalism and empiricism is slowly being dissolved in new approaches.[79] Yet by encompassing in his own philosophical development both ends of this range, and by always stating forthrightly and with eloquence his redefined position, Einstein not only helped us to define our own, but also gave us a virtually unique case study of the interaction of science and epistemology.

REFERENCES AND NOTES

1. [The main results of the study to this point are contained in Essays 5 through 10 in this collection.]

2. These documents are mostly on deposit at the Archives of the Estate of Albert Einstein at Princeton; where not otherwise indicated, citations made here are from those documents. In studying and helping to order for scholarly purposes the materials in the Archives, I am grateful for the help received from the Trustees of the Albert Einstein Estate, and particularly from Miss Helen Dukas.

I thank the Executor of the Estate for permission to quote from the writings of Albert Einstein. I also wish to acknowledge the financial support provided by the Rockefeller Foundation for cataloguing the collection in the Archives at Princeton. The Institute for Advanced Study at Princeton and its director have been most hospitable throughout this continuing work. I am also grateful to M. Vero Besso for permission to quote from the letters of his father, Michelange Besso. All translations here are the author's unless otherwise indicated.

Early drafts of portions of this essay have been presented as invited papers at the *Tagung* of *Eranos* in Ascona (August, 1965), at the International Congress for the History of Science in Warsaw (August, 1965), and at the meeting, *Science et Synthèse*, at UNESCO in Paris (December, 1965), published in SCIENCE ET SYNTHÈSE, Paris: Gallimard, 1967.

3. This letter as well as the next two letters mentioned in the text (those of 3 April 1901 and 13 April 1901) have been published by Hans-Günther Körber, FORSCHUNGEN UND FORTSCHRITTE, 38: 75–76, 1964.

4. Albert Einstein, *Folgerungen aus den Capillaritätserscheinungen*, ANNALEN DER PHYSIK, 4:513–523, 1901.

5. The only other known attempt on Einstein's part to obtain an assistantship at that time was a request to Kammerlingh-Onnes (12 April 1901), to which, incidentally, he also seems to have received no response.

6. J. T. Merz, A HISTORY OF EUROPEAN THOUGHT IN THE NINETEENTH CENTURY, 4 vols. (Edinburgh: William Blackwood & Sons, 1904–1912; reprint ed., New York: Dover Publishing Co., 1965) Volume II, pp. 184, 199.

7. ERINNERUNGEN AN ALBERT EINSTEIN, issued by the Patent Office in Bern, about 1965 (n. d., no pagination).

8. Wilhelm Ostwald, *Chemische Energie*, LEHRBUCH DER ALLGEMEINEN CHEMIE, 2nd ed. (Leipzig: Verlag von Wilhelm Engelmann, 1903), Volume II, Part 1', p. 1014.

9. Albert Einstein, *Autobiographical Notes*, in P. A. Schilpp, ed., ALBERT

EINSTEIN: PHILOSOPHER-SCIENTIST (Evanston, Ill.: Library of Living Philosophers, 1949), p. 15.

10. Anton Reiser, ALBERT EINSTEIN (New York: A. & C. Boni, 1930), pp. 51–52.

11. Ernst Mach, DIE MECHANIK IN IHRER ENTWICKLUNG, HISTORISCH-KRITISCH DARGESTELLT, Leipzig: 1883.

12. Mach, *op. cit.*, preface, 7th ed., 1912.

13. Moritz Schlick, *Ernst Mach, der Philosoph*, in a special supplement on Ernst Mach in the NEUE FREIE PRESSE (Vienna), 12 June 1926. Einstein himself, in a brief and telling analysis, *Zur Enthüllung von Ernst Machs Denkmal*, published in the same issue (the day of the unveiling of a monument to Mach), wrote:

"Ernst Mach's strongest driving force was a philosophical one: the dignity of all scientific concepts and statements rests solely in isolated experiences [*Einzelerlebnisse*] to which the concepts refer. This fundamental proposition exerted mastery over him in all his research and gave him the strength to examine the traditional fundamental concepts of physics (time, space, inertia) with an independence which at that time was unheard of."

14. Among many evidences of Mach's effectiveness, not the least are his five hundred or more publications (counting all editions—for example, seven editions of his THE SCIENCE OF MECHANICS in German alone during his lifetime), as well as his large exchange of letters, books, and reprints (of which many important ones "carry the dedication of their authors," to cite the impressive catalogue of Mach's library by Theodor Ackermann, Munich, No. 634 [1959] and No. 636 [1960]). A glimpse of Mach's effect on those near him was furnished by William James, who in 1882 heard Mach give a "beautiful" lecture in Prague. Mach received James "with open arms. . . . Mach came to my hotel and I spent four hours walking and supping with him at his club, an unforgettable conversation. I don't think anyone ever gave me so strong an impression of pure intellectual genius. He apparently has read everything and thought about everything, and has an absolute simplicity of manner and winningness of smile when his face lights up, that are charming." From James's letter, in Gay Wilson Allen, WILLIAM JAMES, A BIOGRAPHY (New York: Viking Press, 1967), p. 249.

The topicality of Mach's early speculations on what is now part of General Relativity Theory is attested to by the large number of continuing contributions on the Mach Principle. Beyond that, Mach's influence today is still strong in scientific thinking, though few are as explicit and forthright as the distinguished physicist R. H. Dicke of Princeton University, in his recent, technical book THE THEORETICAL SIGNIFICANCE OF EXPERIMENTAL RELATIVITY (London:

Gordon and Breach, 1964), pp. vii–viii: "I was curious to know how many other reasonable theories [in addition to General Relativity] would be supported by the same facts. . . . The reason for limiting the class of theories in this way is to be found in matters of philosophy, not in the observations. Foremost among these considerations was the philosophy of Bishop Berkeley and E. Mach. . . . The philosophy of Berkeley and Mach always lurked in the background and influenced all of my thoughts."

15. Albert Einstein, *Ernst Mach*, PHYSIKALISCHE ZEITSCHRIFT, 17: 101–104, 1916.

16. Einstein, *Autobiographical Notes*, p. 21.

17. Gerald Holton, *On the Origins of the Special Theory of Relativity*, AMERICAN JOURNAL OF PHYSICS, 28:627–636, 1960, and reprinted here as Essay 5.

18. Gerald Holton, *On the Thematic Analysis of Science: The Case of Poincaré and Relativity*, MÉLANGES ALEXANDRE KOYRÉ (Paris: Hermann, 1964), pp. 257–268, and reprinted here as Essay 6.

19. For evidences that this insistence on prior epistemological analysis of conceptions of space and time are Machist rather than primarily derived from Hume and Kant (who had, however, also been influential), see Einstein's detailed rendition of Mach's critique of Newtonian space and time, in note 15; his discussion of Mach in note 9, pp. 27–29; and in note 1.

20. Albert Einstein, *Zur Elecktrodynamik bewegter Körper*, ANNALEN DER PHYSIK, 17:893, 1905.

21. *Ibid*., p. 894.

22. Philipp Frank, *Einstein, Mach, and Logical Positivism*, in Schilpp, *op. cit*., pp. 272–273. "The definition of simultaneity in the special theory of relativity is based on Mach's requirement that every statement in physics has to state relations between observable quantities. . . . There is no doubt that . . . Mach's requirement, the 'positivistic' requirement, was of great heuristic value to Einstein."

23. For example, see Philipp Frank, MODERN SCIENCE AND ITS PHILOSOPHY (New York: George Braziller, 1955), pp. 61–89; Viktor Kraft, THE VIENNA CIRCLE, trans. Arthur Pap (New York: Philosophical Library, 1953); Richard von Mises, ERNST MACH UND DIE EMPIRISTISCHE WISSENSCHAFTSAUFFASSUNG (1938; printed as a fascicule of the series EINHEITSWISSENSCHAFT).

24. Joseph Petzoldt, *Gesellschaft für positivistiche Philosophie*, reprinted in ZEITSCHRIFT FÜR POSITIVISTISCHE PHILOSOPHIE, 1:4, 1913. In that same speech, Petzoldt sounded a theme that became widely favored in the positivistic inter-

pretation of the genesis of relativity theory—namely, that the relativity theory was developed in direct response to the puzzle posed by the results of the Michelson experiment.

In his interesting essay, *Das Verhältnis der Machschen Gedankenwelt zur Relativitätstheorie*, published as an appendix to the eighth German edition of Ernst Mach's DIE MECHANIK IN IHRER ENTWICKLUNG (Leipzig: F. A. Brockhaus, 1921), pp. 490–517, Petzoldt faithfully attempts to identify and discuss several Machist aspects of Einstein's relativity theory:

(1) The theory "in the end is based on the recognition of the coincidence of sensations; and therefore it is fully in accord with Mach's world-view, which may be best characterized as a relativistic positivism" (p. 516).

(2) Mach's works "produced the atmosphere without which Einstein's Relativity Theory would not have been possible" (p. 494), and in particular Mach's analysis of the equivalence of rotating reference objects in Newton's bucket experiment prepared for the next step, Einstein's "equivalence of relatively moving coordinate systems" (p. 495).

(3) Mach's principle of economy is said to be marvelously exhibited in Einstein's succinct and simple statements of the two fundamental hypotheses. The postulate of the equivalence of inertial coordinate systems deals with "the simplest case thinkable, which now also serves as a fundamental pillar for the General Theory. And Einstein chose also with relatively greatest simplicity the other basic postulate [constancy of light velocity]. . . . These are the foundations. Everything else is logical consequence" (pp. 497–498).

25. Letter of Besso to Einstein, 16 February 1939. Among many testimonies to the effect of Einstein on positivistic philosophies of science, see P. W. Bridgman, *Einstein's Theory and the Operational Point of View*, in Schilpp, *op. cit.*, pp. 335–354.

26. Ernst Mach, HISTORY AND ROOT OF THE PRINCIPLE OF THE CONSERVATION OF ENERGY (Chicago: The Open Court Publishing Co., 1911), p. 95, translation by Philip E. B. Jourdain of the second edition (Ernst Mach, DIE GESCHICHTE UND DIE WURZEL DES SATZES VON DER ERHALTUNG DER ARBEIT, Leipzig: J. A. Barth, 1909). For a brief analysis of Mach's various expressions of adherence as well as reservations with respect to the principle of relativity, see Hugo Dingler, DIE GRUNDGEDANKEN DER MACHSCHEN PHILOSOPHIE (Leipzig: J. A. Barth, 1924), pp. 73–86. Friedrich Herneck (*Nochmals über Einstein und Mach*, PHYSIKALISCHE BLÄTTER, 17:275, 1961) reports that Frank wrote him he had the impression during a discussion with Ernst Mach around 1910 that Mach "was fully in accord with Einstein's special relativity theory, and particularly with its philosophical basis."

27. Republished in Max Planck, A SURVEY OF PHYSICAL THEORY (New

York: Dover Publications, 1960), p. 24. We shall read later a reaffirmation of this position, in almost exactly the same words, but from another pen.

After Mach's rejoinder (*Die Leitgedanken meiner naturwissenschaftlichen Erkenntnislehre und ihre Aufnahme durch die Zeitgenossen*, SCIENTIA, 7:225, 1910), Planck wrote a second, much more angry essay, *Zur Machschen Theorie der physikalischen Erkenntnis*, VIERTELJAHRSCHRIFT FÜR WISSENSCHAFTLICHE PHILOSOPHIE, 34:497, 1910. He ends as follows: "If the physicist wishes to further his science, he must be a Realist, not an Economist [in the sense of Mach's principle of economy]; that is, in the flux of appearances he must above all search for and unveil that which persists, is not transient, and is independent of human senses."

28. Philipp Frank, EINSTEIN, HIS LIFE AND TIMES, trans. George Rose, ed. and rev. Suichi Kusaka, New York: Alfred A. Knopf, 1947.

29. Later Einstein found that this procedure did not work; see Albert Einstein, *Notes on the Origin of the General Theory of Relativity*, IDEAS AND OPINIONS, trans. Sonja Bargmann (London: Alvin Redman, 1954), pp. 285–290; and other publications. In a letter of 2 February 1954 to Felix Pirani, Einstein writes: "One shouldn't talk at all any longer of Mach's principle, in my opinion. It arose at a time when one thought that 'ponderable bodies' were the only physical reality and that in a theory all elements that are not fully determined by them should be conscientiously avoided. I am quite aware of the fact that for a long time, I, too, was influenced by this fixed idea."

30. Albert Einstein and Marcel Grossman, *Entwurf einer verallgemeinerten Relativitätstheorie und einer Theorie der Gravitation*, ZEITSCHRIFT FÜR MATHEMATIK UND PHYSIK, 62:225–261, 1913.

31. For a further analysis and the full text of the four letters see Friedrich Herneck, *Zum Briefwechsel Albert Einsteins mit Ernst Mach*, FORSCHUNGEN UND FORTSCHRITTE, 37:239–243, 1963, and *Die Beziehungen zwischen Einstein und Mach, documentarisch Dargestellt*, WISSENSCHAFTLICHE ZEITSCHRIFT DER FRIEDRICH-SCHILLER-UNIVERSITÄT JENA, MATHEMATISCH-NATURWISSENSCHAFTLICHE REIHE, 15:1–14, 1966; and Helmut Hönl, *Ein Brief Albert Einsteins an Ernst Mach*, PHYSIKALISCHE BLÄTTER, 16:571–580, 1960. Many other evidences, direct and indirect, have been published to show Mach's influence on Einstein prior to Mach's death in 1916. For example, recently a document has been found which shows that in 1911 Mach had participated in formulating and signing a manifesto calling for the founding of a society for positivistic philosophy. Among the signers, together with Mach, we find Joseph Petzoldt, David Hilbert, Felix Klein, Georg Helm, Sigmund Freud, and Einstein. (See Herneck, *Nochmals über Einstein und Mach*, p. 276.)

32. Ernst Mach, *Die Leitgedanken meiner naturwissenschaftlichen Erk-*

enntnislehre und ihre Aufnahme durch die Zeitgenossen, PHYSIKALISCHE ZEITSCHRIFT, 11:605, 1910. (Emphasis added.) To be sure, Mach was not always a "Machist" himself.

33. See Einstein, *Time, Space, and Gravitation* (1948), OUT OF MY LATER YEARS (New York: Philosophical Library, 1950), pp. 54–58. Einstein makes the distinction between constructive theories and "theories of principle." Einstein cites, as an example of the latter, the relativity theory, and the laws of thermodynamics. Such theories of principle, Einstein says, start with "empirically observed general properties of phenomena." See also *Autobiographical Notes*, p. 53.

34. Ernst Mach, DIE PRINZIPIEN DER PHYSIKALISCHEN OPTIK, Leipzig: J. A. Barth, 1921. The English edition is THE PRINCIPLES OF PHYSICAL OPTICS, trans. John S. Anderson and A. F. A. Young (London: Methuen, 1926); reprint ed., New York: Dover Publications, 1953), pp. vii–viii.

35. Einstein, in *Séance du 6 avril, 1922: La Théorie de la relativité*, BULLETIN DE LA SOCIÉTÉ FRANÇAISE DE PHILOSOPHIE, 22:91–113, 1922. In his 1913 preface rejecting relativity, Mach expressed himself perhaps more impetuously and irascibly than he may have meant. Some evidence for this possibility is in Mach's letters to Petzoldt. In early 1914, Mach wrote: "I have received the copy of the positivistic ZEITSCHRIFT which contains your article on relativity; I liked it not only because you copiously acknowledge my humble contributions with respect to that theme, but also in general." And within a month, Mach writes—rather more incoherently—to Petzoldt: "The enclosed letter of Einstein [a copy of the last of Einstein's four letters, cited above] proves the penetration of positivistic philosophy into physics; you can be glad about it. A year ago, philosophy was altogether sheer nonsense. The details prove it. The paradox of the clock would not have been noticed by Einstein a year ago."

I thank Dr. John Blackmore for drawing my attention to the Mach-Petzoldt letters, and Dr. H. Müller for providing copies from the Petzoldt Archive in Berlin.

36. R. S. Shankland, *Conversations with Einstein*, AMERICAN JOURNAL OF PHYSICS, 31:56, 1963.

37. A typical example is his letter of 18 September 1930 to Armin Weiner: ". . . I did not have a particularly important exchange of letters with Mach. However, Mach did have a considerable influence upon my development through his writings. Whether or to what extent my life's work was influenced thereby is impossible for me to find out. Mach occupied himself in his last years with the relativity theory, and in a preface to a late edition of one of his works even spoke out in rather sharp refusal against the relativity theory. However, there can be no doubt that this was a consequence of a lessening

ability to take up [new ideas] owing to his age, for the whole direction of thought of this theory conforms with Mach's, so that Mach quite rightly is considered as a forerunner of general relativity theory. . . ."

I thank Colonel Bern Dibner for making a copy of the letter available to me from the Archives of the Burndy Library in Norwalk, Connecticut. Among other hitherto unpublished letters in which Einstein indicated his indebtedness to Mach, we may cite one to Anton Lampa, 9 December 1935: ". . . You speak about Mach as about a man who has gone into oblivion. I cannot believe that this corresponds to the fact since the philosophical orientation of the physicists today is rather close to that of Mach, a circumstance which rests not a little on the influence of Mach's writings."

Moreover, practically everyone else shared Einstein's explicitly expressed opinion of the debt of relativity theory to Mach; thus Hans Reichenbach wrote in 1921, "Einstein's theory signified the accomplishment of Mach's program." (*Der gegenwärtige Stand der Relativitätsdiskussion*, LOGOS, 10:311, 1922.) Even Hugo Dingler agreed: [Mach's] criticism of the Newtonian conceptions of time and space served as a starting point for the relativity theory. . . . Not only Einstein's work, but even more recent developments, such as Heisenberg's quantum mechanics, have been inspired by the Machian philosophy." Hugo Dingler, *Ernst Mach*, ENCYCLOPEDIA OF THE SOCIAL SCIENCES, ed. Edwin R. A. Seligman and Alvin Johnson (New York: Macmillan Co., 1933), Volume 9, p. 653. And H. E. Hering wrote an essay whose title is typical of many others: *Mach als Vorläufer des physikalischen Relativitätsprinzips*, KÖLNER UNIVERSITÄTSZEITUNG, 1:3–4, January 17, 1920. I thank Dr. John Blackmore for a copy of the article.

38. In his article, *Zur Enthüllung von Ernst Machs Denkmal*, in the special supplement of NEUE FREIE PRESSE cited in note 13, Einstein—then already disenchanted for some time with the Machist program—wrote immediately after the portion quoted in note 13:

"Philosophers and scientists have often criticized Mach, and correctly so, because he erased the logical independence of the concepts vis-à-vis the 'sensations,' [and] because he wanted to dissolve the Reality of Being, without whose postulation no physics is possible, in the Reality of Experience . . . ."

There were additional sources, both published and unpublished, on the detailed aspects of the relation between Einstein and Mach, which, for lack of space, cannot be summarized here.

39. For example, by Einstein himself, by Joseph Petzoldt, and by Hugo Dingler. I assign relatively little weight to the possibility that the rift grew out of the difference between Einstein and Mach on atomism. Herneck provides the significant report that according to a letter from Philipp Frank, Mach was personally influenced by Dingler, whom Mach had praised in the 1912

edition of the MECHANIK and who was from the beginning an opponent of relativity theory, becoming one of the most "embittered enemies" of Einstein (Herneck, *Die Beziehungen zwischen Einstein und Mach*, p. 14; see note 31). The copies of letters from Dingler to Mach in the Ernst-Mach-Institute in Freiburg indicate Dingler's intentions; nevertheless, there remains a puzzle about Dingler's role which is worth investigating. It is significant that in his 1921 essay (see note 24) Petzoldt devotes much space to a defense of Einstein's work against Dingler's attacks. See also Joachim Thiele, *Analysis of Mach's Preface*, in NTM, SCHRIFTENREIHE FÜR GESCHICHTE DER NATURWISSENSCHAFTEN, TECHNIK UND MEDIZIN (Leipzig), 2:10–19, 1965.

40. Albert Einstein, *Über Relativitätstheorie*, MEIN WELTBILD (Amsterdam: Querido Verlag, 1934), pp. 214–220; republished as *On the Theory of Relativity*, IDEAS AND OPINIONS, pp. 246–249. Herneck has given the texts of similar discussions on phonographic records by Einstein in 1921 and even in 1924; cf. Herneck's transcriptions of Einstein, *Zwei Tondokumente Einsteins zur Relativitätstheorie*, FORSCHUNGEN UND FORTSCHRITTE, 40:133–134, 1966.

41. Quotations from Einstein, *Zur Methode der theoretischen Physik*, MEIN WELTBILD, pp. 176–187, as reprinted in translation in *On the Method of Theoretical Physics*, IDEAS AND OPINIONS, pp. 270–276, except for correction of mistranslation of one line. There are a number of later lectures and essays in which the same point is made. See, for example, the lecture, *Physics and Reality* (1936, reprinted in IDEAS AND OPINIONS, pp. 290–323), which states that Mach's theory of knowledge is insufficient on account of the relative closeness between experience and the concepts which it uses; Einstein advocates going beyond this "phenomenological physics" to achieve a theory whose basis may be further removed from direct experience, but which in return has more "unity in the foundations." Or see *Autobiographical Notes*, p. 27: "In the choice of theories in the future," he indicates that the basic concepts and axioms will continue to "distance themselves from what is directly observable."

Even as Einstein's views developed to encompass the "*erlebbare, beobachtbare*" facts as well as the "*wild-spekulative*" nature of theory, so did those of many of the philosophers of science who also had earlier started from a strict Machist position. This growing modification of the original position, partly owing to "the growing understanding of the general theory of relativity," has been chronicled by Frank, for example, in *Einstein, Mach, and Logical Positivism*, in Schilpp, *op. cit.*, pp. 296–286.

42. For a discussion of thematic and phenomenic elements in theory construction, see my article, *The Thematic Imagination in Science*, in SCIENCE AND CULTURE, ed. Gerald Holton (Boston: Houghton Mifflin Co., 1965), pp. 88–108, and reprinted here as Essay 1.

43. Einstein, *Autobiographical Notes*, p. 53. Emphasis added. On pp. 9–11, Einstein describes what may be a possible precursor of this attitude in his study of geometry in his childhood.

44. Einstein, *Über das Relativitätsprinzip und die aus demselben gezogenen Folgerungen*, JAHRBUCH DER RADIOAKTIVITÄT UND ELEKTRONIK, 4:411–462, 1907.

45. Walter Kaufmann. *Über die Konstitution des Elektrons*, ANNALEN DER PHYSIK, 19:487–553, 1906. Emphasis in original.

46. *Ibid.*, p. 495.

47. See note 44, p. 439.

48. *Ibid.* Shortly after Kaufmann's article appeared, Max Planck (*Die Kaufmannschen Messungen der Ablenkbarkeit der β-Strahlen in ihrer Bedeutung für die Dynamik der Elektronen*, PHYSIKALISCHE ZEITSCHRIFT, 7:753–761, 1906) took it on himself publicly to defend Einstein's work in an analysis of Kaufmann's claim. He concluded that Kaufmann's data did not have sufficient precision for his claim. Incidentally, Planck tried to coin the term for the new theory that had not yet been named: "*Relativtheorie.*"

49. It should be remembered that Poincaré, with a much longer investment in attempts to fashion a theory of relativity, was quite ready to give in to the experimental "evidence." See note 18.

50. Carl Seelig, ALBERT EINSTEIN (Zurich: Europa Verlag, 1954), p. 195.

51. Einstein, *Einiges über die Entstehung der allgemeinen Relativitätstheorie*, MEIN WELTBILD, pp. 248–256, reprinted in translation as *Notes on the Origin of the General Theory of Relativity*, IDEAS AND OPINIONS, pp. 285–90.

52. "*Da könnt' mir halt der liebe Gott leid tun, die Theorie stimmt doch.*" This semi-serious remark of a person who was anything but sacrilegious indeed illuminates the whole style of a significant group of new physicists. P. A. M. Dirac, *The Evolution of the Physicist's Picture of Nature*, SCIENTIFIC AMERICAN, 208:46–47, 1963 speaks about this, with special attention to the work of Schrödinger, a spirit close to that of his friend, Einstein, despite the ambivalence of the latter to the advances in quantum physics. We can do no better than quote *in extenso* from Dirac's account:

"Schrödinger worked from a more mathematical point of view, trying to find a beautiful theory for describing atomic events, and was helped by deBroglie's ideas of waves associated with particles. He was able to extend deBroglie's ideas and to get a very beautiful equation, known as Schrödinger's wave equation, for describing atomic processes. Schrödinger got this equation by pure thought, looking for some beautiful generalization of deBroglie's ideas and not by keeping close to the experimental development of the subject in the way Heisenberg did.

"I might tell you the story I heard from Schrödinger of how, when he first got the ideas for this equation, he immediately applied it to the behavior of the electron in the hydrogen atom, and then he got results that did not agree with experiment. The disagreement arose because at that time it was not known that the electron has a spin. That, of course, was a great disappointment to Schrödinger, and it caused him to abandon the work for some months. Then he noticed that if he applied the theory in a more approximate way, not taking into account the refinements required by relativity, to this rough approximation, his work was in agreement with observation. He published his first paper with only this rough approximation, and in this way Schrödinger's wave equation was presented to the world. Afterward, of course, when people found out how to take into account correctly the spin of the electron, the discrepancy between the results of applying Schrödinger's relativistic equation and the experiment was completely cleared up.

"I think there is a moral to this story, namely, that it is more important to have beauty in one's equations than to have them fit experiment. If Schrödinger had been more confident of his work, he could have published it some months earlier, and he could have published a more accurate equation. The equation is now known as the Klein-Gordon equation, although it was really discovered by Schrödinger before he discovered his nonrelativistic treatment of the hydrogen atom. It seems that if one is working from the point of view of getting beauty in one's equations, and if one has really a sound insight, one is on a sure line of progress. If there is not complete agreement between the results of one's work and experiment, one should not allow oneself to be too discouraged, because the discrepancy may well be due to minor features that are not properly taken into account and that will get cleared up with further developments of the theory. That is how quantum mechanics was discovered . . ." (pp. 46–47).

53. Published several times—for example, by B. G. Teuber, Leipzig, 1909.

54. Friedrich Herneck, *Zu einem Brief Albert Einsteins an Ernst Mach*, Physikalische Blätter, 15:564, 1959. Frank's remark is reported by Herneck in *Ernst Mach und Albert Einstein*, Symposium aus Anlass des 50. Todestages von Ernst Mach, ed. Frank Kerkhof (Freiburg: Ernst Mach Institut, 1966), pp. 45–61.

55. Philipp Frank, *Das Relativitätsprinzip und die Darstellung der physikalischen Erscheinungen im vierdimensionalen Raum*, Zeitschrift für physikalische Chemie, 74:466–495, 1910.

56. C. B. Weinberg, *Mach's Empirio-Pragmatism in Physical Science* (Thesis, Columbia University, 1937).

57. Mach, *Die Leitgedanken meiner naturwissenschaftlichen Erkenntnislehre und ihre Aufnahme durch die Zeitgenossen*, SCIENTIA, 7:225.

58. Ernst Mach, SPACE AND GEOMETRY (Chicago: Open Court Publishing Co., 1906), p. 138. Mach's attempts to speculate on the use of n-dimensional spaces for representing the configuration of such "mere things of thought"—the derogatory phrase also applied to absolute space and absolute motion in Newton—are found in his first major book, CONSERVATION OF ENERGY (first edition, 1872).

59. Cf. Joseph Petzoldt, *Verbietet die Relativitätstheorie Raum und Zeit als etwas Wirkliches zu denken?* VERHANDLUNGEN DER DEUTSCHEN PHYSIKALISCHEN GESELLSCHAFT, No. 21–24 (1918), pp. 189–201. Here, and again in his 1921 essay (note 24), Petzoldt tries to protect Einstein from the charge—for example, made by Sommerfeld—that space and time no longer "are to be thought of as real."

60. Mach, CONSERVATION OF ENERGY, p. 54.

61. See note 35; also reported in *Einstein and the Philosophies of Kant and Mach*, NATURE, 112:253, 1923.

62. That Einstein did not so misunderstand Newton can be illustrated, for example, in a comment reported by C. B. Weinberg: "Dr. Einstein further maintained that Mach, as well as Newton, tacitly employs hypotheses—not recognizing their non-empirical foundations." (Weinberg, *op. cit.*, p. 55.) Dingler analyzed some of the nonempirical foundations of relativity theory in KRITISCHE BEMERKUNGEN ZU DEN GRUNDLAGEN DER RELATIVITÄTSTHEORIE (Leipzig: S. Hirzel, 1921).

63. Cited by Robert S. Cohen in his very useful essay *Dialectical Materialism and Carnap's Logical Empiricism*, THE PHILOSOPHY OF RUDOLF CARNAP, ed. P. A. Schilpp (La Salle, Ill.: Open Court Publishing Co., 1963), p. 109. I am also grateful to Professor Cohen for a critique of parts of this paper in earlier form.

64. Albert Einstein, *Physics and Reality*, JOURNAL OF THE FRANKLIN INSTITUTE, 221:313–347, 1936.

65. Einstein, *Notes on the Origin of the General Theory of Relativity*, pp. 288, 289.

66. See note 30, pp. 230–231.

67. I am not touching in this essay on the effect of quantum mechanics on Einstein's epistemological development; the chief reason is that while from his "heuristic" announcement of the value of a quantum theory in 1905, Einstein remained consistently skeptical about the "reality" of the quantum

theory of radiation, this opinion only added to the growing realism stemming from his work on general relativity theory. In the end, he reached the same position in quantum physics as in relativity; cf. his letter of 7 September 1944 to Max Born: "In our scientific expectations we have become antipodes. You believe in the dice-playing God, and I in the perfect rule of law in an objectively existing world which I try to capture in a wildly speculative way." (Reported by Max Born, *Erinnerungen an Einstein*, UNIVERSITAS, ZEITSCHRIFT FÜR WISSENSCHAFT, KUNST UND LITERATUR, 20:795–807, 1965.)

68. Max Born, *Physics and Relativity*, PHYSICS IN MY GENERATION (London: Pergamon Press, 1956), p. 205.

69. Albert Einstein, *Über den gegenwärtigen Stand der Feldtheorie*, FESTSCHRIFT PROF. DR. A. STODOLA ZUM 70. GEBURTSTAG, ed. E. Honegger (Zurich and Leipzig: Orell Füssli Verlag, 1929), pp. 126–132. I am grateful to Professor Cornelius Lanczos and Professor John Wheeler for pointing out this reference to me.

70. Dingler, DIE GRUNDGEDANKEN DER MACHSCHEN PHILOSOPHIE, p. 98.

71. Max Planck, THE PHILOSOPHY OF PHYSICS, trans. W. H. Johnson (New York: W. W. Norton & Co. 1936), pp. 122, 125.

72. Albert Einstein, *Religion and Science*, THE NEW YORK TIMES MAGAZINE, 9 November 1930; cf. MEIN WELTBILD, p. 39, and COSMIC RELIGION (New York: Covici-Friede, 1931), p. 48.

Possible reasons for Einstein's growing interest in these matters, partly related to the worsening political situation at the time, are discussed in Frank, EINSTEIN, HIS LIFE AND TIMES. (See note 28). It is noteworthy that while Einstein was quite unconcerned with religious matters during the period of his early scientific publications, he gradually returned later to a position closer to that at a very early age, when he reported he had felt a "deep religiosity. . . . It is quite clear to me that the religious paradise of youth . . . was a first attempt to free myself from the chains of the 'merely personal.'" *Autobiographical Notes*, in Schilpp, *op. cit.*, pp. 3, 5. For a discussion, see Gerald Holton, *Science and New Styles of Thought*, THE GRADUATE JOURNAL, 7:417–420, 1967, and reprinted here as Essay 3.

73. Frank, EINSTEIN, HIS LIFE AND TIMES, p. 215. Einstein's change of mind was, of course, not acceptable to a considerable circle of previously sympathetic scientists and philosophers. See, for example, P. W. Bridgman, *Einstein's Theory and the Operational Point of View* in Schilpp, *op. cit.*, pp. 335–354.

An interesting further confirmation of Einstein's changed epistemological position became available after this paper was first published. Werner Heisenberg, in PHYSICS AND BEYOND (New York, Harper & Row, 1971), pp. 62–66, writes about his conversation with Einstein concerning physics and philosophy. See, for example, p. 63—a portion of a conversation set in 1925–1926:

"But you don't seriously believe," Einstein protested, "that none but observable magnitudes must go into physical theory?"

"Isn't that precisely what you have done with relativity?" I asked in some surprise. "After all, you did stress the fact that it is impermissible to speak of absolute time, simply because absolute time cannot be observed; that only clock readings, be it in the moving reference system or the system at rest, are relevant to the determination of time."

"Possibly I did use this kind of reasoning," Einstein admitted, "but it is nonsense all the same. Perhaps I could put it more diplomatically by saying that it may be heuristically useful to keep in mind what one has actually observed. But in principle, it is quite wrong to try founding a theory on observable magnitudes alone. In reality the very opposite happens. It is the theory which decides what we can observe."

See also Heisenberg's account of Einstein's critique of Mach, *ibid.*, pp. 63–66.

74. Originally entitled *Motiv des Forschens* (in ZU MAX PLANCKS 60. GEBURTSTAG [Karlsruhe: Müller, 1918]), reprinted in translation by James Murphy, as a preface to Max Planck, WHERE IS SCIENCE GOING? (London: Allen & Unwin, 1933), pp. 7–12. In an earlier appreciation of Planck in 1913, Einstein had written only very briefly about his epistemology, merely lauding Planck's essay of 1896 against energetics, and not mentioning Mach. [Einstein's important 1918 essay is analyzed in Essay 10 below.]

75. Planck, *Positivism and External Reality*, INTERNATIONAL FORUM, 1, No. 1:12–16; 1, No. 2:14–19, 1931.

76. Einstein sent his introduction to the editor of the journal on 17 April 1931, but it appears to have come too late for inclusion.

77. Planck, *Positivism and External Reality,* pp. 15–17. Emphasis in original.

78. Albert Einstein, *Bermerkungen zu Bertrand Russells Erkenntnis-Theorie,* in P. A. Schilpp, ed., THE PHILOSOPHY OF BERTRAND RUSSELL (Evanston, Ill.: Library of Living Philosophers, 1944), p. 289.

79. Toward the end, Einstein himself acknowledged a similar point in his *Remarks Concerning the Essays Brought Together in This Co-operative Volume,* in Schilpp, ALBERT EINSTEIN, pp. 679–680: " 'Einstein's position . . . contains features of rationalism and extreme empiricism. . . .' This remark is entirely correct. . . . A wavering between these extremes appears to me unavoidable."

9 EINSTEIN, MICHELSON, AND THE "CRUCIAL" EXPERIMENT

I. Introduction

THE HIGHEST achievements in science are of quite different kinds: the bold theoretical generalization, breathtaking by virtue of its sweeping synthetic power, and the ingenious experiment, sometimes called "crucial," in which the striking character of the result signals a turning point. Albert Einstein's special theory of relativity as first published in 1905 is a supreme example of the first kind, and A. A. Michelson's experiments in the 1880's to find the effect of ether drift on the speed of light are often cited as prototypical examples of the second kind. Even if these two achievements had nothing whatsoever to do with each other, each would continue to be remembered and studied on its own merit. But these two cases have in fact held additional interest for historians and philosophers of science; for, as we shall see, it has been the overwhelming preponderance of opinion over the last half century that Michelson's experiments and Einstein's theory have a close *genetic connection*, one which may be stated most simply in the words of the caption under Michelson's photograph in a recent publication of the Optical Society of America: Michelson "made the measurements on which is based Einstein's Theory of Relativity."

A more detailed account of the experimental origins of relativity

This article was originally published in Isis, Volume 60, 1969, pp. 133–197.

theory is attempted in R. A. Millikan's essay *Albert Einstein on his Seventieth Birthday*. It was the lead article in a special issue in Einstein's honor of the REVIEWS OF MODERN PHYSICS, and the early parts are worth quoting:

> The special theory of relativity may be looked upon as starting essentially in a generalization from Michelson's experiment. And here is where Einstein's characteristic boldness of approach came in, for the distinguishing feature of modern scientific thought lies in the fact that it begins by discarding all *a priori* conceptions about the nature of reality—or about the ultimate nature of the universe—such as had characterized practically all Greek philosophy and all medieval thinking as well, and takes instead, as its starting point, well-authenticated, carefully tested *experimental* facts, no matter whether these facts seem at the moment to be reasonable or not. In a word, modern science is essentially empirical. . . .
>
> But this experiment, after it had been performed with such extraordinary skill and refinement by Michelson and Morley, yielded with great definiteness the answer that there is . . . no observable velocity of the earth with respect to the aether. That unreasonable, apparently inexplicable experimental fact was very bothersome to 19th century physics, and so for almost twenty years after this fact came to light physicists wandered in the wilderness in the disheartening effort to make it seem reasonable. Then Einstein called out to us all, "Let us merely accept this as an established experimental fact and from there proceed to work out its inevitable consequences," and he went at that task himself with an energy and a capacity which very few people on earth possess. Thus was born the special theory of relativity.[1]

The birth of a new theory as the response to a puzzling empirical finding! This sort of thing has happened, but it may also be the stuff of which fairy tales are made. The historian of science senses at once several intriguing problems: How important were experiments to Einstein's formulation of his 1905 paper on relativity? What role did the Michelson experiments play? How good is the evidence on which one is to decide these questions? What light do documents shed on the case, particularly those that appear to provide contradictory evidence? If the Michelson experiments were not of crucial importance, why are there so many who say they were? And if they were, why are there a few who say they were not? What are the philosophical (or other) assumptions made by these two groups? What can this case tell us about the relation between experiment and theory in modern physics? And, above all, what can this case tell us about the rival claims of sensationism and idealism to represent more faithfully the act of modern scientific innovation?

Thus what appears at first to be a limited case opens into the wider field of current scholarship—not the kind of history that uses a wide-angle lens to compose a picture of the rise and fall of major theories, but a case study that focuses a magnifying glass in order to understand a part of modern scientific work. We shall see that there is so much to be observed, so many documents and personae, that even in a full-length essay all the questions cannot be disposed of. Instead, I shall concentrate on those that are particularly illuminated by documents, including some newly found and unpublished ones, and I shall also take the opportunity to gather together and compare previous contributions to this topic that are now widely dispersed.

Partly because of the volume of the resources, many will appear contradictory or ambiguous. Einstein himself made different statements about the influence of the Michelson experiments, ranging from "there is no doubt that Michelson's experiment was of considerable influence on my work . . ." to "the Michelson-Morley experiment had a negligible effect on the discovery of relativity." The initial apparent irreconcilability of the statements need not cause dismay. On the contrary: it is no more comforting to find only unambiguous evidence for one position on a complex issue, for that may indicate that only part of the evidence is in.

Our job will be to weigh incommensurables. And in this act we must try to discern the conceptual framework, motivation, or social mission hiding behind a statement that is asserting to be evidence. Historical statements, like those in physics, have meaning only relative to a specifiable framework. The discovery of the contextual setting will sometimes be as interesting as the use to which a "relativistic" piece of evidence can be put, and thus the light thrown on a specialized problem may help to illuminate a chapter in the history of ideas.

A different purpose of such a study might be simply the correction of popular error. Though it is tempting, this is not my chief aim; nor is it likely to be successful. For the belief that Einstein based his work leading to his 1905 publication of relativity theory on Michelson's result has long been a part of the folklore. It is generally regarded as an important event in the history of science, as widely known and believed as the story of the falling apple in Newton's garden and of the two weights dropped from the leaning tower in Galileo's Pisa—two other cases in which experiential fact is supposed to have provided the genesis of synthetic theory. If Millikan's report and the many others like it are right, then this might be the time when one can still hope to find reliable evidence for them. But if they are not supportable, it is probably in any case too late to stop the spread of a fable which has such inherent appeal.

II. The Symbiosis of Puzzles

At least a brief summary of the essential points of the familiar Michelson experiment may be useful here, even though this will not convey one of the major reasons why it has proved irresistible for so many physics books to give a place of importance to this particular experiment—the fact that it was one of the most fascinating in the history of physics. Its fascination, which has been felt equally by textbook writers and research physicists, derives from its beauty and mystery. Despite the central position of the question of ether drift in late-nineteenth-century physics, nobody before Michelson was able to imagine and construct an apparatus to measure the second-order effect of the presumed ether drift. The interferometer was a lovely thing. Invented by the twenty-eight-year-old Michelson in response to a challenge by Maxwell, it was capable of revealing an effect of the order of one part in ten billion. It is to this day one of the most precise instruments in science, and the experiment is one that carried precision to the extreme limits. Einstein himself later paid warm and sincere tribute to Michelson's experimental genius and artistic sense.[2]

As Michelson recounts in his description of his experiment (in STUDIES IN OPTICS, 1927), among the events that led up to it was first of all George B. Airy's experiment on the angle of aberration of a telescope viewing a star from our moving earth. On the model of light as wave propagation through an ether, the aberration angle was expected to be larger when the observing telescope was filled with water; but on experiment the angle was found to be the same. Augustin Fresnel therefore proposed that the ether is partly carried or dragged along in the motion of a medium (such as water) having a refractive index larger than 1. This hypothesis, in quantitative detail, interpreted Airy's result in a satisfying way and was triumphantly tested in a separate experiment by Armand Fizeau on the effect of moving water on the propagation of a light beam as measured in the laboratory frame of reference. At the same time the experiment implied that a medium of refractive index 1 (such as air) would, when in motion, not carry along any part of the ether.

The hypothesis that the earth moves through an ether which remains unaffected and stagnant all around the earth (later most prominently developed by H. A. Lorentz) invited direct experimental verification. But that required the previously unimaginable feat of looking for the exceedingly small presumed effect of the second order, for the relative earth-ether motion ("ether drift") would show up in a change of the effective light speed by a factor containing the square of the ratio of the speeds of the earth and of light ($v^2/c^2=10^{-8}$).

Michelson's ingenious solution was to let two light beams from the same source simultaneously run a round-trip race along two paths which had effectively the same length in the laboratory but were laid out at 90 degrees, thereby causing the two light beams to be differently affected by their relative motion with respect to the ether. But on bringing the two beams together to compare by their interference pattern these relative effects, Michelson's apparatus (an "interferometer") surprisingly gave what is usually called a negative or null result. More accurately, it gave within experimental error the result that would have been expected on the basis of a quite different hypothesis—namely, that the ether is not stagnant but somehow does get dragged along with the earth and so has no measurable motion or drift with respect to the earth.

The beauty of the design and execution of the experiment was in startling contrast to the mysterious difficulties which attended its interpretation. On one level lay the problem of a detailed understanding of the way the apparatus worked in the context of ether theory, regardless of the meaning of the results. Michelson himself, on presenting in 1882 an account of his first experiment to the Académie des Sciences, acknowledged that he had made an error in his earlier report of 1881 and had neglected the effect of the earth's motion on the path of light in the interferometer arm at right angles to the motion. Alfred Potier, who had pointed out the error to Michelson in 1882, was in error also.[3] On another point—how the moving reflectors in the interferometer affect the angle of reflection—there was a continuing debate for over thirty years. To appreciate the lasting confusion one need only study the record of the summit conference of ether-drift experimenters, held

4–5 February 1927, at the Mt. Wilson Observatory, under the title "Conference on the Michelson-Morley Experiment," with both Michelson and Lorentz in attendance.[4] Although one finds today many simplified accounts of the experiment, in fact a detailed, correct theory of the supposed working of the Michelson interferometer to detect an ether drift is quite complex and is rarely given in full.

But beyond that, on another level the outcome itself was enormously puzzling to everyone at the time, and to many for a long time afterward. The glorious device had yielded a disappointing, even incomprehensible, result in the context of the then-current theory. Michelson himself called his experiment a "failure,"[5] the repeatedly obtained null or nearly null results being contrary to all expectations. Unlike the stereotype of the true scientist who accepts the experimental test that falsifies a theory, he refused to grant the importance of his result, saying, "Since the result of the original experiment was negative, the problem is still demanding a solution."[6] He even tried to console himself with the remarkable observation that "the experiment is to me historically interesting because it was for the solution of this problem that the interferometer was devised. I think it will be admitted that the problem, by leading to the invention of the interferometer, more than compensated for the fact that this particular experiment gave a negative result."[7]

Others were just as mystified and displeased. Lorentz wrote to Rayleigh on 18 August 1892: "I am totally at a loss to clear away this contradiction, and yet I believe that if we were to abandon Fresnel's theory [of the ether], we should have no adequate theory at all. . . . Can there be some point in the theory of Mr. Michelson's experiment which has as yet been overlooked?"[8] Lord Kelvin, who had discerned the result of the experiment as part of a cloud obscuring "the beauty and clearness of the dynamical theory, which asserts heat and light to be modes of motion," could not, even into the 1900's, reconcile himself to the negative findings.[9] Rayleigh, who, like Kelvin, had encouraged Michelson to repeat his first experiment, found the null result obtained by Michelson and Morley to be "a real disappointment."[10] As Loyd S. Swenson has pointed out,[11] Michelson and Morley were so discouraged by the null result of their experiment in 1887 that they disregarded their stated promise that their measurements, which they had taken during only six hours (spread over five days), "will therefore be repeated at intervals of 3 months, and thus all uncertainty will be avoided." Instead, Michelson stopped their work on this experiment and turned to the new use of the interferometer for measuring lengths (which, it turned out, led to his Nobel prize).

In short, to everyone's surprise, including Michelson's, the experiment had turned out to be one of "test" instead of merely "application," to use the terminology of Duhem. Indeed, it was threatening to become for ether theoreticians even, *malgrè lui*, a crucial experiment in the only valid sense of the term, namely as the pivotal occasion causing a significant part of the scientific community to re-examine its previously held basic convictions.

We may gather that for Michelson the experiment was a source of discomfort and perhaps real unhappiness throughout his life, not only because of the null result, but also because of its various explanations. Initially he had felt his findings could only mean that the hypothesis of a stationary ether was incorrect; but the alternatives were no better. The idea that the ether is substantially carried along by the earth was in direct conflict with the well-established results of aberration experiments and Fizeau's measurement of the Fresnel dragging coefficient. And the modification of G. G. Stokes's theory of the ether, which Michelson came to favor, was shown to be untenable by better theoreticians, such as Lorentz, and by the negative result of Oliver Lodge's experiment on the supposed ether drag in the vicinity of rapidly moving discs. Lodge himself confessed impatience with the bothersome Michelson experiments which provided evidence against the existence of a nonviscous ether stagnant in space. Thus Lodge wrote, with only the slight exaggeration that others were to use later: "The one thing in the way of the simple doctrine of an ether undisturbed by motion is Michelson's experiment, viz., the absence of a second-order effect due to terrestrial movement through free ether. This experiment may have to be explained away."[12]

At Chicago in the spring of 1897 Michelson tested the possibility of a differential ether drag at different altitudes, and therefore the applicability of Stokes's hypothesis, which he continued to favor. But the large vertical interferometer also gave negative results. Michelson now was clearly rather exasperated: "One is inclined to return to the hypothesis of Fresnel and try to reconcile in some other way the negative results" of the earlier ether-drift experiments.[13]

Much later, when Michelson came to write the STUDIES IN OPTICS, published in 1927 at the age of seventy-five, he had to end the chapter on *Effects of Motion of the Medium on Velocity of Light*—the subject on which he had spent much of his life—with a question he still could not answer: "It must be admitted, however, that these experiments are not sufficiently conclusive to justify the hypothesis of an ether which is entrained with the earth in its motion. But then how can the negative results be explained?"[14]

By that time two other options had appeared. In his next chapter, Michelson turned first to the proposal by Lorentz and G. F. FitzGerald to explain the "null effect of the Michelson-Morley experiment by assuming a contraction in the material of the support for the interferometer just sufficient to compensate for the theoretical difference in path." But he immediately added, "such a hypothesis seems rather artificial."[15] We note in passing a point that will loom large later—that even to this experimental physicist most sorely in need of an explanation, the Lorentz-FitzGerald hypothesis seemed "artificial," or, to use the terminology of others who expressed the same objection, too patently ad hoc.[16]

As to the other explanation—that implied in Einstein's relativity theory—Michelson, who had long held out against it, now in 1927 proposed a "generous acceptance" of the theory, notwithstanding many "paradoxical" consequences. But it was not a wholehearted acceptance, because "the existence of an ether appears to be inconsistent with the theory," and that seemed to him to be an overwhelming defect: "It is to be hoped that the theory may be reconciled with the existence of a medium, either by modifying the theory, or, more probably by attributing the requisite properties to the ether."[17] At another occasion, also in 1927, Michelson in his last paper published before his death referred to the ether in the following nostalgic words: "Talking in terms of the beloved old ether (which is now abandoned, though I personally still cling a little to it). . . ."[18]

If the result of the Michelson experiment was a mystery for a long time (Swenson has shown that it remained inconclusive into the 1920's[19]), the relativity theory was to most physicists even more mysterious at its announcement in 1905 and for some time afterward. The lag of acceptance of the theory is a major research topic of its own. It took several years before one could say that even among German scientists there was a preponderance of opinion in favor of it, the turning point coming perhaps with the publication in 1909 of Hermann Minkowski's address, *Space and Time*.[20] In fact, the very first response within the scientific community to Einstein's relativity paper, in the same journal in which he had published it, "was a categorical experimental disproof of the theory."[21] For years after Einstein's first publication no new experimental results came forth which could be used to "verify" his theory in the way most physicists were and still are used to look for verification. As Max Planck noted in 1907, Michelson's was then still regarded as the only experimental support.[22] The perceptive physicist Wilhelm Wien had published his earlier disagreement with relativity and was not convinced of the theory until 1909, and then it was not any clear-cut evidence from experiment, but on aesthetic grounds, in words which Einstein must have appreciated: "What speaks for it most of all, however, is the inner consistency which makes it possible to lay a foundation having no self-contradictions, one that applies to the totality of physical appearances, although thereby the customary conceptions experience a transformation."[23]

In retrospect it seems therefore inevitable that during the decade following Einstein's 1905 paper there occurred—especially in the didactic literature—a symbiotic joining of the puzzling Michelson experiment and the all-but-incredible relativity theory. The undoubted result of Michelson's experiments could be thought to provide an experimental basis for the understanding of relativity theory, which otherwise seemed contrary to common sense itself; the relativity theory in turn could provide an explanation of Michelson's experimental result in a manner not as "artificial" or "ad hoc" as reliance on the supposed Lorentz-FitzGerald contraction was widely felt to be. It has proved to be a long-lasting marriage.

III. Implicit History in Didactic Accounts

A look at the secondary material—the frame of reference in which everyone receives his first orientation—will show the degree to which the work of Einstein and Michelson is commonly linked, and the additional pedagogical reasons for this tendency.

Long before we read professional literature of the kind in which Millikan's statement was published, most of us will have been told in our first physics courses what relation supposedly existed between Michelson's experiments and Einstein's work. To be sure, it is not the job of the usual physics textbooks to teach history of science or even to imply it, but with the best intention in the world they do so. As a result, there exists a widely shared, popular, "implicit" history of science. Indeed, since few students take history of science courses, implicit history is the version most widespread; because of its pervasiveness it is also the version that may well shape the judgment of future historians.

On the question under study here (as on so much else) the textbooks are virtually unanimous. Selected practically at random from recent books on my own shelf, the following is a typical quotation, to be found in the excellent text by Robert B. Leighton, PRINCIPLES OF MODERN PHYSICS. The book starts with the theory of relativity in Chapter I, explains the Michelson-Morley experiment in Section 1, and finds, "Einstein finally proposed a radically different approach to the problem posed by the Michelson-Morley experiment. *He explained its null result* simply by returning to the principle of relativity. . . ."[24] Many statements with the same implication can readily be found in other textbooks, including some of my favorite ones.[25]

Although none of the authors actually commits himself unambiguously to a statement of cause and effect, the passages give generally the impression that there was a direct genetic link. Why is this so? The simplest hypothesis would be that it was true. But even before we check this possibility, we must note two suspicious circumstances. In the first place, the Michelson experiments do not *necessarily* entail Einstein's relativity theory. As H. P. Robertson put it in his searching review article *Postulate versus Observation in the Special Theory of Relativity*:

> The kinematical background for this theory, an operational interpretation of the Lorentz transformation, was obtained deductively by Einstein from a general postulate concerning the relativity of motion and a more specific postulate concerning the velocity of light. At the time this work was done an inductive approach could not have led unambiguously to the theory proposed, for the principal relevant observations then available, notably the "ether-drift" experiment of Michelson and Morley (1886), could be accounted for in other, although less appealing ways.[26]

The second point is that in the textbook passages no evidence is ever given to back up the implication of a genetic link; and in the absence of clear evidence either way, the likelihood is a priori great that a pedagogic presentation on any scientific subject will suggest a link from experiment to theory. Almost every science textbook of necessity places a high value on clear, unambiguous inductive reasoning. The norm of behavior in the classroom would seem to be threatened if the text were to allow that correct generalizations have sometimes been made without a base in unambiguous experimental evidence.

Moreover, in the textbook or survey course, where a large amount of ground has to be covered, it is likely (for reasons of space or time if for no other) that one suitable experiment will be selected which can be convincingly presented, rather than a number of different experiments which may be equally good or better candidates. Of course, the dramatic qualities of the Michelson experiment enhance its position as candidate even more.

In the case of relativity theory, the author of a didactic account has an added incentive to foreshorten the period of doubt and uncertainty in the scientific community that followed Einstein's 1905 publication. A student can be expected to accept more easily a theory as non-commonsensical as Einstein's if he can be shown that Einstein, or at least Einstein's readers, became convinced on the basis of some clear-cut experiment.

For such reasons, little is said in textbooks about the dramatic battles that are sometimes required for the gradual acceptance of a new theory. That lack incidentally fits in well with another moralizing function of textbooks—to underplay the scientist's personal involvement and struggle in the pursuit of his scientific work and so to introduce the student to what the textbook author usually, perhaps unconsciously, conceives to be the accepted public norms of professional behavior. Texts do not want to deal with the "private" aspect of science, which can be so different from scientist to scientist and is so far from fully understood in any case. It is simpler to deal with the "public" side of science, on which there is (though perhaps falsely) some consensus. Therefore, the elements that will hold our attention in this essay, the elements that carry the possibility for a classic case study of the difference between private and public science—or for that matter of the relative roles of theory and experiment in modern scientific innovation and of the quasi-aesthetic criteria for decision between rival conceptual systems embracing the same "facts"—give way in textbooks to other, simpler purposes.

The pedagogic usefulness of designating the Michelson experiments as the specific starting point for relativity has on a few occasions been honestly stated. Thus Henri Bergson, in DURATION AND SIMULTANEITY, begins Chapter I, *Half Relativity*, in the following way:

The theory of relativity, even the "special" one, is not exactly founded on the Michelson-Morley experiment, since it expresses in a general way the necessity of preserving a constant form for the laws of electromagnetism when we pass from one system of reference to another. *But the Michelson-Morley experiment has the great advantage of stating the problem in concrete terms and also spreading out the elements of its solution before our very eyes.* It materializes the difficulty, so to speak. From it the philosopher must set forth; to it he will continually have to return, if he wishes to grasp the true meaning of time in the theory of relativity.[27]

It should also be mentioned on behalf of the textbook author that he rarely contradicts what the most prominent scientists themselves are expressing in their own popular and didactic writings. In this case the agreement among physicists has been as striking as among text writers, and its direction is that exhibited by Millikan's opinion. An earlier example—of a physicist who was also the author of the first serious textbook on relativity (1911)—is Max von Laue, who included this estimate:

The negative result of the Michelson experiment, however, forced it [the Lorentz theory of the stagnant ether] to make a new hypothesis which led over to the relativity theory [*zur Relativitätstheorie hinüberleitenden Hypothese*]. In this way, the experiment became, as it were, the fundamental experiment for the relativity theory, just as starting from it [the experiment] one reaches almost directly the derivation of the Lorentz transformation which contains the relativity principle.[28]

It is very significant that Einstein himself, in his frankly didactic publications, has left some of his readers with a similar impression about the relation of his theory to Michelson's work. For example, in his early *gemeinverständliche* book ÜBER DIE SPEZIELLE UND DIE ALLGEMEINE RELATIVITÄTSTHEORIE, a sequence is set forth which was to become so familiar in text presentations:

. . . for a long time the efforts of physicists were devoted to attempts to detect the existence of an ether-drift at the earth's surface. In one of the most notable of these attempts Michelson devised a method which appears as though it must be decisive. . . . But the experiment gave a negative result—a fact very perplexing to physicists. Lorentz and FitzGerald rescued the theory from this difficulty by assuming that the motion of the body relative to the ether produces a contraction of the body in the direction of motion. . . . But on the basis of the theory of relativity the method of interpretation is incomparably more satisfactory.[29]

Without having actually said anything about his own historical route, Einstein's singling out of the Michelson experiment in this and other didactic writings during the first decade of relativity theory cannot have failed to influence and reinforce didactic writings by others—even after the subsequent publication of Einstein's very different and frankly historical accounts, to be discussed later.

One of the most interesting of Einstein's early articles, sometimes cited as an historical document on the influence of Michelson's experiment, is his contribution, *Relativity Theory*, to a collection of thirty-six essays by foremost physicists, intended to convey the "state of physics in our time."[30] Einstein begins: "It is hardly possible to form an independent judgment of the justification of the theory of relativity, if one does not have some acquaintance with the experiences and thought processes which preceded it. Hence, these must be discussed first." There follows a discussion of the Fizeau experiment, leading to Lorentz's theory based on the hypothesis of the stagnant ether. Despite its successes, "the theory had *one* aspect which could not help but make physicists suspicious"[31]: it seemed to contradict the relativity principle, valid in mechanics and "as far as our experience reaches, generally" beyond mechanics also. According to it, all inertial systems are equally justified. Not so in Lorentz's theory: a system at rest with respect to the ether has special properties; for example, with respect to this system alone, the light velocity is constant. "The successes of Lorentz's theory were so significant that the physicists would have abandoned the principle of relativity without qualms, had it not been for the availability of an important experimental result of which we now must speak, namely Michelson's experiment."[32] There follows a description of the experiment and of the contraction hypothesis invoked by Lorentz and FitzGerald. Einstein adds to it sharply, "This manner of theoretically trying to do justice to experiments with negative results through ad hoc contrived hypotheses is highly unsatisfactory."[33] It is preferable to hold on to the relativity principle, and to accept the impossibility-in-principle of discovering relative motion. But how is one to make the principle of constancy of light velocity and the principle of relativity after all compatible? "Whoever has deeply toiled with attempts to replace Lorentz's theory by another one that takes account of the experimental facts will agree that this way of beginning appears to be quite hopeless at the present state of our knowledge."[34]

Rather, Einstein continues, one can attain compatibility of the two apparently contradictory principles through a reformulation of the conception of space and time and by abandoning the ether. The rest of Einstein's short essay is concerned with the introduction of the relativity of simultaneity and of time, the transformation equations, and the length measurement of a rod moving with respect to the observer. "One sees that the above-mentioned hypothesis of H. A. Lorentz and FitzGerald for the explanation of the Michelson experiment is obtained as a consequence of the relativity theory."[35] But this result does not seem to be worthy of listing as one of the achievements of the relativity theory a little later: "We will now briefly enumerate the individual results achieved so far for which we have the relativity theory to thank." The list, as of 1915, was not long: "a simple theory of the Doppler effect, of aberration, of the Fizeau experiment"; applicability of Maxwell's equations to the electrodynamics of moving bodies, and in particular to the motion of electrons (cathode rays, ß-rays) "without invoking special hypotheses"; and "the most important result," the relation between mass and energy, although for that there was then no direct experimental confirmation.[36]

The sequence of ideas in this essay is illuminating. But it is plainly dangerous to quote only the introductory two sentences and the reference to the Michelson experiment, and to call these an "historic account," as some have done in order to imply that Einstein followed this road himself. The whole essay is introduced as dealing with the "justification" of the theory of relativity, not with the genesis. Einstein is saying that "the physicists" would have abandoned the principle of relativity had it not been for the Michelson experiment. Antony Ruhan at the University of Chicago, in an unpublished draft essay, perceptively comments on this passage: "The obvious meaning of this text is that Einstein regarded the experiment of Michelson and Morley as necessary to convince the majority of physicists of the validity of the theory of relativity. This is quite a different point from regarding it as a basis for one's personal discovery of the key to relativity."[37]

To summarize the discussion so far, we have noted strong pressures in the same direction arising from two main sources: (1) the particular history surrounding the difficulties in acceptance of the Michelson results and the Einstein publication, and (2) the particular missions of pedagogic accounts backed up by the popular writings of distinguished physicists. These pressures have tended to the same end—to proclaim the existence of a genetic link between Michelson's and Einstein's work.

To be sure, we have so far not proven whether or not there was such an historical connection. To do that, we shall from Section V on seek the answer in more appropriate documents than didactic writings. But before we turn to such documents and to explicitly historical writings, we must at least briefly note another set of pressures that bore on the question before us: this was the weight of a philosophical view concerning science as a whole, supported by a vocal group of philosophers in the United States and Europe and widely current, particularly after the victories of the empiricist schools around the turn of the century.

IV. The Experimenticist Philosophy of Science

There exists a view of science at the extreme edge of the time-honored tradition of empiricism that will here be called *experimenticism.* It is best recognized by the unquestioned priority assigned to experiments and experimental data in the analysis of how scientists do their own work and how their work is incorporated into the public enterprise of science. A few examples will suffice to indicate the pervasiveness of this attitude. With specific reference to relativity theory, it is well illustrated by the views of Ernst Mach's disciple Joseph Petzoldt, the moving spirit behind the Gesellschaft für positivistische Philosophie of Berlin and its journal, ZEITSCHRIFT FÜR POSITIVISTISCHE PHILOSOPHIE. In the lead article of the inaugural issue (1913), he printed the text of his speech delivered at the opening session of the Gesellschaft on 11 November 1912: with relativity theory had come "the victory over the metaphysics of absolutes in the conceptions of space and time," and a "fusion of mathematics and natural science which at last and finally shall lead beyond the old rationalistic, Platonic-Kantian prejudice."[38] But the fixed hinge on which these desired events turned was, again, the Michelson experiment:

Clarity of thinking is inseparable from knowledge of a sufficient number of individual cases for each of the concepts used in investigation. Therefore, the chief requirement of positivistic philosophy: greatest respect for the facts. The newest phase of theoretical physics gives an exemplary case. There, one does not hesitate, *for the sake of a single experiment,* to undertake a complete reconstruction. The Michelson experiment is the cause and chief support of this reconstruction, namely the electrodynamic theory of relativity. To do justice to this experiment, one has no scruples to submit the foundation of theoretical physics as it has hitherto existed, namely Newtonian mechanics, to a profound transformation.[39]

The full ambitions of the group and their perception of the real enemy were further shown in the next volume, where Petzoldt wrote, "Lorentz's theory is, at its conceptional center, pure metaphysics, nothing else than Schelling's or Hegel's *Naturphilosophie.*" Again, the Michelson experiment, as the one and only experiment cited, is given the credit for ushering in a new era: ". . . the Einsteinian theory is entirely tied to the result of the Michelson experiment, and can be derived from it." Einstein himself "from the beginning conceived the Michelson experiment relativistically. We are dealing here with a principle, a foremost postulate, a particular way of understanding the facts of physics, a view of nature, and finally a *Weltanschauung.* . . . The line Berkeley-Hume-Mach shows us our direction and puts into our hands the epistemological standard."[40]

A few years before, Michelson had been awarded the 1907 Nobel Prize in physics, not for the experiments we have been discussing but "for his optical precision instruments and the search which he has carried out with their help in the fields of precision metrology and spectroscopy." The relativity theory was, of course, still far too new, and regarded as too speculative, to be mentioned in the citations or responses. (Indeed, by the time Petzoldt was writing his eulogies to it, the theory had become too speculative for Mach himself, whereas the Nobel Prize Committee did not award Einstein's prize until 1922, and then, as Einstein was specifically reminded by the Committee, it was for contributions to mathematical physics and especially for his discovery of the [experimentally well-confirmed] Law of the Photoelectric Effect.)[41] In any case, theory was not now of interest at the 1907 award ceremony; the text of the presentation speech (by K. B. Hasselberg) showed the award to Michelson to be clearly motivated by the experimenticist philosophy of science:

As for physics, it has developed remarkably as a precision science, in such

> a way that we can justifiably claim that the majority of all the greatest discoveries in physics are very largely based on the high degree of accuracy which can now be obtained in measurements made during the study of physical phenomena. [Accuracy of measurement] is the very root, the essential condition, of our penetration deeper into the laws of physics—*our only way to new discoveries.* It is an advance of this kind which the Academy wishes to recognize with the Nobel Prize for Physics this year. (Italics supplied.)

Somehow, everyone managed to keep a decorous silence on the experiments which Petzoldt and others of his persuasion were to hail as the crucial turning point for physics and *Weltanschauung*; nobody referred here to Michelson's ether-drift experiments—neither the Swedish hosts nor Michelson himself in his responding lecture ("Recent Advances in Spectroscopy"). These experiments were as embarrassing for experimenticists with ether-theoretic presuppositions as they were welcome for experimenticists with relativistic presuppositions.

The even more extreme view that all scientific advance arises out of the use of instruments was defended by Millikan in his autobiography, where he explained that he moved from the University of Chicago to the California Institute of Technology because there "science and engineering were merged in sane proportions." He set forth his ideological basis as follows:

> Historically, the thesis can be maintained that more fundamental advances have been made as a by-product of instrumental (i.e. engineering) improvement than in the direct and conscious search for new laws. Witness: (1) relativity and the Michelson-Morley experiment, the Michelson interferometer came first, not the reverse; (2) the spectroscope, a new instrument which created spectroscopy; (3) the three-electrode vacuum tube, the invention of which created a dozen new sciences; (4) the cyclotron, a gadget which with Lauritsen's linear accelerator spawned nuclear physics; (5) the Wilson cloud chamber, the parent of most of our knowledge of cosmic rays; (6) the Rowland work with gratings, which suggested the Bohr atom; (7) the magnetron, the progenitor of radar; (8) the counter-tube, the most fertile of all gadgets; (9) the spectroheliograph, the creator of astrophysics; (10) the relations of Carnot's reversible engine to the whole of thermodynamics.[42]

In the work of philosophers of science the discussion of relativity theory is frequently found linked tightly to the Michelson experiment, though rarely more enthusiastically than in Gaston Bachelard's essay *The Philosophical Dialectic of the Concepts of Relativity,* in P. A. Schilpp's volume ALBERT EINSTEIN: PHILOSOPHER-SCIENTIST:

As we know, as has been repeated a thousand times, relativity was born of an epistemological shock; it was born of the "failure" of the Michelson experiment. . . . To paraphrase Kant, we might say that the Michelson experiment roused classical mechanics from its dogmatic slumber. . . . Is so little required to "shake" the universe of spatiality? Can a single experiment of the twentieth [*sic*] century annihilate—a Sartrian would say *néantiser*—two or three centuries of rational thought? Yes, a single decimal sufficed, as our poet Henri de Regnier would say, to "make all nature sing."[43]

And so on and so forth. Einstein chose not to respond to this apotheosis of the Michelson experiment in his replies at the end of the same volume. But he did make a lengthy and subtly devastating reply to another essay in this collection, that by Hans Reichenbach, which has a good deal of the same kind of experimenticist bias.

Reichenbach, who knew Einstein and corresponded with him at certain periods, was, over the years, one of the more persistent and interesting philosophical analysts of the epistemological implications of relativity. (He published, for example, several attempts to cast the theory into axiomatic form. To one of these attempts Einstein himself responded to say he did not find it convincing even on its own grounds; he wrote to Reichenbach on 19 October 1929, "In my view, the logical presentation which you give my theory is, to be sure, possible, but it is not the simplest one.") But Reichenbach's experimenticist conviction never flagged. For example, he wrote that Einstein's work "was suggested by closest adherence to experimental facts. . . . Einstein built his theory on an extraordinary confidence in the exactitude of the art of experimentation."[44] Again, the only historic experiment Reichenbach associated with the genesis of Einstein's theory was, of course, the Michelson experiment; for example, "the theory of relativity makes an assertion about the behavior of rigid rods similar to that about the behavior of clocks. . . . This assertion of the theory of relativity is based mainly on the Michelson experiment. . . ."[45]

In his essay in Schilpp's collection, Reichenbach reverts to the same points,[46] but they are, as it were, only preludes to the conclusion that "it is the philosophy of empiricism, therefore, into which Einstein's relativity belongs. . . . In spite of the enormous mathematical apparatus, Einstein's theory of space and time is the triumph of such a radical empiricism in a field which had always been regarded as a reservation for the discoveries of pure reason."[47] In his reply to this essay Einstein devoted most of his attention to a denial of this claim. He preferred to hold fast to the basic conceptual distinction between "sense impressions" and "mere ideas"—despite the expected reproach "that, in doing so, we are guilty of the metaphysical 'original sin.' "[48] Einstein pleads that one must also accept features not only of empiricism but also of rationalism, indeed that a "wavering between these extremes appears to me unavoidable."[49] Einstein adopts the role of the "non-positivist" in an imaginary dialogue with Reichenbach and urges the useful lesson of Kant that there are concepts "which play a dominating role in our thinking, and which, nevertheless, cannot be deduced by means of a logical process from the empirically given (a fact which several empiricists recognize, it is true, but seem always again to forget)."[50]

The difficulty is, of course, one of relative scientific taste or style. To Reichenbach and his followers, the interest in a scientific theory resides neither in the details of its historical development, nor in the work of an actual person. (As Reichenbach said honestly, "The philosopher of science is not much interested in the thought processes which lead to scientific discoveries; he looks for a logical analysis of the completed theory, including the relationships establishing its validity. That is, he is not interested in the context of discovery, but in the context of justification."[51]) But it is equally understandable that to the originator of the theory the "completed theory" in its public, developed, institutional or textbook form does not have the same exclusive interest.

Unfortunately, neither Reichenbach himself nor his followers always remembered his laudable attempt to distinguish sharply between private and public science, nor have they always adhered to his own disclaimer of interest in the thought processes leading to the discovery. The desire to see a theory as a logical structure, built upon an empirical basis and capable of verification or falsification by more experiment, brings them to discuss presumed historical sequences on the road that led to the discovery; thus, implicit "history" is produced after all (for example, the confident assertion that "Einstein incorporated its [the Michelson experiment's] null result as a physical axiom in his light principle,"[52] and similar attempts at "the unraveling of the history" of relativity theory). When direct evidence, which we shall examine below, against the priority of the Michelson experiment in Einstein's thinking is presented to the experimenticist, the response is this: without the genetic role of this particular experiment, an understanding of the discovery of the theory would become "quite problematic," and one would be left "puzzled concerning the logical, as distinct from psychological grounds which would then originally have motivated Einstein to have confidence in the principle of relativity without the partial support of the Michelson-Morley experiment. . . ."[53]

As a curious postscript to this section, one might mention that the experimenticist interpretation of relativity has also been advocated under quite different circumstances, but none more macabre than the attempt in the 1920's by some German scientists such as W. Wien to point to the supposed experimental origins of relativity in order to remove the theory from the arena of unthinking, inflamed opposition in some quarters in Germany against Einstein personally and against his work.[54] In a small book published in 1921, Wien wrote that he wished

> . . . to give an objective presentation of the theory of the pro and the contra concerning which much is being discussed in a rather unscientific way in public. I hope to have discussed the questions *sine ira et studio*, and I would like to advise everyone who concerns himself with the theory not to give himself out as a follower or an opponent of this theory, but rather to consider the theory in such a manner as is congruous with science, namely as one way to discover peculiarities of nature's laws which may equally well turn out to be right or wrong. The decision concerning this cannot be reached in a dogmatic way, but one must leave it up to the decision of experience.[55]

Wien then reassures his readers that the "relativity theory is, like all physical theories, a result of experience." It is not difficult now to guess the particular experience involved: "The negative result of the Michelson experiment is the fact of experience on which the relativity theory rests. This experiment is for this theory of equal significance as the perpetuum mobile is for the law of the conservation of energy. . . ."[56] This attempt to experimentalize the basis of relativity, incidentally, did not gain acceptance sufficiently to save relativity theory during the Nazi period from the stigma of what was called then (and even as recently as 1954) "the formal rationalism of Jewish thinking."[57]

V. Explicit History

After the nearly complete consensus and confidence implied in textbooks and in experimenticist works with epistemological intentions, we are prepared to find more variety and circumspection in the work of scholars who have undertaken to write explicitly historical accounts. This is indeed the case; there is far less agreement here. The existence of a whole spectrum of differently documented historical studies is itself an interesting problem.

To examine this spectrum and to arrange the items roughly in increasing order of intended seriousness as historical studies, we may start with THE WORLD OF THE ATOM, published in 1966. In short sections based largely on secondary material and inserted between excerpts from original papers, the editors intend to provide some historical guidelines. On the issue of interest to us we find a familiar comment, but now curiously hedged: "The legacy of the Michelson-Morley experiment to atomic theory was tremendous, if indirect. It was the negative result of the experiment that in part led Einstein into one of the fundamental ideas upon which the theory of relativity rests, namely, that the speed of light is the same for all observers, regardless how they may be moving."[58]

Another account that leaves open the extent or directness of the influence is given in a useful biography, MICHELSON AND THE SPEED OF LIGHT, written primarily for high school students by the successful textbook and popular-science author Bernard Jaffe. Speaking of the original work on relativity, Jaffe writes that Einstein "saw the Michelson-Morley ether drift result as perfectly correct, since no ether drift should be expected under the conditions of the experiment."[59] "In this great upheaval in physics, the classic ether-drift experiment of Michelson had been of fundamental significance. To contend, as some have done, that Einstein's Special Theory of Relativity was essentially a generalization of the Michelson experiment, and that it could not have been arrived at without the experiment, is to overstate the case."[60]

Jaffe's biography incidentally does sound, if only faintly, a new note that has some significance for the understanding of the case. He writes, "Unwittingly, Michelson, as it turned out, had supplied the raw material for one of the great structures of science—a synthesis which was to be completed overseas. This was one of the very few instances when a basic discovery was made in America for European exploitation. Almost always it was the other way around."[61] The pride in the American scientist was undoubtedly increased, once relativity theory became palatable, by seeing his work as the root of relativity.[62]

A more detailed analysis, with new documentation, has been provided over the last several years in a series of insightful articles by R. S. Shankland, Professor of Physics at Case-Western Reserve University (formerly Case Institute of Technology, where the Michelson-Morley experiment was repeated before Michelson's departure first for Clark University and then, in 1894, for the University of Chicago). A particular merit of Shankland's work has been his publication of a number of letters from and to Michelson and his report on a series of interviews with Einstein during the period 1950–1954.

On the important question before us, Shankland's writings show a significant development. His earliest article is a short piece on Michelson for the collection LES INVENTEURS CÉLÈBRES—SCIENCES PHYSIQUES ET APPLICATIONS.[63] Subtitled *Expérience de base de la relativité*, the article presents the experiment and the theory as closely coupled as in most of the above versions.[64] Starting with Shankland's next essay on this topic about a dozen years later, *Conversations with Albert Einstein*, another aspect of the case begins to show up. His report (1963) is based on five visits with Einstein at Princeton; their conversations dealt principally with the work of Michelson, particularly the Michelson-Morley and Miller experiments, and with the subsequent studies that led to the clarification of Miller's results (in which Shankland himself took a leading part). This article is a valuable and rich resource which merits mining for all it is worth.

Near the beginning of Shankland's first-hand report we find the main reason for his visit—and a response of Einstein for which nothing we have read above has prepared us:

> The first visit [4 February 1950] to Princeton to meet Professor Einstein was made primarily to learn from him what he really felt about the Michelson-Morley experiment, and to what degree it had influenced him in his development of the Special Theory of Relativity. . . . He began by asking me to remind him of the purpose of my visit and smiled with genuine interest when I told him that I wished to discuss the Michelson-Morley experiment performed at Cleveland in 1887. . . . When I asked him how he had learned of the Michelson-Morley experiment, he told me that he had become aware of it through the writings of H. A. Lorentz [ARCH. NÉERL. 2: 168, 1887, and many later references], but *only after 1905* had it come to his attention! "Otherwise," he said, "I would have mentioned it in my paper." He continued to say the experimental results which had influenced him most were the observations on stellar aberration and Fizeau's measurements on the speed of light in moving water. "They were enough," he said.[65]

Shankland may well have been astonished by Einstein's disclaimer of the direct genetic role of the Michelson-Morley experiment in the creation of the theory of relativity. As if to make sure that he had understood properly, Shankland wisely raised the whole matter again at another visit, two and a half years later, on 24 October 1952.

> I asked Professor Einstein where he had first heard of Michelson and his experiment. He replied, "This is not so easy, I am not sure when I first heard of the Michelson experiment. I was not conscious that it had influenced me directly during the seven years that relativity had been my life. I guess I just took it for granted that it was true." However, Einstein said that in the years 1905–1909, he thought a great deal about Michelson's result, in his discussions with Lorentz and others in his thinking about general relativity. He then realized (so he told me) that he had also been conscious of Michelson's result before 1905 partly through his reading of the papers of Lorentz and more because he had simply assumed this result of Michelson to be true.[66]

These two statements must be faced; their authenticity cannot be minimized. Here and elsewhere in these interviews one is impressed with Einstein's consistent responses and analyses, as well as with the evident readiness and joy with which he thought of the experiment and of Michelson as a person. (Thus Shankland reports he said to him several times, "I really loved Michelson.") To be sure, as Shankland himself emphasizes, many of the events discussed between them had occurred some fifty years before their meetings; Einstein's age was seventy-one and seventy-three years at the time of these two interviews, and during another interview there are two minor episodes where his memory did not serve him absolutely faithfully (Shankland reports that Einstein had forgotten about a speech he gave in Berlin on Michelson's death in 1931, and also about Joos's work). But it would be far too simple to dismiss or play down the repeated, direct response to Shankland on a topic on which Einstein throughout his life had frequent occasion to ponder, write, and lecture, and on which he no doubt had often been asked questions.

The fact that the response is so contrary to practically all other accounts makes it imperative to reexamine the whole question in order to accommodate Einstein's statements: (1) that the Michelson experiment occupied his attention only after 1905 (although he was conscious of the result earlier); (2) that other, earlier ether experiments on stellar aberration and on the Fresnel ether-drag coefficient form the most important experimental bases for his 1905 paper; and (3) that insofar as he was aware of the Michelson result, he was evidently not specially impressed with it upon reading it in Lorentz's paper because he already had assumed it to be true on other grounds.

We shall see to what extent Einstein's response is consistent with a careful reading of his 1905 paper itself and with all other pertinent reports and documents by Einstein which I have been able to find. But it is clear already that when compared to the various confident pronouncements above, Einstein's responses in these interviews seem vague, indefinite, and tentative. It is as if we had been preparing a case for a specific "Yes" or "No," and, at last, encountering the person most involved, we received the unexpected answer, "Neither! That's not the way it happened, and in any case this was for me not really an important consideration." Einstein raises the distinct suspicion that the question may be as trivial or irrelevant to him as it is important to us or to Shankland, who started the series of visits "primarily to learn from him what he really felt about the Michelson-Morley experiment. . . ." It is doubly ironic to find at this point that neither Michelson nor Einstein regarded the famous experiment as decisive for himself, much less "crucial."

In two later articles Shankland adopted the position that the problem that the Michelson-Morley experiment posed "led indirectly to Einstein's Special Theory of Relativity,"[67] and "both postulates [of the 1905 paper] could, of course, be considered as having a close relationship to the Michelson-Morley experiment, but actually Einstein arrived at his theory by a less direct route, becoming aware of the observational material principally through the writings of Lorentz which he began to study as a student in 1895."[68] But perhaps the most significant part of these two articles is the publication in both of a letter which Einstein had supplied, evidently on Shankland's invitation, for a special meeting of the Cleveland Physics Society on 19 December 1952, honoring the centenary of Michelson's birth. This letter, the result of a more careful and reflective response than may have been his *viva voce* answers to the question, is therefore of even greater value. More strongly than in the somewhat sketchy remarks in the interviews, Einstein proposes a point of view for gaining a clearer understanding of his 1905 work on relativity. Following is the document published by Shankland, in which I have inserted alternative readings of some phrases or words in brackets, to indicate translations that would be somewhat more faithful to the copy of the original German text:[69]

I always think of Michelson as the Artist in Science. His greatest joy seemed to come from the beauty of the experiment itself, and the elegance of the method employed. But he has also shown an extraordinary understanding for the baffling fundamental questions of physics. This is evident from the keen interest he has shown from the beginning for the problem of the dependence of light on [upon] motion.

The influence of the crucial [famous] Michelson-Morley experiment upon my own efforts [deliberations] has been rather indirect. I learned of it through H. A. Lorentz's decisive investigation of the electrodynamics of moving bodies (1895) with which I was acquainted before developing [setting forth] the Special Theory of Relativity. Lorentz's basic assumption of an ether at rest seemed to me not convincing in itself and also [*replace* and also *by* precisely] for the reason that it was leading to an interpretation of the result of [*omit* the result of] the Michelson-Morley experiment which seemed to me artificial. What led me more or less [*omit* more or less] directly to the Special Theory of Relativity was the conviction that the electro-motoric [electromotive] force acting on [induced in] a body in motion in a magnetic field was nothing else but an electric field. But I was also guided by the result of the Fizeau experiment and the phenomenon of aberration.

There is, of course, no logical way leading to the establishment of a theory but only groping constructive attempts controlled by careful consideration of factual knowledge.[70]

We first note some differences between the texts at important points. Thus the Michelson-Morley experiment is called "crucial" in the English translation (as it often is in the didactic literature) but "famous" (*berühmt*) in the German original. But the most significant aspect of the document is the definite order of importance which Einstein assigned to four identified experiments. The one "experiment" cited here as having led Einstein "directly" (in the German) or "more or less directly" (in the English text) to the special relativity theory is precisely the thought experiment that appears on the first page of his 1905 paper, namely, the motion of a conductor with respect to a magnetic field. The three other experiments were of some additional importance: the Fizeau and aberration experiments, whose results "also guided" Einstein, and the Michelson-Morley experiment, in its form presented by Lorentz in 1895. But even in this message on the centennial of

Michelson's birth, Einstein assigns Michelson's experiment only the fourth rank as an historic stimulus: he reports that it had only a "rather indirect" role in his own work of 1905; specifically, the experiment (or rather its result) underlined the "artificial" character of the contraction hypothesis that seemed to be needed in order to rescue the conception of a stagnant ether—an artificial character, as we shall see, that was regretted by other physicists, including Lorentz himself.

This document, therefore, provides a likely scenario for the part played by experiments in the genesis of Einstein's 1905 paper. It is believable in itself, considering the authority of the statement. And we shall find that it also fits with all other direct and indirect evidence coming from Einstein, including his letters, his spoken answers to questions, and his original paper.

The final paragraph of the statement is rather startling. Here Einstein, having done his duty by writing this response to a specific question, goes beyond it and volunteers a glimpse of the methodological position that had also surfaced in his interviews with Shankland: "There is, of course, no logical way leading to the establishment of a theory but only groping constructive attempts controlled by careful consideration of factual knowledge." Entirely in accord with the honest self-appraisal of an original scientist, Einstein's forthright confession is yet so contrary to the widely current myths which present scientific work as the inexorable pursuit of logically sound conclusions from experimentally indubitable premises. Systematizers, axiomatizers, text writers, and others may yearn for linearized sequences both in scientific work itself and in accounts of it; the truth, alas, is different. Einstein had often mentioned this, for example in speaking to Shankland about the origins of his work of 1905. Shankland reports in *Conversations*: "This led him to comment at some length on the nature of mental processes in that they do not seem at all to move step by step to a solution, and he emphasized how devious a route our minds take through a problem. 'It is only at the last that order seems at all possible in a problem.' "[71] Similarly, in commenting on the correct view an historian should take of the work of physicists, Einstein told him: "The struggle with their problems, their trying everything to find a solution which came at last often by very indirect means, is the correct picture."[72]

This view Einstein had expressed repeatedly—though not previously in this context—explicitly from about 1918 on and more emphatically from the early 1930's on. Examples may be found in his essay in honor of Max Planck in 1918 ("there is no logical way to the discovery of these elementary laws. There is only the way of intuition" based on *Einfühlung* in experience[73]); in his Herbert Spencer lecture of 1933 (concerning the "purely fictitious character of the fundamentals of scientific theory"[74]); in his *Autobiographical Notes* written in 1946 ("A theory can be tested by experience, but there is no way from experience to the setting up of a theory"[75]); in his reply to Jacques Hadamard, who had asked Einstein for a self-analysis of his thought processes ("The words or the language, as they are written or spoken, do not seem to play any role in my mechanism of thought. The psychical entities which seem to serve as elements in thought are certain signs and more or less clear images which can be 'voluntarily' reproduced and combined"[76]); and on many other occasions.[77]

In believing that there can be an essential abyss between experience and logically structured theory, and in believing also in the related "distinction between 'sense impressions' on the one hand and mere ideas on the other," which, as we noted, he confessed made him "guilty of the metaphysical 'original sin,' "[78] Einstein separated himself from his earlier positivistic allegiances and from most of the prominent philosophies of science of his time. That he did not do this lightly is indicated by the frequency with which he kept reiterating these points over the years. In this connection Einstein's document for the Michelson centenary will be a key to the final evaluation of the problem under discussion.

To round out this section on explicitly historical accounts, we turn to the relatively few other sources that have serious historical intentions. Here we find a great variety of opinions among authors. At one end Sir Edmund Whittaker refers to "the Michelson-Morley experiment and the other evidence which had given rise to relativity theory" in his famous and amply discussed chapter, entitled significantly *The Relativity Theory of Poincaré and Lorentz*, in A HISTORY OF THE THEORIES OF AETHER AND ELECTRICITY.[79] At the other end is T. W. Chalmers's HISTORIC RESEARCHES: "It should be made clear that, in spite of frequent statements to the contrary, the theory for relativity *did not* owe its inspiration and origin to the null result of the ether drift experiments. . . . Had the ether drift experiments never been performed, the theory for relativity would have arisen in the way it did, but it would have lacked one of its several sources of experimental confirmation."[80]

The most comprehensive historical study of the Michelson experiment has recently been carried out by Loyd Swenson in his Ph.D. dissertation mentioned earlier—THE ETHEREAL AETHER: A HISTORY OF THE MICHELSON-MORLEY AETHER-DRIFT EXPERIMENTS, 1880–1930. Swenson is not primarily concerned with the special question of this essay, but rather analyzes historically the idea of the *experimentum crucis* and specifically the way in which Michelson's experiments were discussed before and after 1905. A number of points useful for our examination have already been noted in Section II. In particular, Swenson gives additional evidence of the consternation which Einstein's relativity paper caused in the early days. Thus he cites William F. Magie's address of 28 December 1911 to the American Physical Society, of which he had just been elected president: Magie stated that relativity "may fairly be said to be based on the necessity of explaining the negative result of the famous Michelson-Morley experiment, and on the convenience of being able to apply Maxwell's equations of the electromagnetic field without change of form to a system referred to moving axes."[81] Swenson continues, "Magie insisted on the counter-evidence that the principle of relativity was not essential to obtain explanations for the experiments by Fizeau, Mascart, Brace, nor those of Kaufmann and Bucherer. Why then, he [Magie] asked, should we allow the Michelson-Morley experiment to upset all our primary concepts of physics?" And quoting Magie again, "The principle of relativity accounts for the negative result of the experiment of Michelson and Morley, but without an ether how do we account for the interference phenomena which made that experiment possible?"[82] Clearly, like Oliver Lodge, Magie did not think much of the attempt to make relativity and Michelson's work support each other.[83]

In summary, what emerges from explicit historical accounts, including Einstein's interviews with Shankland, is that the story we found earlier in didactic or philosophical resources is, at best, suspect and needs a serious critique. The study so far has served to clarify some of the options that will emerge from direct documentary evidence, first of all from the basic paper by Einstein in 1905.

VI. Direct Evidence in Einstein's 1905 Paper

Einstein's paper on relativity of his *annus mirabilis* of 1905—still so fresh, so clearly the inspired work of genius—has been discussed so often that we need here only a reminder of some chief points.[84] The purpose of the paper, as implied in the title and the first lines of the introduction, is to provide an electrodynamics of moving bodies, based on the laws previously formulated in Maxwell's electrodynamics for bodies at rest. As Einstein said about forty years later in his *Autobiographical Notes*: "The special theory of relativity owes its origin to Maxwell's equations of the electromagnetic field. Inversely the latter can be grasped formally in satisfactory fashion only by way of the special theory of relativity."[85]

The first reason for doing this work is indicated in Einstein's first sentence: "It is known that Maxwell's electrodynamics—as usually understood at the present time—when applied to moving bodies, leads to asymmetries which do not appear to be inherent in the phenomena." This is a dissatisfaction of an aesthetic kind, and incidentally one which had not been thought of by other physicists as a major defect that must be remedied. At any rate, we note that Einstein does not start by pointing to some conflict between theory and the known facts. The well-known example he appends to the first sentence is that "the customary view draws a sharp distinction" between the reasons why a current is induced in a conductor when, on the one hand, a conductor is at rest while in the field of a moving magnet (here the current is said to be due to an electric field in the neighborhood of the magnet) and, on the other hand, the conductor is moving in the field of a resting magnet (here no electric field exists in the neighborhood of the magnet, but following Hertz, an electromotive force appears in the conductor and is responsible for the electric current). Yet the magnitude and direction of the observed currents in the two cases are found to be the same, given the same relative motion. The young author implies that the theory "as usually understood at the present time" is deficient because of the asymmetry of presumed causes, and therefore that a reformulation of electrodynamics is needed to change the "understanding" by removing the asymmetry (as is done later in the paper[86]).

The example Einstein has chosen is, on the surface, rather pedestrian and not at all novel, going back to Faraday's work. But that is of course the mark of his originality. In leading up to the reformulation of the most fundamental notions of space and time, Einstein does not have to depend on a sophisticated effect or a new or even ancient experimental puzzle. He refers to observations long known and believed to be well understood by everyone. This was also how Galileo argued, in the DIALOGUE CONCERNING THE TWO CHIEF WORLD SYSTEMS. So, too, neither Copernicus's DE REVOLUTIONIBUS nor Newton's PRINCIPIA was based on newly available experimental facts; nor was either designed to explain observations which previous theory had failed to accommodate. To be sure, the more ordinary scientific paper, particularly in the modern period, is apt to start from some new experimental result of an observation that has been recalcitrant, resisting absorption into the existing theoretical structure.

After the details Einstein gave for the case of the induced-current experiment with conductors and magnets, his next two sentences are surprising in their unspecific language and in their combination of apparently unrelated matters:

> Examples of this sort, together with the unsuccessful attempts to discover any motion of the earth relatively to the "light medium," suggest that the phenomena of electrodynamics as well as of mechanics possess no properties corresponding to the idea of absolute rest. They suggest rather that, as has already been shown to the first order of small quantities, the same laws of electrodynamics and optics will be valid for all frames of reference for which the equations of mechanics hold good.[87]

We stop briefly at this significant place: it is precisely here that the Michelson experiment might have been referred to. It is, after all, foremost in our textbooks among the experiments on the effects of the earth's motion upon the observed speed of light. But however we may speculate on whether or not Einstein thought about it, his paper does not support such speculation. Again, when Einstein later spoke specifically of the origins of the relativity theory in his *Autobiographical Notes*, he failed to mention Michelson's experiment anywhere. Some have found this frustrating, but it would be presumptuous to think that Einstein had any obligation to explain himself on this point, either in 1905 or later. As it will turn out, he did leave more than enough material for the historian to deal adequately with the problem.

What is clear, at any rate, is that the famous experiment is not referred to by name, nor are any other experiments on the presumed effect of the earth's motion relative to the ether. Also, in the sentence referring to the unnamed unsuccessful attempts to discover an ether drift, the experiments are not labeled as crucial. They seem to play a supporting role of the following kind: the results of the specified magnet-and-conductor experiments and of the unspecified optical experiments are in conflict with the notions of absolute space and with other ideas of absolutistic physics. Also, in terms of Maxwell's electromagnetic theory, they are closely related experiments. These facts make it that much more reasonable to fasten upon Maxwell's theory as the place where, through its relativization, both optics and electrodynamics must be reinterpreted together.

The suggestion (conjecture, *Vermutung*) referred to in Einstein's last sentence, whose purport he calls the "principle of relativity," is then, without further discussion, raised in status to become the first of two postulates which form the basis of the rest of the paper. The second postulate (constancy of the speed of light in empty space) is added in the same sentence without citing any evidence which might increase its plausibility. The reader is to find its warrant in the success of the theory based on the postulates.

Neither here nor later in the paper are the fundamental statements set forth in a logical way, connected to a well-marshalled set of facts and experiments and bolstered by detailed reasons and examples. (See Table I for an attempt to set out schematically the structure underlying these pages of the 1905 paper.) On the contrary the paper has the freshness of an outpouring of genius that makes it believable that it was written within "five or six weeks" (as Einstein wrote to one of his biographers, Carl Seelig, on 11 March 1952), and in a year which saw him send three basic papers off at intervals of less than eight weeks while also doing his job at the Bern Patent Office.[88] It is also quite in accord with Einstein's statement that he had then recently seen "there is no logical way to the establishment of a theory," but that he had to leap over the gap "to the discovery of a universal formal principle."[89]

We can thus relate these passages of 1905 with a well-known portion of Einstein's *Autobiographical Notes* (written in 1946, published in 1949), in which he reported on what he regarded to be the origins of relativity theory in his own speculations:

Reflections of this type [on the limitations of Maxwell's theory for the description of pressure fluctuations on a light reflector] made it clear to me as long ago as shortly after 1900, i.e., shortly after Planck's trailblazing work, that neither mechanics nor thermodynamics could (except in limiting cases) claim exact validity. By and by I despaired of the possibility of discovering the true laws by means of constructive efforts based on known facts. The

Table I. Brief schematic structure of the introductory section of Einstein's 1905 paper on relativity

	Statements in sequence	Examples or reasons given
A	Maxwell's electrodynamics for bodies at rest leads to asymmetries that do not belong to the phenomena.	1. Thought experiments à la Faraday on electrodynamic interaction between magnet and conductor. 2. Other "examples of a similar kind" (not specified).
B	Attempts to discover a motion of the earth relative to the "light aether" have failed.	(No explicit reasons given.)
C	Postulation of "principle of relativity" for mechanics, optics, and electrodynamics.	1. A + B "lead to" conjecture (the postulate) C (not shown how). 2. It has already proved useful for first order (v/c) theory.
D	Postulation of principle of constancy of light speed.	(No explicit reasons given.)
E	C + D will have consequences:	1. A simple, noncontradictory electrodynamics will result, based on Maxwell's theory. 2. The "light aether" will be superfluous.

longer and the more despairingly I tried, the more I came to the conviction that only the discovery of a universal formal principle could lead us to assured results. The example I saw before me was thermodynamics. The general principle was there given in the theorem: the laws of nature are such that it is impossible to construct a *perpetuum mobile* (of the first and second kind). How, then, could such a universal principle be found? After ten years of reflection such a principle resulted from a paradox upon which I had already hit at the age of sixteen: If I pursue a beam of light with the velocity c (velocity of light in a vacuum), I should observe such a beam of light as a spatially oscillatory electromagnetic field at rest. However, there seems to be no such thing, whether on the basis of experience or according to Maxwell's equations. From the very beginning it appeared to me intuitively clear that, judged from the standpoint of such an observer, everything would have to happen according to the same laws as for an observer who, relative to the earth, was at rest. For how, otherwise, should the first observer know, i.e., be able to determine that he is in a state of fast uniform motion?

One sees that in this paradox the germ of the special relativity theory is already contained.[90]

This passage has its exact parallel in the 1905 paper, in the conceptual leap from a simple experiment (indeed, also a kind of *Gedanken*-experiment—the relative motion of conductor and magnet) to the general principle from which the content of the relativity theory will derive. Moreover, it appears not to have been noticed before that the seminal paradox pondered in his youth and the experiment in the beginning of the 1905 paper are physically of precisely the same kind: in one case the question concerns the electric and magnetic fields a moving observer finds to be associated with a light beam; in the other case, it concerns the electric and magnetic fields experienced by a moving conductor; and the solutions in both cases follow from the same transformation equations. It seems therefore possible that Einstein may have had this youthful thought-experiment with the light beam in mind when he wrote in 1905 the otherwise rather obscure phrase, "examples of this sort."[87] Indeed, the paradox of the light beam would have been a natural bridge to the reference immediately following in the 1905 paper, namely to the experiment attempting "to discover any motion of the earth relative to the 'light medium.' "

The next sentence of the 1905 paper is one in which Einstein, now near the end of his introduction, almost gratuitously notes a result to be expected from his approach. "The introduction of a 'luminiferous ether' will prove to be superfluous inasmuch as the views here to be developed will not require an 'absolutely stationary space' provided with special properties. . . ." This quickly brings him to the "Kinematical Part," with its examination of the concepts of space and time. The Lorentz transformation equations are derived from the postulates and lead to the transformation of the Maxwell-Hertz equations that encompass all electrodynamic phenomena, including moving magnets at one end and moving light beams at the other. All the rest follows from this: the relativistic Doppler effect, aberration, and the pressure of radiation exerted on a reflector. It is significant that aberration and change in frequency of a light wave are the two oldest known optical effects due to the motion of the earth relative to the stars—known long before the Michelson experiment—and that the problem of radiation pressure is one which Einstein reported later to have intrigued him very early as evidence of the limits of applicability of Maxwell's theory.

But at this point Einstein stops himself: "All problems in the optics of moving bodies can be solved by the method here employed. What is essential is that the electric and magnetic force of the light which is influenced by a moving body be transformed into a system of co-ordinates at rest relatively to the body." (Here the experiment of the magnet and induced current connects again with the experiment on the light beam.) "By this means all problems in the optics of moving bodies will be reduced to a series of problems in the optics of stationary bodies."[91]

It would be incredible for a more ordinary physicist, particularly for a young man proud to show the power of his new theory, to do what Einstein now does (or does not do): he fails to explain that these sentences contain a reinterpretation of the null result of the Michelson experiment, the relativistic equivalent of the Lorentz-FitzGerald contraction, and the solution of other problems that had preoccupied many of the best physicists during the previous two decades. Depending on one's predisposition, the last sentence cited above can be interpreted as a statement of elegance, of arrogance, of ignorance of the detailed experiments in the "optics of moving bodies," of lack of serious concern with the messy details of experimental physics, or even of mere lack of time to go into further detail in a thirty-one-page paper, written rapidly during an immensely productive period.

I suspect that all of these elements were at work, but the first is the most prominent. For example, in forgoing for the second time the opportunity to mention the Michelson experiment, Einstein is only facing the fact that from the point of view of relativistic physics, *nothing important happens at all* in the experiment. The result is "natural," fully expected, and trivially true. The abandonment of the ether and the acceptance of the transformation equations meant the disappearance of both the objective and the very vocabulary for discussing the ether-theoreticians' interests in the null result and in the possible "causes" of the "contraction." The two views of the experiment were thus different to the point of being unbridgeable, which accounts for the inconclusiveness of debate between the two factions long after the paper was published, for example at the 1927 meeting on the Michelson experiment. The relativists simply could not see the complex problems that were seriously evident to the ether theorists to whom, in Dugas's happy phrase, the ether still formed "the substratum of thought in physics."

This recognition should prepare us for the possibility that analogous philosophical differences separate the two points of view on the overriding question—how to understand the history of the development of relativity theory. On one hand is the largely intuitive, holistic approach of Einstein himself, with some empiricist and rationalist elements; on the other hand, there are the axiomatic and experimenticist approaches we noted earlier. To the first group it appears natural that an experiment that played no decisive role should not be mentioned; for the second group, however, the absence in Einstein's paper of specific acknowledgments to the role of Michelson's experiment becomes a problem, and Einstein is seen as having shirked an "obligation" to explain himself.[92]

Returning to Einstein's paper of 1905, we find at the end a short section on the dynamics of moving electrons—that particularly famous case of moving bodies discovered only eight years earlier, and concurrently the subject of widely discussed experiments, such as those by Kaufmann. None of the experiments, however, is mentioned. The paper ends with three statements, one of which "may be tested experimentally," and which together are boldly summarized as "the laws according to which, by the theory here advanced, the electron must move."[93] This is in fact the only place in the paper where new experimental results are predicted explicitly, although elsewhere there are a few implications, for example, that an object moving at high relative speeds "will appear shortened" and that a moving clock will appear to a stationary observer to keep time more slowly. In a few instances a result is said to be in accord with established experiments (the case of the pressure of light, for instance). Yet, while Einstein develops the equations which can explain with striking ease such historic experiments as the Fizeau test of Fresnel's theory of ether drag or the aberration observations, the equations are not explicitly applied to them.

In a paper meant to be read by a community of scientists used to being presented with experimental verification of new theories, it would have greatly helped the establishment of relativity theory if there had been, so to speak, a pedagogic section elaborating on the consequences of the basic work. One can only speculate that this remarkable omission may be additional evidence that the author perceived as obvious what to others would be a delighted discovery, and that he was single-mindedly attending to his announced main purpose—the reformulation of electrodynamics based on Maxwell's equations, together with reform of the notions of space and time which this work necessitates.[94]

By his own, later criteria for a sound theory—"internal perfection" and "external confirmation"[95]—one might have expected a little more detailed discussion of both these criteria. However, the paucity of cited "facts" and experiments conforms to Einstein's dictum of what constitutes "external confirmation": not that a theory must be *built* on clearly visible empirical facts; not that the theory must be *verified* by decisive experimentation; but rather, "the theory must not *contradict empirical facts.*"[96] This criterion is even coupled with a warning against trying to secure "the adaptation of the theory to the facts by means of artificial additional assumptions" such as are "often, perhaps even always, possible."[97] This remark led Einstein to discuss the criteria concerning the premises of a theory, "with what may briefly but vaguely be characterized as the 'naturalness' or 'logical simplicity' of the premises." He admits that "an exact formulation . . . meets with great difficulties," for it involves "a kind of reciprocal weighing of incommensurable quantities." Though he cannot be more precise, "it turns out that among the 'augurs' there usually is agreement in judging the 'inner perfection' of theories. . . ."[98] Here we face again the role of what can only be called scientific taste in deciding which theory or hypothesis to accept and which to reject. We shall come back to this important point later.

To summarize: the style of thought that emerges from the direct examination of the 1905 paper is fully consonant with Einstein's report half a century later on the occasion of the centennial of Michelson's birth. It is not a theory of the more usual kind, designed to save one or even a few phenomena, but a theory "whose object is the *totality* of all physical appearances."[99] There is nothing in the paper which gives support to the idea that Einstein must have considered Michelson's experiment as crucial or even of primary importance, or even that he did know or had to know of its existence. Michelson's null result is evident "on other grounds" if one accepts the general sway of Maxwell's theory over all electrodynamics and optics and applies the relativity principle to it. Such basic experimental results as are assumed in the paper could well be those of Faraday, Fizeau, and the aberration experiments. From first to last, Einstein's paper is a work on a grandiose scale, specifically intended to transform the electrodynamic theory as then understood; and in the process it implies its own methodology and its own metaphysics to serve as the philosophical basis of a renewed science. This is of course why we think of it still as so powerful a work.

VII. Indirect Evidence: From 1905 and Earlier

We have seen no direct evidence in the 1905 paper itself to support the common textbook story, and on the contrary we have found more plausible evidence in the opposite direction. However, this cannot by any means complete our search, for there are a number of different types of indirect evidence to be examined, including evidence from contemporaneous work of Einstein in 1905 and from his early work, comments, and letters.

Reading other work by Einstein in 1905, we may reasonably expect to find additional insight into his assessment of the relation between experimental fact and theory. As I have discussed elsewhere,[100] the three epochal papers of Einstein published in 1905, though on such very different fields of physics—quantum theory of light, Brownian motion, relativity—all shared two significant properties. They arose from the same general problem: fluctuations in radiation pressure, for which, in the case of a mirror suspended in a radiation cavity, Einstein knew by 1905 that Maxwell's theory leads to wrong predictions. And they shared the same style of construction: "Each begins with the statement of formal asymmetries or other incongruities of a predominantly aesthetic nature . . . , then proposes a principle—preferably one of the generality of, say, the Second Law of Thermodynamics, to cite Einstein's repeated analogy—which removes the asymmetries as one of the deduced consequences, and at the end produces one or more experimentally verifiable predictions."[101]

The experimental part in each of the three papers was by far the least developed and for many the most unconvincing part. Thus, Millikan, whose Nobel Prize in physics was awarded in part for his experimental confirmation of Einstein's theory for the photoelectric effect, remarked later that Einstein's explanation of the effect in 1905

> at the time ignored and indeed seemed to contradict all the manifold facts of interference and thus to be a straight return to the corpuscular theory of light which had been completely abandoned since the time of Young and Fresnel around 1800 A.D. I spent 10 years of my life testing the 1905 equation of Einstein's, and, contrary to all my expectations, I was compelled in

1915 to assert its unambiguous experimental verification in spite of all its unreasonableness since it seemed to violate everything that we knew about the interference of light.[102]

We conclude that Einstein's other work of 1905 was quite consistent with the relativity paper, and that in particular the attitude toward experiments was the same: reliance on a very small number of experiments either as the basis of the theory or as a support to its claim to serious attention.

Elsewhere I have shown that at least from 1907 on we have evidence that even in the case of "disconfirming facts," such as Kaufmann's confident "experimental disproof" of relativity in 1906, Einstein continued to believe in a theory that seemed to him to have a "greater probability" because it embraced a "greater complex of phenomena."[103] I shall now review, in approximately chronological order, the other documents I have been able to find that have a bearing on Einstein's attitude on the importance of experimentation, particularly with respect to the Michelson experiment.

There is good evidence that as a young student Einstein regarded himself as an empiricist. The later attainment of a more complex position was definite enough to be dated by Einstein himself in his *Autobiographical Notes* as "shortly after 1900," when he found after Planck's publications of 1900 that neither mechanics nor thermodynamics could claim exact validity. "By and by I despaired of the possibility of discovering the true laws by means of constructive efforts based on known facts."[104] But prior to that, while a student at the Polytechnic Institute of Zurich from 1896 to 1900, he tended toward practical experimentation: "There I had excellent teachers (for example, Hurwitz, Minkowski), so that I really could have gotten a sound mathematical education. However, I worked most of the time in the physical laboratory, fascinated by the direct contact with experience."[105] This dating fits with evidence provided in 1930 by Einstein's biographer Anton Reiser,[106] who, as I have pointed out elsewhere,[107] was in fact Einstein's son-in-law, Rudolf Kayser, who worked on the biography with Einstein's acquiescence. An interesting passage in the biography refers to young Einstein's early interest in constructing by himself an experiment of the type we have been here discussing, while under the sway of "pure empiricism":

He encountered at once, in his second year of college, the problem of light, ether and the earth's movement. This problem never left him. He wanted to construct an apparatus which would accurately measure the earth's movement against the ether. That his intention was that of other important theorists, Einstein did not yet know. He was at that time unacquainted with the positive contributions, of some years back, of the great Dutch physicist Hendrik Lorentz, and with the subsequently famous attempt of Michelson. He wanted to proceed quite empirically, to suit his scientific feeling of the time, and believed that an apparatus such as he sought would lead him to the solution of a problem, whose far-reaching perspectives he already sensed.

But there was no chance to build this apparatus. The skepticism of his teachers was too great, the spirit of enterprise too small. Albert had thus to turn aside from his plan, but not to give it up forever. He still expected to approach the major questions of physics by observation and experiment. His thought was most intensely bound up with reality. As a natural scientist he was a pure empiricist. He did not entirely believe in the searching power of the mathematical symbol. After several years this state of affairs changed completely.[108]

The story fixes a date (the academic year 1897–1898) before which Einstein would not have heard of the Michelson experiment, and it helps us also to understand what otherwise may seem to be puzzles and contradictions in Einstein's references to his state of knowledge of the Michelson experiment prior to the publication of the 1905 paper. It makes it that much more believable that Einstein could have accepted without surprise the result of the Michelson experiment when he did become aware of it, for he had pondered about an ether-drift experiment of his own; and at any rate his empiricist type of approach had later "changed completely." As we noted from Shankland's first interview with Einstein in 1950, "when I asked him how he had learned of the Michelson-Morley experiment, he told me that he had become aware of it through the writings of H. A. Lorentz, but *only after 1905* had it come to his attention!"[109] And in 1952 Einstein told Shankland, "I am not sure when I first heard of the Michelson experiment. I was not conscious that it had influenced me directly during the seven years that relativity had been my life. I guess I just took it for granted that it was true." Shankland immediately adds, "However, Einstein said that in the years 1905–1909, he thought a great deal about Michelson's result, in his discussions with Lorentz and others in his thinking about general relativity. He then realized (so he told me) that he had also been conscious of Michelson's result before 1905 partly through his reading of the papers of Lorentz and more because he had simply assumed this result of Michelson to be true."[110]

Shankland's report of Einstein having read Lorentz's papers before 1905 must now be clarified. We know such reading in fact did not include the famous 1904 paper of Lorentz,[111] in which he gave his theory of electromagnetism for moving bodies to quantities of the second order ($v^2 : c^2$), nor probably most of the other papers that partly had prepared for it.[112] We have positive evidence of Einstein having read only one paper and one book by Lorentz—the paper of 1892 (published in French) and the book of 1895 (published in German), in which the theory given is to the first order of the quantity ($v : c$). This fits entirely with Einstein's remark in the 1905 paper, "as has already been shown to the first order of small quantities. . . ." Also, Einstein wrote specifically on this point to his last biographer, Carl Seelig: "As for me, I knew only of Lorentz's important work of 1895—*La théorie électromagnétique de Maxwell* [actually published in 1892], and VERSUCH EINER THEORIE DER ELEKTRISCHEN UND OPTISCHEN ERSCHEINUNGEN IN BEWEGTEN KÖRPERN [1895]—but not Lorentz's later publication, nor the consecutive investigations by Poincaré. In this sense my work of 1905 was independent."[113] We may well take Einstein's quoted word to Seelig; and even if he had read other papers by Lorentz prior to Lorentz's 1904 paper, he would not have found in them a more extensive discussion of the Michelson experiment than Lorentz gave in his 1895 book. It will therefore be profitable if we look with a little care at Lorentz's 1895 work.[114]

In this work Lorentz is still satisfied with the construction of a theory which will explain first-order effects, that is, quantities in $v : c$. Therefore the Fresnel ether theory and the aberration experiment are prominently mentioned on page 1 and are further dealt with in the text. But the Michelson ether-drift experiments are only briefly mentioned (on p. 2) with the casual warning that they will find interpretation in the framework of a first-order theory only by means of what he calls a *Hülfshypothese.* The matter is not brought up again until page 120, in a short section toward the end of the 139-page book. This, however, has become that part of the book which everyone in the field knows, because the particular excerpt—with its subtitle "Michelson's Interference Experiment" now serving as the main title—was put at the beginning of the famous reprint collection of Lorentz, Einstein, Minkowski, and Weyl, published by Teubner in 1913.[115] (Thus do scissors-wielding and extract-prone editors distort the appearance of history!)

In Lorentz's book this section appears in the last chapter, one devoted to residual difficulties in Lorentz's otherwise quite successful first-order theory, and is entitled "Experiments whose Results do not allow Explanation without Further Ado." He discusses three such embarrassing experiments. The first takes about five pages and is on an unexpectedly absent rotation of the plane of polarization (Mascart, 1872), which Lorentz attempts to explain by a separate ad hoc hypothesis without much conviction. Then come the 1881 and 1887 ether-drift experiments of Michelson (also in about five pages). There, on "the basis of Fresnel's [ether] theory," a $v^2 : c^2$ effect was predicted. To explain why it had not been observed, Lorentz first rejects Stokes's original aberration theory as one that raises too many difficulties; as an alternative he will try to obtain the removal of the puzzle "by means of a hypothesis which I have already announced some time ago [in 1892–1893] and which, as I later found, FitzGerald reached also." This saving *Hülfshypothese* is introduced completely ad hoc: "If one assumes that the arm [of the interferometer] lying in the direction of the earth's motion is shorter than the other by $L(v^2 : c^2)$ . . . , the result of the Michelson experiment is fully explained." He acknowledges that "this hypothesis at first glance may appear estranging [*befremdend*], but one must nevertheless admit that it is not so far-fetched" if one associates it with yet another assumption, namely that the molecular forces that determine the dimensions of a body are modified by relative ether motion, "similar to what may now be certainly stated for electric and magnetic forces."[116]

No explicit comment is made which connects this assumed shrinkage with the Lorentz transformations in their still primitive form, as published earlier in the book. And throughout this section Lorentz stresses the ad hoc nature of the argument. Thus when he comes back to the equality of transformation properties he has assumed for both the molecular forces and electrostatic forces, he confesses there is "no ground" for the assumption.[117] The reader is left with the impression that the results of this experiment indeed "*sich nicht ohne Weiteres erklären lassen,*" as the title of the chapter warned. It appears, therefore, that the work would not have been adversely affected if the Michelson experiment had not been done at all!

The third experiment Lorentz takes up in this 1895 book is the polarization experiment of Fizeau on glass columns, in which Fizeau had thought to have detected an effect of the earth's motion on the plane of polarization of light. These are the final thirteen pages of the book and therefore the longest of the three items. This experiment is singled out by Lorentz's remarks that its results "deserve our attention in high measure."[118]

So, when Einstein mentions in 1905, after the opening paragraph of his relativity paper, "the unsuccessful attempts to discover any motion of the earth relatively to the 'light medium,' " without naming specifically any one of them, he could have had in mind any two or more of at least *seven* experiments—the first two which were cited in the last chapter of Lorentz's work of 1895, and five additional ones that became known by 1905: the experiments by Rayleigh of 1902 on rotary polarization and double refraction; by Brace on double refraction in 1904 and the repetition, with negative effect, of the Fizeau experiment on glass columns in 1905; and by Trouton and Noble in 1903 on the turning couple of a condenser. (One might also add other ether experiments which had a bearing: e.g., Arago, 1810; Fizeau, 1851; Lodge, 1892.) Pointing out that Einstein had all these choices, as is clear to anyone who has read the available literature, does not in itself deny the possibility of an inspirational role for the Michelson experiment. On the other hand, there is also no warrant for believing that Einstein's phrase "the unsuccessful attempts" must point directly to any one of these experiments, including Michelson's. It is in fact entirely possible that he had heard and read about any two of these "unsuccessful attempts" and had drawn the correct conclusion that all of them would be unsuccessful.

We turn now briefly to Lorentz's basic 1904 paper, which everybody, including Einstein after 1905, read and cited. Lorentz there forges ahead with the improvement of the first-order theory of 1895 and specifically tries to reach a theory to cover second-order phenomena. The Michelson experiment is therefore much more important to Lorentz's argument than it was in 1895. It now appears at the very beginning of the paper, followed by some of the other more recent, embarrassing experiments (Rayleigh, Brace, Trouton and Noble, and Kaufmann), and the explanation of Michelson's results is given at two different places.[119] It is a masterful paper. In my opinion it is, together with Poincaré's paper of 1905, the best work prior to Einstein's relativity theory that electrodynamics was capable of achieving. But it is significant that in this peak achievement of classical physics we find two striking flaws, both later acknowledged by Lorentz himself.

The first is that the transformation equations Lorentz developed here do not after all achieve what he had hoped: Maxwell's equations are not completely invariant even at small speeds. In 1912 Lorentz generously added a footnote to the 1913 republication of his 1904 paper in the Teubner collection, one that was unfortunately left out in the English-language reprint of the essay:

> One will notice that in this work the transformation equations of Einstein's Relativity Theory have not quite been attained. Neither equation (7) nor formula (8) has the form given by Einstein, and as a result I was unable to make the term $-wu'_x/c^2$ in equation (9) disappear and to put equation (9) exactly in the form which holds for a system at rest. On this circumstance depends the clumsiness [*Unbeholfene*] of many of the further considerations in this work. It is owing to Einstein that the relativity principle was first announced as a general, strictly and exactly valid law. . . .[120]

In Lorentz's THE THEORY OF ELECTRONS we find a similar statement. Lorentz points out that "besides the fascinating boldness of its starting point, Einstein's theory has another advantage over mine." Einstein has attained exact covariance "by means of a system of new variables slightly different from those which I have introduced." And Lorentz adds significantly: "I have not availed myself of his substitutions, only because the formulae are rather complicated and look somewhat artificial, unless one deduces them from the principle of relativity itself."[121] This remark incidentally serves to announce the theme of the relativity of "artificial" assumptions, to be discussed below in more detail.[122]

The recognition of a second flaw in Lorentz's work, one that now strikes us as even more serious than the first, is implied in another typically generous comment by Lorentz in 1909 in THE THEORY OF ELECTRONS. There he hints at a further aspect of the difference between the "clumsiness" of his own hypothesis-ridden, constructive theory and the "fascinating boldness" of Einstein's approach, which sweeps away the whole complex machinery of ether-electrodynamics, a theory in which Einstein started from general principles and jettisoned wherever possible any assumption, even one that was as precious to contemporary physics as the light ether. Einstein's results concerning electromagnetic and optical phenomena, Lorentz writes,

> . . . agree in the main with those which we have obtained in the preceding pages, the chief difference being that Einstein simply postulates what we have deduced, with some difficulty and not altogether satisfactorily, from the fundamental equations of the electromagnetic field. By doing so, he may certainly take credit for making us see in the negative results of experiments like those of Michelson, Rayleigh, and Brace, not a fortuitous compensation of opposing effects, but the manifestation of a general and fundamental principle.[123]

In a note added in the edition of 1915 Lorentz went further: ". . . Einstein's theory . . . gains a simplicity that I had not been able to attain."[124] Einstein agreed in a publication soon afterward, stating that the relativity theory grew out of the Maxwell-Lorentz theory of electrodynamics "as an amazingly simple summary and generalization of hypotheses which previously had been independent from one another. . . ."[125] This feeling of attaining a view in which previously separate processes, phenomena, or mechanisms now appear interdependent has always been a quasi-aesthetic experience of the most treasured kind in the sciences. One is reminded of Copernicus' triumphant boast in the DE REVOLUTIONIBUS that in his heliocentric system "not only do all their [the planets] phenomena follow from that, but also this correlation binds together so closely the order and magnitudes of all the planets and of their spheres or orbital circles, and the heavens themselves, that nothing can be shifted around in any part of them without disrupting the remaining parts and the universe as a whole."

VIII. Against an Ad Hoc Physics

That feeling of having unveiled a grand scheme, one so beautifully interconnected that all of it can be entered through a very small set of postulates, is precisely what seems lacking in Lorentz's theory. Devoted though Poincaré was to Lorentz, he reserved some of his most caustic criticism for the way more and more new hypotheses appeared in this work:

On a fait des expériences qui auraient dû déceler les termes du premier ordre; les résultats ont été négatif; cela pouvait-il être par hasard? Personne ne l'a admis; on a cherché une explication générale, et Lorentz l'a trouvée; il a montré que les termes du premier ordre devaient se détruire, mais il n'en était pas de même de ceux du second. Alors on a fait des expériences plus précises; elles ont aussi été négatives; ce ne pouvait non plus être l'effet du hasard; il fallait une explication; on l'a trouvée; on en trouve toujours; les hypothèses, c'est le fonds qui manque le moins.

Mais ce n'est pas assez. . . .[126]

Lorentz surely did not enjoy the ad hoc character of his theory. The matter was evidently much on his mind when he wrote the 1904 paper. On the second page Lorentz gives the chief reasons for publishing a new treatment:

The experiments of which I have spoken are not the only reason for which a new examination of the problems connected with the motion of the Earth is desirable. Poincaré has objected to the existing theory of electric and optical phenomena in moving bodies that, in order to explain Michelson's negative result, the introduction of a new hypothesis has been required, and that the same necessity may occur each time new facts will be brought to light. Surely this course of inventing special hypotheses for each new experimental result is somewhat artificial. It would be more satisfactory if it were possible to show by means of certain fundamental assumptions and without neglecting terms of one order of magnitude or another, that many electromagnetic actions are entirely independent of the motion of the system. Some years ago [1899], I already sought to frame a theory of this kind. I believe it is now possible to treat the subject with a better result. . . .[127]

There is little doubt that Lorentz's hopes for a more satisfactory theory were not entirely in vain. Thus he could say "the only restriction as regards the velocity will be that it be less than that of light."[128] But the work is far from dispensing with the need to set up "special hypotheses" to explain new experimental results, and Lorentz has to introduce explicitly or implicitly at least eleven different ad hoc assumptions or hypotheses (the terms are used interchangeably in his paper), as I have pointed out previously.[129] For a paper that is to deal with physics from a fundamental point of view it is veritably obsessed with hypotheses. I have made a quick count of the number of times the term "hypothesis" or "assumption" and their direct equivalents ("I shall now suppose") appear: it happens at least thirty times in these pages, and even more if one adds circumlocutions and implicit references (e.g., "the velocity will be subjected to a limitation . . ."). Whether all the hypotheses introduced are ad hoc in the same sense or to the same degree, whether the criteria of ad hocness are different for different types of theory, or whether such distinctions are significantly meaningful for physicists or philosophers are not issues at this point. What is more relevant for an understanding of Einstein's work is to read Lorentz's paper through the eyes of someone such as Einstein whose style of work is to minimize the making of hypotheses.

On the other hand, to appreciate Lorentz's purpose and problem fully we must remember that he saw the crisis of physics in the 1890's and 1900's in a rather different way from Einstein. With his immense knowledge in virtually every part of physics, Lorentz was then particularly deeply involved in constructing, step by step, a viable theory for electrodynamics, based as far as possible on existing principles and mechanisms, relying on experimental results as a guide to the detailed construction of a modification of existing theory. The result, as a careful reading of Lorentz's papers shows, was that Poincaré's accusation was all too justified. But Lorentz was attending to a very difficult problem with conventional tools, and today we also know how to live in some areas of contemporary physics with far less elegant work and more blatantly ad hoc hypotheses. Nothing better could probably be done as long as the ether was kept at the heart of physics; and that seemed to Lorentz, as to Poincaré, Michelson, and many others, a thematic necessity to the end of his life.[130]

The simplification introduced by Einstein's approach is far easier to discern with hindsight. The decisions at this point in the origins of contemporary science were quite analogous to the decisions at the fork in the road to modern science—in accepting or rejecting the Copernican system in the sixteenth century. A famous passage in Herbert Butterfield's THE ORIGINS OF MODERN SCIENCE is as appropriate for the twentieth-century turning point as for the earlier one:

> . . . at least some of the economy of the Copernican system is rather an optical illusion of more recent centuries. We nowadays may say that it requires smaller effort to move the earth round upon its axis than to swing the whole universe in a twenty-four hour revolution about the earth; but in the Aristotelian physics it required something colossal to shift the heavy and sluggish earth, while all the skies were made of a subtle substance that was supposed

> to have no weight, and they were comparatively easy to turn, since turning was concordant with their nature. Above all, if you grant Copernicus a certain advantage in respect of geometrical simplicity, the sacrifice that had to be made for the sake of this was tremendous. You lost the whole cosmology associated with Aristotelianism—the whole intricately dovetailed system in which the nobility of the various elements and the hierarchical arrangement of these had been so beautifully interlocked. In fact, you had to throw overboard the very framework of existing science, and it was here that Copernicus clearly failed to discover a satisfactory alternative. He provided a neater geometry of the heavens, but it was one which made nonsense of the reasons and explanations that had previously been given to account for the movement in the sky.[131]

Here again a similar agonizing choice had to be made: in order to extend the principle of relativity from mechanics (where it had worked) to all of physics, and at the same time to explain the null results of all optical and electrical ether-drift experiments, one needed "only" to abandon the notion of the absolute frame of reference and, with it, the ether. But without these the familiar landscape changed suddenly, drastically, and in every detail. Physics was left without its old hope, already partly and sometimes gratifyingly fulfilled, namely to explain all phenomena by means of one consistent, mechanistic theory.

In the compass of this work we cannot do full justice to the question of what constitutes an ad hoc hypothesis as it affects the work of scientists such as Lorentz and Einstein; but we must at least include some comment, for it is an essential issue for two related reasons. Somewhat along the line of the textbook story, a few philosophers of the experimenticist persuasion appear to think that Lorentz's and Poincaré's theory was on the whole satisfactory (e.g., they argue that not all the hypotheses were ad hoc in the same sense or that Einstein's theory did not drastically reduce the number and artificiality of the needed hypotheses), so that the Michelson experiment becomes more plausibly the "crucial" event that forces a radical reconsideration. More important for our purposes is another reason for clarifying the meanings of ad hoc in the actual work of scientists: to be able to appreciate the differences of style between the chief protagonists and indeed the differences between classical nineteenth-century physics and modern twentieth-century physics.

To begin with, operationally it does not matter whether an ad hoc hypothesis appears to have some support in theory (thus it is easily seen from Lorentz's papers that he tried to link the contraction hypothesis with the supposedly analogous behavior of "molecular forces" on the one hand and electric and magnetic forces on the other, though without being able to say anything revealing about these presumed molecular forces). It also does not much matter that the contraction hypothesis was not completely self-contained; Lorentz himself made clear that his ad hoc assumptions might have uses beyond merely explaining the experiments that forced their invention (e.g., the prediction of a negative result for the Michelson experiment for light rays traversing transparent objects).[132] These other applications of the ad hoc hypothesis are not of real interest in any case; they were not urged as tests that would decide on its acceptability, and even if such tests had been carried out successfully, it is unlikely that they would have increased the appeal of the hypothesis to Einstein.

A refinement of the concept of ad hoc, including divisions into such categories as "logically ad hoc" and "psychologically ad hoc," may perhaps be of interest for epistemological discussions. Such work, however, has yet to make a really important point, and at any rate we must not draw attention away from what *is* an important point, namely the scientist's feeling of ad hocness about an hypothesis whether his own or not—for example, the distaste for the contraction hypothesis expressed by Einstein, Poincaré, Michelson, and even Lorentz himself. To understand what almost any working scientist feels when he has to evaluate an hypothesis seems to be difficult for those who are not actually engaged in creative scientific work. Hence it will be useful to develop a field that can fairly be called the aesthetics of science. For these matters are still decided in practice on the basis on which Copernicus confessed, apparently so vaguely yet so correctly, to have made his decision against accepting "the planetary theory of Ptolemy and most other astronomers, although consistent with the numerical data": one of his chief objections was that he found their sort of system not "sufficiently pleasing to the mind." This criterion of choice is familiar to every working scientist.[133]

The essential difference between the scientist's use of ad hoc and the logician's is that the former regards it largely as a matter of private science, or science-in-the-making (which may be designated by S_1), whereas the latter regards it as a matter of public science (S_2). It would be wrong to think of S_1 and S_2 as always sharply separated, but it is much more erroneous to overlook the deep distinctions between these two meanings of science. In the case at hand they refer to the differences between two legitimate but different uses of ad hoc that may be termed ad hoc S_1 and ad hoc S_2. For a scientist engaged in original activity, his designation ad hoc (or its equivalent term) is an essentially aesthetic judgment which he makes within S_1 while he imagines, considers, introduces, or rejects an hypothesis. Ad hoc S_1, in this sense of an act of individual, initially private judgment by a scientist who may very well be deeply and not only rationally involved, differs fundamentally from ad hoc S_2 in the sense of a public statement with permanent, more or less clear epistemological properties, one that has been published and has become part of science-as-an-institution.[134]

There is no doubt that an hypothesis that is ad hoc S_1 has logical properties; the point is that they are not ruling in the actual use of such hypotheses. The scientist who adopts somebody's hypothesis or creates his own for a specific purpose, "in order to account" for a bothersome result or feature of the theory, regards it as ad hoc—not necessarily in a derogatory sense— *regardless* of its "logical" status. This helps to explain the significance of the passionate and personal "unscientific" language generally used to describe such hypotheses, and incidentally points to gradations of aesthetic or "psychological" acceptability. Thus we have found in the scientific literature characterizations of the following kinds for acceptable ad hoc hypotheses: "not inconceivable," "reasonable," "plausible," "fundamental," "natural," "appealing," "elegant," "likely," "assumed a priori to get the desired results," "auxiliary" or "working hypothesis." On the other hand, when an ad hoc hypothesis in rejected, we see it described in the following way: "artificial," "complex," "contrived," "implausible," "bothersome," "unreasonable," "improbable," "unlikely," "unnecessary," "ugly." Sometimes ad hoc itself is applied in the pejorative sense, and then it has such meanings. Or the individual's judgment is transferred to nature itself, as in the ancient motif which Newton phrased, "Nature is pleased with simplicity, and affects not the pomp of superfluous causes." (Note that in all these cases we do not consider now the quite separate question whether the hypothesis will later turn out to be "right" or "wrong.")

The derogatory characterizations evoke strong behavioral responses, which the author of an ad hoc hypothesis can sometimes foresee and try to ameliorate. Thus in 1892, in the French-language publication which

Einstein said he read, Lorentz announced the contraction hypothesis with considerably hedged qualifications: "Now, some such changes in the arms of Michelson's first experiment, and in the dimensions of the slab in the second one, is so far as I can see, not inconceivable. . . . An influence of the order of $\rho^2:v^2$ *is not excluded and that is precisely what we need*."[135] And, in 1899, writing on the effect of motion on the masses of charged particles, Lorentz said: "Such a hypothesis seems very startling at first sight. Nevertheless, we need not wholly reject it."

The language used when the contraction hypothesis was introduced will give us an important clue to some properties of ad hoc hypotheses which cannot be accommodated in an axiomatic treatment of the theory. Writing on 10 November 1894, Lorentz sent a letter to FitzGerald: "My dear Sir, In his *Aberration Problems* Prof. Oliver Lodge mentioned a hypothesis *which you have imagined in order to account* for the negative result of Mr. Michelson's experiment."[136] Similarly, when Lodge mentioned the idea of FitzGerald's in public on 27 May 1892 he said, "Professor FitzGerald has suggested *a way out of the difficulty* by supposing the size of bodies to be a function of their velocity through the ether."[137]

It is true that the contraction hypothesis, initially so welcome to ether theoreticians because it "explained" the Michelson result, later turned out to be unacceptable because it predicted an unsymmetrical change of dimensions for different inertial systems. But our attention should be fixed on the fact illustrated by these examples that the contraction hypothesis *when it was made* was clearly and quite blatantly ad hoc—or, if one prefers to use the patois of the laboratory, ingeniously "cooked up," for the narrow purpose which it was to serve. Indeed, in A. M. Bork's article[138] there are two quotations from Lodge's later reminiscences which provide the plausible setting of a casual chat between Lodge and FitzGerald during which the hypothesis seems to have been first discussed as a nice, wild idea among friends. It is precisely from such situations that some of the finest advances develop—for example, the postulation of the spin of the electron by Goudsmit and Uhlenbeck.[139] But in the case of this particular hypothesis, it always retained a casual, improbable character, even when Lorentz tried to propose it independently in a somewhat different form, relying lightly on an explanatory scheme based on the presumed analogy between molecular forces (whatever they may be) and electric and magnetic forces. I do not find it at all surprising that FitzGerald himself seems to have been satisfied to let others discuss the hypothesis, that Einstein and others called it ad hoc in a clearly derogatory sense, or that it has traditionally been called the very paradigm of an ad hoc hypothesis.

This working sense of "ad hoc" is not, to be sure, the only meaning of the phrase, but it is one which at this stage of the understanding of hypotheses is the important one for any historical analysis that claims to deal with the actual contribution to science by an individual person. And it is a sense that cannot be dismissed as "merely psychological" or "only psychological." Whether or not epistemological analysis can establish other, perhaps "largely unnoticed," senses for the phrase remains to be seen. C. G. Hempel has shown clearly that "there is, in fact, no precise criterion for ad hoc hypotheses."[140] Then, too, the epistemological distinctions between various meanings of ad hoc hypotheses seem to be more difficult than was once thought: the epistemologist Grünbaum's recent analysis that had confidently announced clear distinctions between "logically ad hoc" and "psychologically ad hoc" soon had to be withdrawn by its author, as the result of the demonstration by Hempel that there were serious inadequacies in the work.[141] One may hope for eventual progress, but one need not wait for it to show the operative sense of ad hoc hypotheses that does exist among scientists. C. C. Gillispie has correctly said: "The special theory of relativity was rather a restriction upon science than an induction from positive phenomena. In his taste for 'inner perfection' in theory, Einstein answered to an aesthetic which logicians of science have not yet reduced to empirical terms, or to inter-subjective agreement."[142]

Some questions arise which we should deal with at another occasion at length and here at least cursorily. Are not *all* hypotheses ad hoc? The answer, in brief, is No. In the special example of the contraction hypothesis, there were alternatives of at least two kinds. One was to explain the negative result of the Michelson experiment by showing that conclusions equivalent to the Lorentz-FitzGerald contraction could be derived from hypotheses or postulates that were *not* proposed specifically in order to account for the phenomenon—for example, from Lorentz's transformation equations, although the latter may have initially been completely ad hoc *with respect to their own original purpose* of providing for the invariance of Maxwell's equations.[143] Indeed, when Lorentz proposed the transformation equations he had the sense of ad hocness with respect to their purpose; he therefore refused to acknowledge one of the possible consequences of the Lorentz transformation equations, namely a physical meaning of the concept *Ortszeit* that he felt had been invented for a narrower purpose. The other possibility was to derive a statement, whose import would be equivalent to the contraction hypothesis, from statements that were even more distant from any specific worry about the Michelson experimental results—for example, from the two basic postulates in Einstein's relativity paper of 1905. This was so simple to do that Einstein, as we saw, did not deign to go through the derivation in his 1905 paper and simply hinted that the reader could derive it for himself for all optical experiments.

The important point here is that "ad hoc" is not an absolute but a relativistic term. Postulates 1 and 2 may be said to have been introduced ad hoc with respect to the relativity theory of 1905 as a whole; Einstein cites little support even for the *Vermutung* [conjecture, suspicion] of postulate 1 and virtually nothing to support the *Vermutung* of postulate 2. But these two principles were not ad hoc with respect to the Michelson experiment, for they were not specifically "imagined in order to account" for its result.

Thus a statement may be ad hoc relative to one context but not ad hoc relative to another. The relativity of ad hocness in this sense is beautifully illustrated by another episode from the history of the FitzGerald proposal itself. As S. G. Brush has shown,[144] FitzGerald did intend to publish his contraction hypothesis in a letter to the journal Science in 1889, but he never saw a printed copy and thought it had not appeared because the journal stopped publishing at about that time. This initial proposal was entirely ad hoc with respect to the ether drift experiment:

> I would suggest that almost the only hypothesis that can reconcile this opposition [between the Michelson experiment and the Fresnel ether] is that the length of material bodies changes, according as they are moving through the ether or across it, by an amount depending on the square of the ratio of their velocities to that of light.[145]

To add plausibility, he referred to the "not improbable supposition that the molecular forces are affected by the motion, and that the size of a body alters consequently." Rather like Lorentz, FitzGerald based himself here on the qualitative analogy: "We know that electric forces are affected by the motions of the electrified bodies relative to the ether." Later, however, FitzGerald did not seem to regard his proposal any longer as an ad hoc hypothesis, and his reply to Lorentz on 14 November 1894 emphasizes the distinction from the view Lorentz expressed in his own first letter. FitzGerald wrote, "My dear Sir, I have been preaching and lecturing on the doctrine that Michelson's experiment proves, and is one of the only ways of proving, that the length of a body depends on how it is moving through the ether. . . . Now that I hear you as an advocate and authority I shall begin to jeer at others for holding any other view." What FitzGerald had published in 1889 rather modestly as "almost the only hypothesis that can reconcile this opposition," supported by "a not improbable supposition"—a useful idea, as Lorentz's work showed when he used his own very similar one—had become in 1894 a "doctrine" for FitzGerald, "proved" by the Michelson experiment, and in that context no longer ad hoc. Hence we see "ad hoc" to be a concept that is relativistic in more senses than we may have thought: relativistic for one person with respect to time ad hoc for FitzGerald in 1899 but not in 1894), and relativistic for different persons at the same time (ad hoc for Lorentz in 1894 but not for FitzGerald in 1894).

How is one to decide whether an hypothesis is ad hoc or not? And, moreover, whether it is repulsively ad hoc or acceptably so? It is here that we connect with Einstein's criterion of the "inner perfection" of the theory. The criterion is the feeling for the "naturalness" or "logical simplicity" of the premises. And we recall again that Einstein makes immediately two points: one is that this point of view "has played an important rôle in the selection and evaluation of theories since time immemorial," and secondly that "an exact formulation of [it] meets with great difficulties," because "the problem here is not simply one of a kind of enumeration of the logically independent premises (if anything like this were at all unequivocally possible), but that of a kind of reciprocal weighing of incommensurable qualities."[146]

As in the pursuit of other aspects of science itself, one's best guide to a decision about the ad hocness of an hypothesis is not logical analysis alone. Here, for better or worse, Einstein's dictum applies: "There is of course no logical way leading to the establishment of a theory, but only groping constructive attempts. . . ." Nor is the scientist likely to be much helped by a criterion such as Karl Popper's: "As regards *auxiliary hypotheses* we decide to lay down the rule that only those are acceptable whose introduction does not diminish the degree of falsifiability or testability of the system in question, but, on the contrary, increases it."[147] Rather, what is required is a feeling for the state of affairs, a *Fingerspitzengefühl*, which not only distinguishes the quality of insight of different scientists but also forces them to take often quite different views on the same hypothesis. I use the dangerous word "feeling" deliberately, because it indicates accurately the difficult-to-define locus where judgments of the scientific relevance of hypotheses are made. Thus Max Wertheimer recorded Einstein's feelings about Lorentz's contraction hypothesis: "He felt the auxiliary hypothesis to be an hypothesis ad hoc, which did not go to the heart of the matter. . . . He felt the trouble went deeper than the contradiction between Michelson's actual and the expected results."[148]

A chief difficulty with the more abstract discussions of the matter which I have seen so far is that in some cases this essential *Fingerspitzengefühl* is lacking. In the absence of a firsthand feeling for scientific taste, historical or philosophical scholarship, particularly that directed to a case at the very front of a major scientific advance, is endangered because it is likely to be uninterested in or impotent before the personal dimension, the private (S_1) aspect of science, the essential judgment whether or not some approach does "go to the heart of the matter." So-called philosophical mastery must be supplemented by an understanding of matters of scientific taste and feeling. Otherwise it may be brought to bear on an empty case, or worse, one that exists only as a visible model constructed to reflect the maker's own hidden theory of science. It can lead such a person to scold an Einstein for not having behaved like an obedient student in the classroom of a logician, for not having used the "right" terminology, and for not having shouldered an "obligation" to his philosophical masters.[149] It may also lead him to overlook the usefulness of the (partial) definition of an ad hoc hypothesis implied in Newton's exceedingly liberal and permissive Fourth Rule of Reasoning: that an hypothesis is to be accepted as "accurately or very nearly true, notwithstanding any contrary hypotheses that may be imagined," until we have additional evidence by which the hypotheses may be revised or made more accurate.[150]

I suspect that the task of the epistemologist is made difficult because the very nature of his work is different from the necessarily more lax epistemological attitude of the working scientist. Einstein warned of this in a famous passage in the *Remarks* which we have cited, where he compares the attitudes of the practitioners of the disciplines and notes that the scientist cannot let himself be "restricted in the construction of his conceptual world by the adherence to an epistemological system."[151] Hans Reichenbach wisely recognized the danger when he wrote in the same volume a passage we have already touched on:

> The physicist who is looking for new discoveries must not be too critical; in the initial stages he is dependent on guessing, and he will find his way only if he is carried along by a certain faith which serves as a directive for his guesses. When I, on a certain occasion, asked Professor Einstein how he found his theory of relativity, he answered that he found it because he was so strongly convinced of the harmony of the universe. No doubt his theory supplies a most successful demonstration of the usefulness of such a conviction. But a creed is not a philosophy; it carries this name only in the popular interpretation of the term.[152]

Returning from this general overview of aspects of the status of ad hoc hypotheses, we can now refine and summarize our assessment of the probable role of the Michelson experiment in Einstein's work leading to the 1905 paper. In reading Lorentz's book of 1895, Einstein will have found that the experiment was not thought to be the crucial event upon which a new physics must be built: it was only one of several second-order experiments that at the time could be explained only by invoking yet another unhappy ad hoc hypothesis to add to all the others on which current theory was built. Lorentz himself explicitly called the contraction hypothesis a *Hülfshypothese*, and later felt compelled to make an (essentially vain) attempt to explain the Michelson result with more appealing assumptions. Einstein characterized the contraction hypothesis as unsatisfactory on several occasions after 1905. Michelson agreed ("such an hypothesis seems rather artificial"[153]). So did Poincaré. So did others who had to face whether or not to work with it. And that is what counts in the characterization of ad hocness.

We conclude that the chief lesson of the Michelson experiment for Einstein was a secondary one: that the then-current contraction explanation of the result of the experiment, by what he felt to be its unappealing ad hoc character, compromised further the ether-committed theory of electrodynamics which Einstein already knew for many other, also largely aesthetic, reasons to be inadequate. The problem Einstein saw was not the logical status of the contraction hypothesis, not Michelson's experimental result itself (for it *could* be accommodated, even if not *ohne Weiteres*), but the inability of Lorentz's theory to fulfill the criterion of "inner perfection" of a theory.

I have elsewhere drawn attention to the distinctions Einstein made between "constructive theories" and "theories of principle."[154] Theories of the latter type, such as the relativity theory and thermodynamics, start with "empirically observed general properties of phenomena." The accent is not on any one property or phenomenon but on a creative digest or synthesis of *die gesammten Erfahrungstatsachen*, the totality of physical experience in a field. It is thus an unhappy caricature to think that any one experiment would be a chief reason for restructuring all of electrodynamics.

This completes our examination of the evidence during the earlier period of Einstein's work. We now turn to the later period, from the 1920's into the 1950's.

IX. Indirect Evidence: After 1905

First we return briefly to Shankland's interviews to calibrate an endpoint in the long development of Einstein's attitude toward experimentation in general. Thus, Shankland asked Einstein in 1952 about J. L. Synge's recently published approach to relativity which predicted a small positive effect in a Michelson-type experiment. "Einstein stated strongly that he felt Synge's approach could have no significance. He felt that even if Synge devised an experiment *and found a positive result*, this would be completely irrelevant. . . . [Later] he again said that more experiments were not necessary, and results such as Synge might find would be 'irrelevant.' He told me not to do any experiments of this kind."[155]

This attitude was characteristic not only at the endpoint. In reading through the documents in the Einstein Archive one finds abundant evidence from the earliest period on that Einstein felt there was a necessity for order in natural law, the perception of which, once he had obtained it even on the basis of the hints furnished by a few chosen experimental facts, allowed him to judge the significance of further experiments. Einstein's responses to the result of D. C. Miller's repetition of the Michelson-Morley experiment is rather typical.

On Christmas Day, 1925, Einstein received a cable from a newspaper service in the United States: "President Miller, American Physical Society, announces the discovery of aether drift. Says 'my work annuls second postulate Einstein theory.' Please cable collect 200-word opinion, press rates. . . ." Apparently he gave no answer, but on the same day Einstein wrote to his oldest friend, Michele Besso, "I think that the Miller experiments rest on an error in temperature. I have not taken them seriously for a minute." Again, on 14 March 1926, in a letter to A. Piccard, Einstein wrote, "I believe that in the case of Miller, the whole spook is caused by temperature influences (air)."[156] As it turned out, Einstein's intuitive response was right. In fact it was Shankland and his colleagues[157] who found in a painstaking and beautiful analysis of Miller's data that the "ether-drift" differences at different altitudes which Miller had reported "were in fact due to the greatly differing temperature conditions in the basement laboratory at Case and at Mt. Wilson."[158]

We now come to a very significant document that needs to be understood in its setting: the speech given early in 1931 by Einstein on his visit to Pasadena, California, when he came for the first and last time face to face with Michelson. The occasion must have been moving. Michelson, twenty-seven years his senior, was much beloved by Einstein from a distance, as we have noted. Einstein told Shankland he particularly appreciated "Michelson's artistic sense and approach to science, especially his feeling for symmetry and form. Einstein smiled with pleasure as he recalled Michelson's artistic nature—here there was a kindred bond."[159]

But Michelson was well known to be no friend of relativity, the destroyer of the ether. Like so many, he was convinced that his own ill-fated experiments were the basis for the theory, and he explained the route in a talk given in 1927: the Michelson experiment had led Lorentz to propose the transformations, and "these contained the gist of the whole relativity theory."[160] Einstein reminisced later, "He told me more than once that he did not like the theories that had followed from his work,"[161] and he said he was a little sorry that his own work started this "monster."[162] Michelson was now seventy-nine years old, weak after a serious stroke that had first forced him to his sickbed two years earlier. The picture taken on that occasion shows the frail old man, standing next to Einstein, with his usual erect dignity on this last public appearance, but he was marked for the death that came three months later.

Among others present at a grand dinner in the new Athenaeum on 15 January 1931[163] were the physicists and astronomers C. E. St. John, W. W. Campbell, R. A. Millikan, W. S. Adams, R. C. Tolman, G. E. Hale, and E. P. Hubble, as well as Mrs. Einstein and two hundred members of the California Institute Associates. Millikan set the stage with his opinion concerning the characteristic features of modern scientific thought (it takes, "as its starting point, well-authenticated, carefully tested experimental facts, no matter whether these facts seem to fit into any general philosophical theme. . . . In a word, modern science is essentially empirical, and no one has done more to make it so than the theoretical physicist, Albert Einstein"). It is, in fact, largely the same material Millikan republished eighteen years later as part of the introduction to the Einstein issue of the REVIEWS OF MODERN PHYSICS. But to the sentence "Thus was born the special theory of relativity" Millikan in 1931 added, "I now wish to introduce the man who laid its experimental foundations, Professor Albert A. Michelson. . . ."

Michelson kept his short response in the channel laid out by Millikan:

> I consider it particularly fortunate for myself to be able to express to Dr. Einstein my appreciation of the honor and distinction he has conferred upon me for the result which he so generously attributes to the experiments made half a century ago in connection with Professor Morley, and which he is so generous as to acknowledge as being a contribution on the experimental side which led to his famous theory of relativity.

Einstein had not yet responded. Millikan next called on Campbell, representing the splendid group of experimental astronomers present, saying, "I am herewith assigning him the task of sketching the development of the experimental credentials of the general theory of relativity." Campbell recounted the success of the three chief tests, in which the California astronomers had played leading roles.

Millikan then started to introduce Einstein, but prefaced it with a last reinforcement of the philosophical message that had been building up, this time by referring to Millikan's own "experimental verification of predictions" contained in the early papers of Einstein. Seen from his perspective, his evaluation of Einstein's paper on the quantization of light energy (1905) was not surprising: "The extraordinary penetration and boldness which Einstein showed in 1905 in accepting a new group of experimental facts and following them to what seemed to him to be their inevitable consequences, whether they were reasonable or not as gauged by the conceptions prevalent at the time, has never been more strikingly demonstrated."

At last, the stage and the expectations were fully set for Einstein's response. What happened next—or rather, what is supposed to have happened—is widely known from the account given in Jaffe's biography of Michelson:

> Einstein made a little speech. Seated near him were Michelson, Millikan, Hale, and other eminent men of science. "I have come among men," began Einstein, "who for many years have been true comrades with me in my labors." Then, turning to the measurer of light, he continued, "You, my honored Dr. Michelson, began with this work when I was only a little youngster, hardly three feet high. It was you who led the physicists into new paths, and through your marvelous experimental work paved the way for the development of the Theory of Relativity. You uncovered an insidious defect in the ether theory of light, as it then existed, and stimulated the ideas of H. A. Lorentz and FitzGerald, out of which the Special Theory of Relativity developed. Without your work this theory would today be scarcely more than an interesting speculation; it was your verifications which first set the theory on a real basis."
>
> Michelson was deeply moved. There could be no higher praise for any man.[164]

It is the kind of response for which the developing occasion had prepared, and Jaffe gives a natural and clear-cut evaluation of the problem of a possible genetic connection between Michelson's experiment and Einstein's work, one entirely in accord with all the textbook versions we cited earlier: "In 1931, just before the death of Michelson, Einstein publicly attributed his theory to the experiment of Michelson."[165]

However, reading Jaffe's passage carefully, we need not go so far. Michelson's "stimulating the ideas" of Lorentz and FitzGerald out of which in turn the special theory of relativity "developed" is not a scenario in contradiction with the likely chain of events discussed above: the then-current Lorentz-FitzGerald contraction explanation of the Michelson experiment, as found in the two works of Lorentz (of 1892 and 1895) which we know Einstein had read, by its unappealing ad hoc character compromised further the ether-committed theory of electrodynamics which Einstein already knew for many other reasons to be inadequate. What is missing from Einstein's short response was, to be sure, an elaboration of inputs other than those he was here mentioning; but this clearly was not the occasion to do it.

It is more difficult for us to match the last sentence given above to the ideas we have been developing. As is true for similar remarks at other times, perhaps these too referred to the public acceptance of relativity rather than to Einstein's own developing thoughts leading to his 1905 paper. But the remarks "without your work . . . it was your verifications . . ." more likely would be a personal acknowledgment to Michelson, a public attribution of the kind that Jaffe clearly saw in it. And in that case, as Kepler put it half-way through the ASTRONOMIA NOVA, "our hypothesis goes up in smoke."

But it turns out that Jaffe's widely read version of Einstein's talk has fallen into the trap of preconceptions that had unwittingly been set up for Einstein. A heading, a little sentence, and a long ending from Einstein's talk were omitted by Jaffe, and they make a lot of difference. The text of Einstein's talk, in the original German, was published in 1949[166] together with the rather inadequate English translation used in the SCIENCE account and, with omissions, in Jaffe's book. The talk starts with "Liebe Freunde!"—it is, of course, addressed to the whole company, among whom there were so many whose scientific lives were closely linked to Einstein's. And just between the last two sentences quoted by Jaffe, we find a sentence omitted from his passage that switches the discussion away from Michelson and special relativity toward the assembled astronomers and general relativity. The correct passage reads: "You uncovered an insidious defect in the ether theory of light, as it then existed, and stimulated the ideas of H. A. Lorentz and FitzGerald, out of which the special theory of relativity developed. *These in turn led the way to the general theory of relativity, and to the theory of gravitation.* Without your work this theory would today be scarcely more than an interesting speculation; it was your verifications which first set the theory on a real basis" (italics supplied). Then follows immediately an acknowledgment of each of the experimental contributions that had "set the [general] theory on a real basis": those by Campbell, St. John, Adams, and Hubble.[167]

What remains is still a fine compliment to Michelson. Yet even standing before him, and under the accumulated pressures of the dramatic affair, Einstein agreed neither with Millikan's nor with Michelson's version of the genetic connection (nor, of course, with Jaffe's). He did not avail himself of the occasion to say straight out what everyone seemed to have come to hear him say: "Michelson's is the crucial experiment that was the basis for my own work." Rather he seemed to see Michelson as one of the figures on the continuous, long way leading to relativity theory. For even while working on the theory in the spring of 1905, Einstein claimed that he saw his work not as a violent break with the past, but more modestly, as a continuation and improvement of existing trends.[168]

As to the unfortunate omission of the sentence in Jaffe's book, one knows such things happen, and at the most awkward moment. The significance lies in this: mistakes of this sort favor the prepared mind. And worse, through no fault of Jaffe's, his evaluation has been repeatedly republished by others who, apparently without a scholarly examination of the available original text, found comfort in his evaluation for their purpose of forging a tight genetic link from Michelson to Einstein.

The next document to be examined was published after Einstein's return from Pasadena. It is a summary, in third person, of Einstein's remarks, presented on 17 July 1931, to the Physikalische Gesellschaft of Berlin, in memory of Michelson who had died on 9 May, 1931. Once more, the chance offered itself to Einstein to say what all textbooks had long been saying, and to do so under the most natural circumstances. But, as at Pasadena, this did not happen. Einstein is reported to have said (without direct attribution) that Michelson's greatest idea "was the invention of his famous interference apparatus, which came to be of greater significance both for relativity theory as well as for the observation of spectral lines," and "this negative result [of the Michelson experiment] greatly advanced the belief in the validity of the general relativity theory."[169] The last phrase casts some doubt on the accuracy of the report, since the Michelson experiment could be interpreted by the special rather than the general theory. But even in this form the report is most consistent with Einstein's earlier statements pointing out the usefulness of the experiment in convincing other physicists of the value of relativity theory.

There ensued a lull for some years during which the question of a possible debt to Michelson was apparently not raised. Then, in a letter of reply on 17 March 1942 to Jaffe, Einstein again had to make a statement on the matter. It has most of the characteristics of the replies to Shankland eight to twelve years later, that is, the influence would have been to strengthen a previously obtained conviction and to remove doubts (on the part of others?):

It is no doubt that Michelson's experiment was of considerable influence upon my work insofar as it strengthened my conviction concerning the validity of the principle of the special theory of relativity. On the other side, I was pretty much convinced of the validity of the principle before I did know this experiment and its results. In any case, Michelson's experiment removed practically any doubt about the validity of the principle in optics and showed that a profound change of the basic concepts of physics was inevitable.[170]

In 1946 Einstein wrote at Schilpp's request the *Autobiographical Notes*, from which we have already quoted all sections relevant to the question before us. It has often been remarked, particularly by those who believe in the "missing link," that they found it "frustrating" that even on that occasion Einstein did not provide what they so desired. But he mentioned neither the name nor the experiment.

From 1950 on, for the remaining five years of Einstein's life, the question somehow began to be asked much more frequently than ever before. The various requests fall during the period of Shankland's interviews, which we have already mined, and the answers overlap on the whole quite consistently with them.

On 8 July 1953 Einstein was interviewed by the physicist N. L. Balazs, whose account was then published by Michael Polanyi in 1958. Balazs reported:

. . . The Michelson-Morley experiment had no role in the foundation of the theory. He got acquainted with it while reading Lorentz's paper about the theory of this experiment (he of course does not remember exactly when, though prior to his papers), but it had no further influence on Einstein's considerations and the theory of relativity was not founded to explain its outcome at all.[171]

Polanyi, a prominent physical chemist and long-term acquaintance of Einstein's, also published a second statement, "approved for publication by Einstein early in 1954": "The Michelson-Morley experiment had a negligible effect on the discovery of relativity."[172]

At about the same time the pious duty fell on Einstein to acknowledge a debt to one of his foremost heroes and mentors. This was done in the short essay *H. A. Lorentz als Schöpfer und als Persönlichkeit,* dated 27 February 1953 and composed for delivery at Leyden that year during the commemoration of the centenary of Lorentz's birth. After a fine tribute to Lorentz's leadership and eminence in physics at the turn of the century, Einstein gives a (partial) list of chief hypotheses on which Lorentz had based his reconstruction of electrodynamics, adding, "It is a work of such consistency, lucidity, and beauty as has only rarely been attained in an empirical science."

But this empirically based, constructive theory has its limits, and, in describing them, Einstein points out two quite different features of electrodynamics around the turn of the century which, he implies, set the stage for Einsteinian relativity theory. One of these is primarily aesthetic:

> To him [Lorentz], Maxwell's equations in empty space held only for a particular coordinate system distinguished from all other coordinate systems by its state of rest. This was a truly paradoxical situation because the theory seemed to restrict the inertial system more strongly than did classical mechanics. This circumstance, which from the empirical point of view appeared completely unmotivated, was bound to lead to the theory of special relativity.[173]

This remark is entirely consistent with the long tradition that the primary impetus for Einstein was the essential requirement of finding symmetry and universality in the operations of nature.

The other problem with Lorentz's theory to which Einstein draws attention in the essay concerns the experimental side: the inability of the theory to encompass all the relevant phenomena in an elegant manner. In the half-paragraph devoted to this failure, Einstein sketches a severely abbreviated version of the state of affairs around the turn of the century, far less adequate than Lorentz's own confession. Instead of listing all the awkward experiments treated in Lorentz's VERSUCH of 1895 or the larger number available by 1905, Einstein mentions only the one for which Lorentz, with FitzGerald, had fashioned the famous saving hypothesis of contraction, namely the Michelson experiment; and then Einstein adds two more sentences, more obscure than most others we have now read on this topic and seemingly more at variance with them:

> The only phenomenon whose explanation in this manner did not succeed fully—that is, without additional assumptions—was the famous Michelson-Morley experiment. That this experiment led [or brought, or guided; *hinführte*]

to the special relativity theory would have been inconceivable without the localization of the electromagnetic field in empty space. Indeed, the essential step was in any case the tracing back [of the phenomenon?] to Maxwell's equations in empty space or—as one called it then—in the aether.[174]

The middle sentence is, of course, of particular interest to us here, but its doubly negative and rather obscure construction gives us no warrant to change it to some other statement such as "The experiment led me to the special relativity theory." During nearly fifty years of responding on this question, Einstein had shown no hesitation to use the first person singular when he wished to; whenever he did so, he coupled the reminiscence of his own progress and the experiment by saying at most, as in the reply to Jaffe in 1942, the experiment "was of considerable influence upon my work insofar as it strengthened my conviction concerning the validity of the principle of the special theory of relativity."

It seems to me closest to Einstein's intention that in the passage for Lorentz's memorial celebration he wanted to provide a brief indication of an experimental counterpart to the aesthetic-theoretical guidelines we noted a little earlier, and that he hit upon the most frequently used experimental illustration for the purpose of all didactic expositions (including Einstein's own[175]). We learned earlier that when Einstein wrote in the passive voice in answering questions or obligations, he spoke of the importance the Michelson experiment had for the further development and acceptance of the theory by *other* physicists. When he mentioned the influence of the experiment on himself explicitly and in first person, he said the effect was "negligible," "indirect," "rather indirect," "not decisive," or at most "considerable" in the limited sense of the reply to Jaffe. We have, therefore, learned to distinguish between Einstein's evaluations of the effects on public science and on private science.

These considerations lead directly to the last document in this case, the last of the replies of Einstein, given about a year before his death. On 2 February 1954, F. G. Davenport of the Department of History of Monmouth College, Illinois, wrote to Einstein that in connection with a study on "Scientific Interests in Illinois 1865–1900" he was looking into evidence that Michelson had "influenced your thinking and perhaps helped you to work out your theory of relativity." Not being a scientist, he asked for "a brief statement in non-technical terms, indicating how Michelson helped to pave the way, if he did, for your theory."

In a hitherto unpublished letter,[176] Einstein answered very soon after receipt, on 9 February 1954. Perhaps having benefited from the repeated questions during the previous few years and by having pondered again over his answers, he now seemed very willing to respond in detail to a stranger and to offer the letter for publication, as well as to invite continued correspondence. It is a thoughtfully composed reply which we can let stand by itself as the summation of what we have learned from other documents—including the need to differentiate sharply between the effect of the experiment on the development of physics and its effect on the development of Einstein's own thought, between the beauty of the immortal experiment and its subsidiary place in theory, between the statements a scientist may make in direct response to repeated questions and the statements he volunteers (in the latter case Einstein preferred not to speak about specific experiments except Fizeau's and the aberration experiments), and between the large interest the whole question has held for many people and the small interest it seemed to have held for Einstein.

Dear Mr. Davenport:

Before Michelson's work it was already known that within the limits of the precision of the experiments there was no influence of the state of motion of the coordinating system on the phenomena, resp. their laws. H. A. Lorentz has shown that this can be understood on the basis of his formulation of Maxwell's theory for all cases where the second power of the velocity of the system could be neglected (effects of the first order).

According to the status of the theory, it was, however, natural to expect that this independence would not hold for effects of second and higher orders. To have shown that such expected effect of the second order was de facto absent in one decisive case was Michelson's greatest merit. This work of Michelson, equally great through the bold and clear formulation of the problem as through the ingenious way by which he reached the very great required precision of measurement, is his immortal contribution to scientific knowledge. This contribution was a new strong argument for the non-existence of "absolute motion," resp. the principle of special relativity which, since Newton, was never doubted in Mechanics but *seemed* incompatible with electro-dynamics.

In my own development Michelson's result has not had a considerable influence. I even do not remember if I knew of it at all when I wrote my first paper on the subject (1905). The explanation is that I was, for general reasons, firmly convinced how this could be reconciled with our knowledge of electro-dynamics. One can therefore understand why in my personal struggle Michelson's experiment played no role or at least no decisive role.

You have my permission to quote this letter. I am also willing to give you further explanations if required.

Sincerely yours,
ALBERT EINSTEIN

But if he truly felt this way, why did Einstein not make any voluntary statements to stop the myth he must have seen spreading all around him? Many opportunities offered themselves. Why did he wait for interrogators such as Jaffe, Balazs, Shankland, Polanyi, and Davenport to raise the issue before denying what almost everyone else seemed to be affirming? The answer, of course, can be found in Einstein's general pattern of response. It would have been most uncharacteristic for Einstein to take pen in hand to attack a myth of this kind. Even on the purely scientific issues he only very rarely published a correction of (not to speak of an attack on) the many erroneous interpretations of his work, and it is even less conceivable that he would, of his own will, publish anything that would seem to increase the degree of originality of his own work or imply a diminished status of another scientist. It is also relevant that he tolerated even the most vicious printed attacks on his work and person by Nazi scientists (and nonscientists) with astonishing humor.

In fact, from the point of view of the historian, Einstein's characteristic fault was to be too tolerant. A notable episode will illustrate the point. When Sir Edmund Whittaker was composing his second volume (1953) of A HISTORY OF THE THEORIES OF AETHER AND ELECTRICITY,[177] in which he explicitly ascribes the chief original work on special relativity to Lorentz and Poincaré, Einstein's old friend Max Born, then in Edinburgh, saw Whittaker's manuscript. Having seen the rise of relativity practically from its beginning, Born was astonished and somewhat angry about this misleading version. He wrote to Einstein in dismay that Whittaker had persisted in the plan to publish his version despite the contrary evidence which Born had submitted (including translations from the German originals of some relevant articles he had prepared for Whittaker). Though Einstein was probably somewhat wounded, he wrote on 12 October 1953 to reassure Born:

Don't give any thought to your friend's book. Everyone behaves as seems to him right, or, expressed in deterministic language, as he has to. If he convinces others, that's their problem. At any rate, I found satisfaction in my efforts, and I don't think it is sensible business to defend my few results as "property," like an old miser who has laboriously gathered a few coins for himself. I don't think ill of him. . . . And I don't have to read the thing.[178]

X. Concluding Remarks

Historians are quite used to the large discrepancy we find here between documentable history of science on the one hand and, on the other, the popular history found in texts and in the writings of eminent scientists and some philosophical analysts. Putting together all extant first-hand documents, including the 1905 paper, the Shankland interviews, the *Autobiographical Notes*, and the letters, we see they fit together and tell a story for which the secondary sources had not prepared us. It is a scenario of which we cannot, in the nature of the case, be absolutely certain, but one which is highly probable. Indeed, the role of the Michelson experiment in the genesis of Einstein's theory appears to have been so small and indirect that one may speculate that it would have made no difference to Einstein's work if the experiment had never been made at all. To be sure, the public acceptance of the theory might well have been delayed, but through his reading in Lorentz's works Einstein in 1905 had available enough other "unsuccessful attempts to discover any motion of the earth relatively to the 'light medium,' " and enough other evidences of what Lorentz himself called a "clumsiness" in the then-current theory.

This special case may yield some more widely applicable conclusions. Above all, it forces us to ask anew what are the most appropriate styles and functions of historical scholarship today. On this point, Einstein's own opinion is illuminating. Shankland had asked Einstein during their first conversation in 1950 whether "he felt that writing out the history of the Michelson-Morley experiment would be worthwhile":

He said, "Yes, by all means, but you must write it as Mach wrote his SCIENCE OF MECHANICS." Then he gave me his ideas on historical writing of science. "Nearly all historians of science are philologists and do not comprehend what physicists were aiming at, how they thought and wrestled with their problems. Even most of the work on Galileo is poorly done." A means of writing must be found which conveys the thought processes that lead to discoveries. Physicists have been of little help in this, because most of them have no "historical sense." Mach's SCIENCE OF MECHANICS, however, he considered one of the truly great books and a model for scientific historical writing. He said, "Mach did not *know* the real facts of how the early workers considered their problems," but Einstein felt that Mach had sufficient insight so that what he said is very likely correct anyway. The struggle with their problems, their trying everything to find a solution which came at last often by very indirect means, is the correct picture.[179]

In discussing the approach of "nearly all historians" (perhaps somewhat too brusquely) Einstein accentuates the need to deal with the private phase of scientific effort—how a man thinks and wrestles with a problem. In discussing the physicists themselves (perhaps also too brusquely) Einstein accentuates the need for a particular kind of historical sense, one that largely intuits how a scientist may have proceeded, even in the absence of "the real facts" about the creative phase. It is a challenging statement, a recommendation to adopt for research in the history of science a lesson Einstein had learned from his research in physics: *just as in doing physics itself*, Einstein here advises the historian of science to leap across the unavoidable gap between the necessarily too limited "facts" and the mental construct that must be formed to handle the facts. And in such an historical study, *as in physics itself*, the solution comes often "by very indirect means"; the best outcome that can be hoped for is not certainty but only a good probability of being "correct anyway."

One can well agree with this call for new ways of writing about the thought processes that lead to major discoveries, without having to agree at this late date with the particular model of Mach's SCIENCE OF MECHANICS. The most obvious difficulty with following Einstein's advice is of course the unspecifiability of "sufficient insight." Another is that any study of the processes of discovery—that evanescent, partly unconscious, unobserved, unverbalized activity—is by definition going to yield a report with apparently vague and contradictory elements. Yet another is that the invitation to leap courageously may cause the historian to slight even some of the most pertinent and easily available documents. And a fourth trouble is that there are some problems which now seem largely unsolvable by any method and may remain so for a long time: the problem of genius, of reasons for thematic and aesthetic choices, of interaction between private and public science, not to speak of the problem of induction.

Ernst Mach himself would perhaps have objected to Einstein's characterization of his work on the history of science, laudatory though it was intended to be. But Einstein was right nevertheless in ascribing to Mach, and recommending to others, an unconventional method, despite the difficulties and dangers it may pose. For in this way one can at least hope to penetrate beyond the more pedestrian or trivial aspects on an historic case of such magnitude, to recognize more fully the feat of intellectual daring and superb taste that was needed to create the theory.

Of course, experiments are essential for the progress of science. Of course, the chain from a puzzling new experiment to a theoretical scheme that explains it is the more usual process, particularly in the everyday—as Einstein later acknowledged by referring to three specific examples—accomplishments of most scientists. Of course, experiments did influence his developing thought processes while he was struggling with the problem of understanding electrodynamics in a new way, to get at the "heart of the matter." Of course, Michelson's experiment played an (indirect) role in this, if only because Einstein found Lorentz's theory of electrodynamics to be inadequate precisely because "it was leading to an interpretation of the result of the Michelson-Morley experiment which seemed to me artificial," as Einstein wrote in the Michelson centenary message, or as too patently "ad hoc," as he wrote on other occasions.

Yet, the experimenticist fallacy of imposing a logical sequence must be resisted. Not only is it false to the actual historic development of thought processes that may have led to major scientific discoveries; not only might the doctrine inhibit creative work in science if it were taken too seriously; but also, by drawing attention primarily to the externally visible clay that provides factual support and operational usefulness for the developed theory, it does not do adequate justice to the full grandeur of the theory. Nor need we be deterred by the experimenticist's predictable outcry that we have a choice only between his own view at one extreme and private intuition at the other. The basic achievement of Einstein's theory was not to preserve hallowed traditional conceptions or mechanisms; it was not to produce a logically and tightly structured sequence of thoughts; it was not to build on a beautiful and pedagogically persuasive experiment. Rather, the basic achievement of the theory was that at the cost of sacrificing all these, it gave us a new unity in the understanding of nature.

NOTES

1. R. A. Millikan, *Albert Einstein on his Seventieth Birthday*, Reviews of Modern Physics, 21:343–344, 1949; italics in original.

2. Quoted in R. S. Shankland, *Conversations with Albert Einstein*, American Journal of Physics, 31: 47–57, 1963. From time to time, current research papers in physics come back to the Michelson experiment to view it in a new light; e.g., P. P. Phillips, *Is the Graviton a Goldstone Boson?* Physical Review, 146: 966–973, 1966.

It should be noted that in this paper, as in most of the physics literature, the term "Michelson experiment" is used more or less interchangeably with "Michelson experiments," "Michelson-Morley experiments," etc. A useful summary of the whole sequence of closely related experiments by Michelson (1881), by Michelson, together with his colleague Edward W. Morley (published 1887), by Morley and D. C. Miller (1902–1904), etc., will be found in R. S. Shankland *et al.*, *New Analysis of the Interferometer Observations of Dayton C. Miller*, Reviews of Modern Physics, 27: 167–178, 1955; T. W. Chalmers, Historic Researches, London: Morgan Brothers, 1949; and most compactly in W. K. H. Panofsky and Melba Phillips, Classical Electricity and Magnetism (Reading, Massachusetts: Addison-Wesley, 1955), pp. 233–235.

3. R. S. Shankland, *Michelson-Morley Experiment*, American Journal of Physics, 32: 23, 1964. In 1886 (Archives Néerlandaises des Sciences Exactes et Naturelles, 21: 104–176, 1886), Lorentz showed also that Michelson's analysis in 1882 of the action of the interferometer was in error, *i.e.*, that it predicted double the expected fringe shift.

4. *Conference on the Michelson-Morley Experiment*, Astrophysical Journal, 68: 341–402, 1928.

5. Bernard Jaffe, Michelson and the Speed of Light (Garden City, New York: Doubleday & Co., 1960), p. 89.

6. *Ibid.*, p. 90.

7. A. A. Michelson, Light Waves and Their Uses (Chicago: University of Chicago Press, 1903), p. 159.

8. Shankland, *Michelson-Morley Experiment*, p. 32.

9. *Ibid.*

10. R. S. Shankland, *Rayleigh and Michelson*, Isis, 58: 87, 1967.

11. Loyd S. Swenson, Jr., *The Ethereal Aether: A Descriptive History of the Michelson-Morley Aether-Drift Experiments, 1880–1930*, dissertation, Claremont Graduate School, 1962. I am grateful to Dr. Swenson for letting me read his thesis and for several useful discussions. [Dr. Swenson's work has subsequently been rewritten and published as The Ethereal Aether: A History of the Michelson-Morley-Miller Aether-Drift Experiments, 1880–1930, Austin, Texas: University of Texas Press, 1972.]

12. Oliver Lodge, *Aberration Problems*, PHILOSOPHICAL TRANSACTIONS OF THE ROYAL SOCIETY, 184: 753, 1893. As we shall see, two years later Lorentz could cite two other long-known measurements that also were not in accord with the predictions of the ether theory.

13. Swenson, *op. cit.*, p. 205, quoting Michelson, AMERICAN JOURNAL OF SCIENCE, 3: 478, 1897.

14. A. A. Michelson, STUDIES IN OPTICS (Chicago: University of Chicago Press, 1927), p. 155.

15. *Ibid.*, p. 156.

16. Moreover, as was realized by and by, the Lorentz-FitzGerald contraction was by no means sufficient to provide the needed relativistic basis for electromagnetic phenomena. We shall come back to this point in due course.

17. Michelson, STUDIES IN OPTICS, p. 161.

18. *Conference on the Michelson-Morley Experiment, op. cit.*, p. 342.

19. See Swenson, *op. cit.*

20. Hermann Minkowski, *Space and Time*, reprinted in H. A. Lorentz *et al.*, THE PRINCIPLE OF RELATIVITY, trans. W. Perrett and G. B. Jeffery (London: Methuen & Co., 1923), pp. 75–91.

21. Gerald Holton, *On the Origins of the Special Theory of Relativity*, AMERICAN JOURNAL OF PHYSICS, 28:634, 1960, and reprinted here as Essay 5.

22. Max Planck, *Zur Dynamik bewegter Systeme*, SITZUNGSBERICHTE DER AKADEMIE DER WISSENSCHAFTEN IN BERLIN, 29:542, 1907.

23. Wilhelm Wien, ÜBER ELEKTRONEN, 2nd ed. (Leipzig: B. G. Teubner, 1909), p. 32. I thank Stanley Goldberg for bringing this and the previous reference to my attention.

Similar references are also given in the useful paper by K. F. Schaffner, *The Lorentz Electron Theory and Relativity* (in press), whose author I thank for a preprint of his paper, received shortly before completing this study. [A shorter version was published subsequently in AMERICAN JOURNAL OF PHYSICS, 37: 498 ff, 1969.]

Max von Laue, in his text of 1911 on relativity theory, still had to confess that "a really experimental decision between the theory of Lorentz and the theory of relativity is indeed not to be gained; and that the former, in spite of this, has receded into the background is chiefly due to the fact that, close as it comes to the theory of relativity, it still lacks the great simple universal principle, the possession of which lends the theory of relativity from the start an imposing appearance" (DAS RELATIVITÄTSPRINZIP, Braunschweig: Friedrich Vieweg & Son, 1911, pp. 19–20). And we shall soon see how short the list was that Einstein himself gave for the results achieved by the relativity theory up to 1915.

What made special relativity theory at last a widely accepted basic part of physics were developments far from the scope of Einstein's 1905 paper itself —foremost among them the *experimental* successes such as the eclipse expedition of 1919 with its successful test of a prediction of the general theory of relativity, the use of relativistic calculations to explain the fine structure of spectral lines, and the Compton effect. In the meantime the interested public and indeed many physicists had to look for support for relativity theory, particulary in the face of its challenging paradoxes and iconoclastic demands, in the ease with which it explained Michelson's results.

24. R. B. Leighton, PRINCIPLES OF MODERN PHYSICS (New York: McGraw-Hill, 1959), p. 5; italics supplied.

25. "[The Michelson-Morley Experiment] was one of the most remarkable experiments in the nineteenth century. Simple in principle, *the experiment led to a scientific revolution* with far-reaching consequences." Charles Kittel, Walter D. Knight, and Melvin A. Ruderman, MECHANICS: THE BERKELEY PHYSICS COURSE, Volume I (New York: McGraw-Hill, 1965), p. 332, footnote; italics supplied.

"As mentioned above, attempts were made to determine the absolute velocity of the earth through the hypothetical 'ether' that was supposed to pervade all space. The most famous of these experiments is one performed by Michelson and Morley in 1887. It was 18 years later before the negative results of the experiment were finally explained, by Einstein." Richard Feynman, Robert B. Leighton, and Matthew Sands, THE FEYNMAN LECTURES ON PHYSICS, Volume I (Reading, Massachusetts: Addison-Wesley, 1963), p. 15.

"Michelson and Morley found that the speed of the earth through space made no difference in the speed of light relative to them. The inference is clear, either that the earth moves in some way through the ether space more slowly than it moves about the sun, or that all observers must find that their motion through space makes no difference in the speed of light relative to them. *The above inference was clear, at least to Einstein*, who knew of the 'unsuccessful attempts to discover any motion of the earth relative to the "light medium."' " James A. Richards, Jr., Frances W. Sears, M. Russel Wehr, and Mark W. Zemansky, MODERN COLLEGE PHYSICS (Reading, Massachusetts: Addison-Wesley, 1962), p. 769; italics supplied.

"After a period of adjustment and speculation, with varying degrees of success, *this famous experimental result* [of Michelson and Morley] *led to a more far-reaching postulate*, one of the basic pillars of *Einstein's Relativity Theory*." Gerald Holton, INTRODUCTION TO CONCEPTS AND THEORIES IN PHY-

SICAL SCIENCE (Cambridge, Massachusetts: Addison-Wesley, 1952), p. 506; italics supplied.

Though they are very much in the minority, we must not fail to mention that a few textbook authors do not imply a genetic connection between the Michelson experiments and Einstein's relativity theory, and a few even specifically disclaim such a connection. Examples of this minority are R. B. Lindsay and Henry Margenau, FOUNDATIONS OF PHYSICS: A SURVEY OF CERTAIN VIEWS, New York: John Wiley & Sons, 1936; R. A. R. Tricker, THE ASSESSMENT OF SCIENTIFIC SPECULATION, London: Mills & Boon, 1965; A. P. French, SPECIAL RELATIVITY, New York: W. W. Norton & Co., 1968; and Robert Resnick, INTRODUCTION TO SPECIAL RELATIVITY, New York: John Wiley & Sons, 1968.

26. H. P. Robertson, *Postulate versus Observations in the Special Theory of Relativity*, REVIEWS OF MODERN PHYSICS, 21: 378, 1949.

27. Henri L. Bergson, DURATION AND SIMULTANEITY (Originally published as DURÉE ET SIMULTANÉITÉ, Paris: Alcan, 1922) reprinted in translation (Indianapolis: Bobbs-Merrill, 1965), p. 9; italics supplied. We note that here, as in the quotation of von Laue below, too much power is assigned to the conclusions that can be deduced from the experiment concerning the transformation equations. But these passages were of course written before the results of the Kennedy-Thorndike experiment (1932) and the Ives-Stilwell experiment (1938).

Another honest assessment of the usefulness of the approach Bergson outlined is given in Mary B. Hesse, FORCES AND FIELDS, THE CONCEPT OF ACTION AT A DISTANCE IN THE HISTORY OF PHYSICS (London: Nelson & Sons, 1961), p. 226:

"The experimental basis of the special theory of relativity is generally taken to be the Michelson-Morley experiment. This was not the only experiment which found its most convenient explanation in terms of that theory, for there were others concerning the optical and electromagnetic properties of moving bodies which led in the same direction, but since the Michelson-Morley experiment is familiar and comparatively simple, it will be convenient to use it in an analysis of the logical status of the theory."

See also Emile Meyerson, LA DÉDUCTION RELATIVISTE (Paris: Payot, 1925), pp. 111–113, and Ernst Cassirer, *Einstein's Theory of Relativity*, SUBSTANCE AND FUNCTION AND EINSTEIN'S THEORY OF RELATIVITY, trans. William C. Swabey and Marie C. Swabey (Chicago: Open Court Publishing Co., 1923), p. 375.

28. von Laue, DAS RELATIVITÄTSPRINZIP, *op. cit.*, p. 13 (Unless indicated to

the contrary, all translations from German sources have been prepared by the author.)

Another distinguished example is provided by Arthur Holly Compton, a Nobelist ex-colleague of Michelson: "This experiment [Michelson-Morley's] more than anything else was the occasion for the development of the theory of relativity. . . ." Quoted in Marjorie Johnston, ed., THE COSMOS OF ARTHUR HOLLY COMPTON (New York: Alfred A. Knopf, 1967), p. 196, from an essay published originally in 1931.

29. Albert Einstein, ÜBER DIE SPEZIELLE UND DIE ALLGEMEINE RELATIVITÄTSTHEORIE (Braunschweig: Friedrich Vieweg & Son, 1917); republished in later editions and translations, e.g., New York: Crown Publishers, 1961, pp. 52–53.

An earlier discussion along these lines is contained in Einstein's *Über das Relativitätsprinzip und die aus demselben gezogenen Folgerungen*, JAHRBUCH DER RADIOAKTIVITÄT UND ELEKTRONIK, 4:411–462, 1907. The article had originally been commissioned as a review essay by the journal's editor, Johannes Stark. It is explicitly not an historical review; Einstein says, "In what follows the attempt is made to pull together into a coherent whole [*zu einem Ganzen*] the publications which so far have come out of the unifications of H. A. Lorentz's theory and the principle of relativity. . . . In this I follow the publications of H. A. Lorentz [1904] and A. Einstein [1905]." Here again we find a sequence of sentences which can be considered implicit history:

"However, the negative results of the experiment of Michelson and Morley show that in a certain case also an effect of second order is absent, although it should have been noticeable on the basis of the Lorentz theory. It is known that this contradiction between theory and experiment was formally removed through the assumption by Lorentz and FitzGerald, according to which moving bodies experience a certain contraction in the direction of their motion. This assumption, introduced ad hoc, appeared however to be an artificial means to rescue the theory." Einstein continues, still in a neutral and passive voice, to say that to overcome the difficulty it turned out "surprisingly" that all one had to do was to understand the concept of time sharply enough, that is, to recognize that the auxiliary quantity *Ortszeit*, introduced by Lorentz, should be defined as "time" itself. Only the idea of a light ether had to be given up.

30. Einstein, *Relativity Theory*, DIE PHYSIK, ed. Emil Warburg (Leipzig: B. G. Teubner, 1915), pp. 703–713.

31. *Ibid.*, p. 705.

32. *Ibid.* p. 706.

33. *Ibid.*, p. 707.

34. *Ibid.*, pp. 707–708.

35. *Ibid.*, p. 712.

36. *Ibid.*, pp. 712–713.

37. Antony Ruhan, personal communication, draft, p. 138. There are other discussions of this type to be found in Einstein's manuscripts and articles. An interesting glimpse of Einstein's classroom lecture on relativity, with special attention to his use of experiments, is furnished by Einstein's manuscript notebook which he used in preparation for his lecture in Berlin. Entitled *Relativitäts Vorlesungen, Winter 1914–15* (now kept at the Einstein Archive at the Institute for Advanced Study in Princeton), it was composed about the same time as his article in DIE PHYSIK and shows a similar structure. The Fizeau and aberration experiments are mentioned on the first page of the notebook and later are developed, with the characterization "Erfahrungs-Resultat." The Michelson experiment is mentioned next, with the notation: "erwies sich als nicht zutreffend. Verkürzung in Bewegungsrichtung eingeführt. Unbefriedigend, weil Hypothesen ad hoc."

On another pedagogic occasion, in a lecture *On the Theory of Relativity* at King's College, London, in 1921, Einstein said: ". . . I am anxious to draw attention to the fact that this theory is not speculative in origin; it owes its invention entirely to the desire to make physical theory fit observed fact as well as possible. . . . The law of the constant velocity of light in empty space, which has been confirmed by the development of electrodynamics and optics, and the equal legitimacy of all inertial systems (special principle of relativity), which was proved in a particularly incisive manner by Michelson's famous experiment, between them made it necessary, to begin with, that the concept of time should be made relative, each inertial system being given its own special time." Albert Einstein, IDEAS AND OPINIONS, trans. Sonja Bargmann (New York: Crown Publishers, 1954), p. 246.

This talk was delivered toward the end of the period in which Einstein still used the language of the empiricist interpretation of science and just before the posthumous publication of Mach's attack on relativity theory, which formed a kind of turning point in Einstein's writings. I have discussed this change, including the position taken in his talk in *Mach, Einstein, and the Search for Reality*, DAEDALUS, Spring, 1968, and reprinted here as Essay 8.

38. Joseph Petzoldt, *Positivistische Philosophie*, ZEITSCHRIFT FÜR POSITIVISTISCHE PHILOSOPHIE, 1:3–4, 1913.

39. *Ibid.*, p. 8; italics added.

40. Joseph Petzoldt, *Die Relativitätstheorie der Physik*, ZEITSCHRIFT FÜR POSITIVISTISCHE PHILOSOPHIE, 2:10–11, 1914.

41. The official document, dated 10 December 1922, of the Royal Swedish Academy of Sciences, now kept at the Einstein Archive, stated specifically that the Academy, "independent of the value that may be credited to the relativity and gravitation theory after eventual confirmation, bestows the prize . . . to Albert Einstein, being most highly deserving in the field of theoretical physics, particularly his discovery of the law pertaining to the photoelectric effect." See also LES PRIX NOBEL EN 1921–1922: 64–71, 1923.

42. THE AUTOBIOGRAPHY OF ROBERT A. MILLIKAN (New York: Prentice-Hall, 1950), p. 219.

43. Gaston Bachelard, *The Philosophical Dialectic of the Concepts of Relativity*, in P. A. Schilpp, ed., ALBERT EINSTEIN: PHILOSOPHER-SCIENTIST (Evanston, Illinois: The Library of Living Philosophers, 1949), pp. 566–568.

44. Hans Reichenbach, FROM COPERNICUS TO EINSTEIN, trans. Ralph B. Winn (New York: Philosophical Library, 1942), p. 51.

45. Hans Reichenbach, THE PHILOSOPHY OF SPACE AND TIME, trans. Maria Reichenbach and John Freund (New York: Dover Publications, 1957), p. 195; (a translation of DIE PHILOSOPHIE DER RAUM ZEIT LEHRE [Berlin: Walter de Gruyter, 1928.]) In a footnote, Reichenbach adds that the assertion "does not follow from this experiment alone," but no others are indicated.

Reichenbach and his followers are forced to give high prominence to the Michelson experiment in the supposed development of relativity theory, partly because of the claim that the experiment logically is not dependent on the theory:

"The opinion has been expressed that the contraction of one arm of the apparatus is an 'ad hoc hypothesis,' while Einstein's hypothesis [that both arms are equally long in every inertial system] is a natural explanation that is a consequence of the relativity of simultaneity. Both of these explanations are wrong. The relativity of simultaneity has nothing to do with the contraction in Michelson's experiment, and Einstein's theory explains the experiment as little as does that of Lorentz" (pp. 195–196). "It would be mistaken to argue that Einstein's theory gives an explanation of Michelson's experiment, since it does not do so. Michelson's experiment is simply taken over as an axiom" (p. 201). The same point is echoed by Adolf Grünbaum, who writes, "far from explaining the outcome of the Michelson-Morley experiment as a consequence of more fundamental principles, Einstein incorporated its null result as a physical axiom in his light principle." *Logical and Philosophical Foundations of the Special Theory of Relativity*, in Arthur Danto and Sidney Morgenbesser eds., PHILOSOPHY OF SCIENCE (New York: Meridian Books, 1960), p. 419.

46. Reichenbach, *The Philosophical Significance of the Theory of Relativity*, in Schilpp, *op. cit.*, e.g., p. 301.

47. *Ibid.*, pp. 309–310.

48. Albert Einstein, *Remarks Concerning the Essays Brought Together in This Cooperative Volume*, in Schilpp, *op. cit.*, p. 673 (hereinafter referred to as *Remarks*).

49. *Ibid.*, p. 680.

50. *Ibid.*, p. 678. A few pages later, Einstein summarizes his eclectic approach in a memorable passage (pp. 683–684): "The reciprocal relationship of epistemology and science is of noteworthy kind. They are dependent upon each other. Epistemology without contact with science becomes an empty scheme. Science without epistemology is—insofar as it is thinkable at all—primitive and muddled. However, no sooner has the epistemologist, who is seeking a clear system, fought his way through to such a system, than he is inclined to interpret the thought-content of science in the sense of his system and to reject whatever does not fit into his system. The scientist, however, cannot afford to carry his striving for epistemological systematic that far. He accepts gratefully the epistemological conceptual analysis; but the external conditions, which are set for him by the facts of experience, do not permit him to let himself be too much restricted in the construction of his conceptual world by the adherence to an epistemological system. He therefore must appear to the systematic epistemologist as a type of unscrupulous opportunist: he appears as *realist* insofar as he seeks to describe a world independent of the acts of perception; as *idealist* insofar as he looks upon the concepts and theories as the free inventions of the human spirit (not logically derivable from what is empirically given); as *positivist* insofar as he considers his concepts and theories justified only to the extent to which they furnish a logical representation of relations among sensory experiences. He may even appear as *Platonist* or *Pythagorean* insofar as he considers the viewpoint of logical simplicity as an indispensable and effective tool of his research."

51. Reichenbach, *The Philosophical Significance of the Theory of Relativity*, in Schilpp, *op. cit.*, p. 292.

52. Grünbaum, *Logical and Philosophical Foundations of the Special Theory of Relativity*, in Danto and Morgenbesser, *op. cit.*, p. 419.

53. Adolf Grünbaum, PHILOSOPHICAL PROBLEMS OF SPACE AND TIME (New York: Alfred A. Knopf, 1963), p. 381.

54. For details of this turbulent period (1921–1922), see Philipp Frank, EINSTEIN: SEIN LEBEN UND SEINE ZEIT (Munich: Paul List Verlag, 1949), especially Chapters 7 and 8. (The German-language version of the book is preferable to the much less detailed English version.) Right-wing protest manifestos against relativity were being published even as Nazi terrorist activities were

leading up to the attempted *Putsch* of 1923. Einstein's friend Rathenau was among those assassinated; Einstein was told he was in danger, too, and in fact he left Germany for a time.

55. Wilhelm Wien, DIE RELATIVITÄTSTHEORIE VOM STANDPUNKT DER PHYSIK UND ERKENNTNISLEHRE (Berlin: J. A. Barth, 1921), p. 3.

56. *Ibid.*, p. 7.

57. Heinrich Lange, GESCHICHTE DER GRUNDLAGEN DER PHYSIK (Freiburg and Munich: Karl Alber, 1954), Volume I, p. 301.

58. Henry A. Boorse and Lloyd Motz, eds., *The Luminiferous Ether Receives a Mortal Blow*, in THE WORLD OF THE ATOM (New York: Basic Books, 1966), Volume I, p. 373.

59. Jaffe, *op. cit.*, p. 99.

60. *Ibid.*, p. 100. Jaffe also reprints an interesting letter from Einstein which we shall take up at a later point together with Einstein's other direct responses to the same question.

61. *Ibid.*, p. 90.

62. Michelson was America's first Nobel prize scientist (and, in 1907, the only American Nobelist other than Theodore Roosevelt, who had received the 1906 peace prize); the next American physicist so honored was Millikan (1923). There is some evidence that the Nobel award to Michelson had for American science some of the same significance in terms of increased national self-esteem as did Yukawa's Nobel prize in physics (1949) for Japanese science.

63. R. S. Shankland, *Michelson 1852–1931, Expérience de base de la relativité*, in LES INVENTEURS CÉLÈBRES—SCIENCES PHYSIQUES ET APPLICATIONS (Paris: Lucien Mazenod, 1950), pp. 254–255.

64. *Ibid.*, p. 255: "Contre toute attente, l'observation finale, faite en juillet 1887, ne permit de constater aucun déplacement sensible dans les franges d'interférence. Ce résultat surprenant ne fût pas entièrement apprécié à l'époque, mais après: le travail de pionnier de FitzGerald et H. A. Lorentz, Einstein le généralisa dans sa grande théorie de la Relativité restreinte de 1905, et l'expérience Michelson-Morley acquit sa juste place, en tant que l'une des expériences cruciales de l'histoire des sciences. Ses répétitions ultérieures ne firent que confirmer ce résultat et son importance pour la théorie de la Relativité."

65. Shankland, *Conversations with Albert Einstein*, pp. 47–48; italics in the original.

66. *Ibid.*, p. 55.

67. R. S. Shankland, *The Michelson-Morley Experiment*, SCIENTIFIC AMERICAN, 211, No. 5:107, 1964.

68. Shankland, *Michelson-Morley Experiment*, AMERICAN JOURNAL OF PHYSICS, p. 34.

69. Shankland published the English translation that was sent to him by Einstein, and he reports that he also received the German original; in both the author's holograph draft and the carbon copy of the latter (both these versions are now in the Einstein Archive at Princeton) the first two sentences are in English and appear to represent the acceptance by Einstein of a draft submitted for the occasion by Shankland on the basis of the interviews. Because of the significant differences between the German and English versions after the first two sentences, it would be of interest to find who made the English translation of Einstein's German original. For many years Einstein's friends and assistants helped him with English translations, but in this instance, according to Miss Helen Dukas, we may believe it was Einstein's own. She informed me that in his last years he sometimes made the translations himself.

70. Shankland, *Michelson, Morley Experiment*, p. 35. Einstein's original text appears as follows in copies in the Einstein Archive:

"I always think of Michelson as the Artist in Science. His greatest joy seemed to come from the beauty of the experiment itself, and the elegance of the method employed.

"Aber er hat auch ein aussergewöhnliches Verständnis gezeigt für die fundamentalen Rätsel in der Physik. Dies sieht man aus dem Interesse, dass er von Anfang an dem Problem der Abhängigkeit des Lichtes von der Bewegung entgegenbrachte.

"Mein eigenes Nachdenken wurde mehr indirekt durch dass berühmte Michelson-Morley Experiment beeinflusst. Ich erfuhr von diesem durch Lorentz' bahnbrechende Untersuchung über die Elektrodynamik bewegter Körper (1895), von der ich vor Aufstellung der speziellen Relativität Kenntnis hatte. Lorentz' Grundannahme vom ruhenden Aether schien mir gerade deshalb nicht überzeugend, weil sie zu einer Interpretation des Michelson-Morley Experimentes führte, die mir unnatürlich erschien. Mein direkter Weg zur speziellen Relativitäts-Theorie wurde hauptsächlich durch die Überzeugung bestimmt, dass die in einem im Magnetfelde bewegten Leiter induzierte elektromotorische Kraft nichts anderes sei als ein elektrisches Feld. Aber auch das Ergebnis des Fizeau'schen Versuches und das Phänomen der Aberration führten mich.

"Es führt ja kein logischer Weg zur Aufstellung einer Theorie, sondern nur tastendes Konstruieren mit sorgfältiger Berücksichtigung des Thatsachen-Wissens."

71. Shankland, *Conversations with Albert Einstein*, p. 48.

72. *Ibid.*, p. 50.

73. Albert Einstein, *Motiv des Forschens*, in ZU MAX PLANCKS 60. GEBURTSTAG, Karlsruhe: C. F. Müller, 1918; reprinted in MEIN WELTBILD.

74. *On the Method of Theoretical Physics*, The Herbert Spencer Lecture delivered at Oxford, June 10, 1933. It has been reprinted in MEIN WELTBILD and THE WORLD AS I SEE IT.

75. Einstein, *Autobiographical Notes*, in Schilpp, *op. cit.*, p. 89.

76. Jacques Hadamard, AN ESSAY ON THE PSYCHOLOGY OF INVENTION IN THE MATHEMATICAL FIELD (New York: Dover Publications, 1954; originally, Princeton: Princeton University Press, 1945), pp. 142–143.

77. See Holton, *Mach, Einstein, and the Search for Reality*, pp. 636–637; pp. 219–220 in this volume.

78. Einstein, *Remarks*, in Schilpp, *op. cit.*, p. 673.

79. Sir Edmund Whittaker, A HISTORY OF THE THEORIES OF AETHER AND ELECTRICITY, Volume II: THE MODERN THEORIES, 1900–1926 (London: Nelson & Sons, 1953), p. 38. [For a discussion, see the essay referenced in note 21.]

80. Chalmers, *op. cit.*, p. 81; italics in original.

81. Swenson, *The Ethereal Aether*, pp. 280–281 (citing William F. Magie, *The Primary Concepts of Physics*, SCIENCE, 35:287, 1912).

82. *Ibid.*, p. 281.

83. Swenson had also further evidence of Michelson's own dismay: "In reply to a request for his own estimates of his most significant achievement, Michelson said: 'I think most people would say that it was the experiment which started the Einstein theory of relativity. That experiment is the basis of Einstein. But I should think of it only as one of a dozen of my experiments in the interference of light waves,'" pp. 317–318 (citing James O'Donnell Bennett's article on A. A. Michelson at the age of seventy, THE CHICAGO TRIBUNE, Rotogravure Section, 1923, p. 22). In this connection a passage in Shankland's *Conversations* is relevant: "Michelson said to Einstein that he was a little sorry that his own work had started this 'monster.'" *Conversations with Albert Einstein*, p. 56.

84. Translations which follow are taken largely from Albert Einstein, *On the Electrodynamics of Moving Bodies*, pp. 37–65 in H. A. Lorentz, *et al.*, THE PRINCIPLE OF RELATIVITY, trans. W. Perrett and G. B. Jeffery (London: Methuen & Co., 1923; reprint ed., New York: Dover Publications [1951]). Einstein's original article is *Zur Elektrodynamik bewegter Körper*, ANNALEN DER PHYSIK, 17:1905. Although I have corrected the translation at several

points, none of these corrections changes materially the sense of the translations by Perrett and Jeffery. Note that several of the footnotes that appear in reprints of Einstein's original paper were actually added later, supplied by Arnold Sommerfeld.

85. Einstein, *Autobiographical Notes*, in Schilpp, p. 63.

86. Einstein, *On the Electrodynamics of Moving Bodies*, pp. 51–55. By requiring that Maxwell's equations be covariant, Einstein shows that a purely electric (or purely magnetic) force field in one system is experienced as an electromagnetic field in another system moving with respect to the first. Hence, "If a unit electric point charge is in motion in an electromagnetic field, the force acting upon it is equal to the electric force which is present at the locality of the charge, and which we ascertain by transformation of the field to a system of coordinates at rest relative to the electrical charge. . . . We see that the electromotive force plays in the developed theory merely the part of an auxiliary concept, which owes its introduction to the circumstance that electric and magnetic forces do not exist independently of the state of motion of the system of coordinates."

We recall Einstein had said in his Michelson centennial message, "What led me more or less directly to the Special Theory of Relativity was the conviction that the electromotive force acting on a body in motion in a magnetic field was nothing else but an electric field."

87. *Ibid.*, p. 37.

88. Martin J. Klein, *Thermodynamics in Einstein's Thought*, SCIENCE, 157: 513, 1967, writes about this job and quotes Einstein as follows: ". . . contrary to what is sometimes suggested, this job kept him busy—'eight hours of exacting work every day.' "

89. Einstein, *Autobiographical Notes*, in Schilpp, *op. cit.*, p. 53.

90. *Ibid.*, pp. 51–53.

91. Einstein, *On the Electrodynamics of Moving Bodies*, p. 59.

92. Grünbaum, PHILOSOPHICAL PROBLEMS OF SPACE AND TIME, p. 380.

If the history of ether theory itself is a guide, one may predict that on this issue, as on others, few of the protagonists can be expected to change their minds, owing to the stability of the "substratum of thought," the thematic presuppositions.

93. Einstein, *On the Electrodynamics of Moving Bodies*, pp. 64–65.

94. As Einstein stated (cited by Max Born, *Physics and Relativity*, in HELVETICA PHYSICA ACTA, Supplementum IV, 1956, p. 249): "The new feature of it [the work of 1905] was the realization of the fact that the bearing of the

Lorentz-transformations transcended their connection with Maxwell's equations and was concerned with the nature of space and time in general. A further new result was that 'Lorentz invariance' is a general condition for any physical theory. This was for me of particular importance because I had already previously found that Maxwell's theory did not account for the micro-structure of radiation and could therefore have no general validity. . . ."

95. Einstein, *Autobiographical Notes*, in Schilpp, *op. cit.*, p. 23.

96. *Ibid.*, p. 21; italics added.

97. *Ibid.*, pp. 21–23.

98. *Ibid.*, pp. 23–25.

99. *Ibid.*, p. 23.

100. Holton, *On the Origins of the Special Theory of Relativity*, pp. 629–630; pp. 167–169 in this volume.

101. *Ibid.*, p. 629; p. 168 in this volume.

102. Millikan, *Albert Einstein on His Seventieth Birthday*, p. 344.

103. Holton, *Mach, Einstein, and the Search for Reality*, esp. pp. 651–653; pp. 234–237 in this volume.

104. Einstein, *Autobiographical Notes*, in Schilpp, *op. cit.*, p. 53.

105. *Ibid.*, p. 15.

106. Anton Reiser, ALBERT EINSTEIN, A BIOGRAPHICAL PORTRAIT, New York: A. & C. Boni, 1930.

107. Gerald Holton, *Influences on Einstein's Early Work in Relativity Theory*, AMERICAN SCHOLAR, 37:59–79, 1968–1969, and reprinted here as Essay 7.

108. Reiser, *op. cit.*, pp. 52–53.

109. Shankland, *Conversations with Albert Einstein*, p. 48; italics in original.

110. *Ibid.*, p. 55.

111. H. A. Lorentz, *Electromagnetic Phenomena in a System Moving with Any Velocity Smaller Than That of Light*, PROCEEDINGS OF THE ACADEMY OF SCIENCES, AMSTERDAM, 6:809–831, 1904.

112. The fact that most of these papers seem to have been difficult to obtain and were published in Dutch or English may play a role here.

113. Cited by Born, *Physics and Relativity*, p. 248. The book of Lorentz's is VERSUCH EINER THEORIE DER ELEKTRISCHEN UND OPTISCHEN ERSCHEINUNGEN IN BEWEGTEN KÖRPERN, Leiden: E. J. Brill, 1895; reprinted without change by B. G. Teubner, Leipzig, 1906.

As I have shown elsewhere (*Influences on Einstein's Early Work in Relativity Theory*) in almost all of the books from which Einstein may have learned Maxwell's theory, the Michelson experiment was not even mentioned—neither in the VORLESUNGEN ÜBER THEORETISCHE PHYSIK, Leipzig: J. A. Barth, 1897–1907 by Helmholtz, although it was in his own laboratory that Michelsons's first experiment had been begun, nor in Hertz's essays, nor in August Föppl's EINFÜHRUNG IN DIE MAXWELLSCHE THEORIE DER ELEKTRIZITÄT, Leipzig, B. G. Teubner, 1904. In fact, the most significant point about the German treatises is that there is a remarkable paucity of references to actual experimental situations, not only to Michelson's, but also to all other ether-drift experiments. One clear exception is Paul Drude's LEHRBUCH DER OPTIK (Leipzig: S. Hirzel, 1900; issued in 1902 as THE THEORY OF OPTICS, trans. C. R. Mann and R. A. Millikan, preface by A. A. Michelson, London: Longmans Green & Co., 1902.) But the discussion of the optical properties of moving bodies is based entirely on Lorentz's VERSUCH of 1895, which has "developed a complete and elegant theory. It is essentially this theory which is here presented" (p. 457, English translation). Thus in discussing an explanation of the Michelson experiment in terms of the contraction hypothesis, Drude copies almost word for word the corresponding sentences from Lorentz's VERSUCH.

114. A brief draft of the next few pages was presented in the PROCEEDINGS OF THE INTERNATIONAL CONFERENCE ON RELATIVISTIC THEORIES OF GRAVITATION, (London: 1965), Volume I, pp. 14–18 (mimeographed edition).

115. H. A. Lorentz, *et al.*, DAS RELATIVITÄTSPRINZIP, Leipzig and Berlin: B. G. Teubner, 1913. The collection has been often reprinted and exists in English translation as THE PRINCIPLE OF RELATIVITY.

116. Lorentz, VERSUCH EINER THEORIE DER ELEKTRISCHEN UND OPTISCHEN ERSCHEINUNGEN IN BEWEGTEN KÖRPERN, p. 123.

117. "Wozu freilich kein Grund vorliegt." *Ibid.*, p. 124.

118. *Ibid.*, pp. 2, 127. It turned out that Fizeau's experimental results were erroneous, as Lorentz had in effect begun to suspect.

119. Lorentz, THE PRINCIPLE OF RELATIVITY, pp. 22, 29.

120. Lorentz, *et al.*, DAS RELATIVITÄTSPRINZIP, p. 10.

121. H. A. Lorentz, THE THEORY OF ELECTRONS (New York: G. E. Stechert, 1909; 2nd ed. [1915]), p. 230.

122. Professor Charles Kittel has pointed out to me that Joseph Larmor's book AETHER AND MATTER, Cambridge: Cambridge University Press, 1900, which neither Einstein nor Lorentz cites, sets forth the complete and exact Lorentz transformation equations for x, t, E, and B. Larmor claimed it only to the order of $(v/c)^2$, which is all he needed and used to explain the null re-

sult of the Michelson-Morley experiment; but since he was trying to make $x^2 - c^2t^2$ invariant, as we do today, Larmor got the whole transformation exactly.

One of Larmor's erstwhile students, Ebenezer Cunningham, in a letter of 14 December 1963 to Kittel, explained that he did not notice the exactness of the transformation in 1903–1904, but thought, like Lorentz, he had it only to second order. Therefore he spoke of it as the Larmor-Lorentz transformation. In 1904 he moved from Cambridge to Liverpool, discovered there the exactness, and wrote to Larmor, who replied briefly he knew it though he had not referred to it in publication or lecture. Cunningham adds simply: "Larmor did not seem at all enthusiastic about the idea that an algebraic transformation happened to be exact." One is reminded of Lorentz's disinclination to assign a more general meaning to the concept of *Ortszeit.* It is an interesting sidelight on the limitation which can be put on a theory if it is seen as serving immediate purposes.

123. Lorentz, THE THEORY OF ELECTRONS, (1909), p. 230. Lorentz and Einstein came to be close friends, each admiring the scientific contributions of the other despite their fundamentally different approaches to electrodynamics.

124. *Ibid.* (1915), p. 321.

125. Einstein, ÜBER DIE SPEZIELLE UND DIE ALLGEMEINE RELATIVITÄTSTHEORIE, p. 28.

126. Henri Poincaré, CONGRÈS DE PHYSIQUE DE 1900, 1:22, 1900.

127. Lorentz, THE PRINCIPLE OF RELATIVITY, pp. 12–13. An example of Lorentz's sensitivity to this issue was shown in *Deux Mémoires de Henri Poincaré sur la physique mathématique*, ACTA MATHEMATICA, 38:293–308, 1921; reprinted at the end of Volume IX of Poincaré's ŒUVRES (Paris: Gauthier-Villars, 1954), in which Lorentz gives an assessment of the contribution of Poincaré to physics, particularly to relativity theory and quantum theory. And on that occasion Lorentz allowed himself a word of regret about his own method of procedure (p. 684): "In order to explain Michelson's 1881 experiment, the hypothesis of an immobile ether was not sufficient. I was obliged to make a new supposition which had the effect of admitting that the translation of a body through the ether produced a slight contraction of the body in the direction of motion. This hypothesis was the only one possible. It had been imagined by FitzGerald, and it found acceptance by Poincaré, but Poincaré nevertheless did not hide how little satisfaction was given to him by theories in which one multiplies special hypotheses invented for particular phenomena. This criticism was for me more reason to search for a general theory. . . ."

For more on Poincaré's dissatisfaction with the hypotheses then needed, see

Charles Scribner, Jr., *Henri Poincaré and the Principle of Relativity*, AMERICAN JOURNAL OF PHYSICS, 32: 672–678, 1964, and Stanley Goldberg, *Henri Poincaré and Einstein's Theory of Relativity*, AMERICAN JOURNAL OF PHYSICS, 35:934–944, 1967.

128. Lorentz, THE PRINCIPLE OF RELATIVITY, p. 13.

129. Holton, *On the Origins of the Special Theory of Relativity*, p. 630; p. 170 in this volume.

130. See Gerald Holton, *On the Thematic Analysis of Science: The Case of Poincaré and Relativity*, MÉLANGES ALEXANDRE KOYRÉ (Paris: Hermann, 1964), pp. 257–258.

131. Herbert Butterfield, THE ORIGINS OF MODERN SCIENCE: 1300–1800 (New York: Macmillan Co., 1958), pp. 29–30.

132. Nor did others think that the contraction hypothesis might not have other uses also. Several, such as the application to the Trouton-Noble experiment, are well known. Others are not. Thus, when Fritz Hasenöhrl found that these computations yielded a net increase in the temperature of a radiation-filled cavity carried through a closed cycle of velocity under adiabatic conditions he noted that "our contradiction [with the Second Law of Thermodynamics] is solved when the density of the true radiation does not remain constant. . . . The simple assumption is that perhaps the dimensions" are changed by the factor $(1 - \frac{1}{2} \beta^2)$ in "complete agreement with the assumption of Lorentz and Fitzgerald." *Zur Theorie der Strahlungen in bewegten Körpern*, ANNALEN DER PHYSIK, 15:369, 1904. Hasenöhrl also made other uses of the Lorentz-FitzGerald contraction, for example in SITZUNGSBERICHTE DER AKADEMIE DER WISSENSCHAFTEN IN WIEN, 116:1391, 1907, and 117:207, 1908. I thank Stanley Goldberg for bringing these passages to my attention.

133. See, for example, the letter of Heisenberg to Pauli, in which he vented his feelings about Schrödinger's approach to quantum mechanics: "The more I ponder the physical part of Schrödinger's theory, the more disgusting [*desto abscheulicher*] it appears to me." Schrödinger was no less candid about his feelings concerning Heisenberg's theory when he wrote: "I was frightened away [*abgeschrecht*], if not repelled [*abgestossen*], by what appeared to me a rather difficult method of transcendental algebra. . . ." *Cf.* Max Jammer, THE CONCEPTUAL DEVELOPMENT OF QUANTUM MECHANICS (New York: McGraw-Hill, 1966), p. 272. Similar criteria guide the personal decision by which a scientist chooses the more probable theory either when the evidence for and against two theories seems equally balanced or when in fact the evidence seems to be against the theory that is nevertheless preferred. A case of the first kind applied to Galileo, who decided in the DIALOGUES CONCERNING THE TWO

Chief World Systems that "it is much more probable that the diurnal motion belongs to the earth alone than to the rest of the universe excepting the earth." A case of the second kind was Einstein's response to an apparently authoritative experimental disproof of his 1905 paper (discussed in the essay referenced in note 103 above).

134. I have discussed some differences in the meanings of S_1 and S_2 in *On the Duality and Growth of Physical Science*, American Scientist, 41:89–99, 1953, and reprinted here as Essay 11.

135. H. A. Lorentz, *The Relative Motion of the Earth and the Aether,* Verhandelingen der Konink. Akademie van Wetenschappen, Amsterdam, 1:74, 1892; also in H. A. Lorentz, Collected Papers (The Hague: Martinus Nijhoff, 1937), Volume IV, pp. 219–223; italics added. Some of these and similar quotations are found in Alfred M. Bork, *The "FitzGerald" Contraction,* Isis, 57:199–207, 1966. See also the earlier discussion centering on notes 116 and 117 above.

136. Draft copy in Algemeen Rijksarchief, The Hague, published by Stephen G. Brush, in *Note on the History of the FitzGerald-Lorentz Contraction,* Isis, 58:231, 1967; italics added.

137. Oliver Lodge, *On the Present State of Knowledge of the Connection between Ether and Matter: An Historical Summary,* Nature, 46:164–165, 1892; italics added.

138. Bork, *The "FitzGerald" Contraction, loc. cit.*

139. The story is told briefly in Jammer, *op. cit.*, pp. 149–150, and at greater length in the sources he cites there.

140. C. G. Hempel, Philosophy of Natural Science (Englewood Cliffs, New Jersey: Prentice-Hall, 1966), p. 30.

141. Adolf Grünbaum, *The Bearing of Philosophy on the History of Science,* Science, 143:1406, 1410, 1412, 1964.

142. C. C. Gillispie, The Edge of Objectivity (Princeton, New Jersey: Princeton University Press, 1960), p. 516.

143. That the covariance of Maxwell's equations rather than the explanation of ether experiments was the primary motivation for the Lorentz transformation equations is not always granted; e.g., S. J. Prokhovnik in The Logic of Special Relativity (Cambridge: Cambridge University Press, 1967), p. 6, writes: "However, their [Lorentz's and Poincaré's] manner of saving the aether concept had a certain artificial character. Their transformation was devised solely to explain a null effect associated with an undetectable medium. It was the shadow of a phantom of zero dimensions."

144. Brush, *op. cit.*

145. G. F. FitzGerald, *The Ether and the Earth's Atmosphere*, SCIENCE, 13:390, 1889.

146. Einstein, *Autobiographical Notes*, in Schilpp, *op. cit.*, p. 23.

147. Karl Popper, THE LOGIC OF SCIENTIFIC DISCOVERY (New York: Basic Books, 1959), pp. 82–83.

148. Max Wertheimer, PRODUCTIVE THINKING (New York: Harper & Brothers, 1945), pp. 173–174. This work has to be used cautiously.

149. Grünbaum, PHILOSOPHICAL PROBLEMS OF SPACE AND TIME, pp. 380–381; Grünbaum, *The Bearing of Philosophy on the History of Science.*

150. It is in this spirit that Planck, in *Zur Dynamik bewegter Systeme*, SITZUNGSBERICHTE DER AKADEMIE DER WISSENSCHAFTEN IN WIEN, 29:542–570, 1907, argued that one should accept the principle of relativity, since there was nothing as yet which forced one *not* to accept it as exact. And it is in the same sense that Einstein's first criterion for a good theory asked that "the theory must not *contradict* empirical facts."

A similar point has been correctly raised in several publications by Michael Polanyi, e.g., in Herbert Feigl and Grover Maxwell, eds., CURRENT ISSUES IN THE PHILOSOPHY OF SCIENCE (New York: Holt, Rinehart & Winston, 1961), pp. 53–55; and also by Stillman Drake, e.g., in *The Scientific Personality of Galileo*, GALILEO STUDIES (Ann Arbor, Mich.: University of Michigan Press, 1970) pp. 63–78.

"In recent years, I have read many scholarly discussions—I might better say 'scholastic' discussions—on the question whether Galileo had any right, as a scientist, to conclude in favor of Copernicus on the basis of the evidence in his possession. . . . The crucial point is not whether Galileo had or did not have ocular evidence decisively in favor of Copernicus [by later evidence]; it is how he behaved when he considered that he did have such evidence."

There are, of course, useful discussions of ad hoc hypotheses by philosophers, e.g., by Mary B. Hesse, *op. cit.*, pp. 226–235. She also raises the significant point, not discussed above, that the contraction hypothesis was "somewhat ad hoc because it entailed that motion in the aether is in principle unobservable" (p. 228).

We may take this occasion to remark that an ad hoc hypothesis, in particular a poor one, leaves the feeling that the operations of nature are constricted or restricted by arbitrary human intervention. On the other hand, a large-scale generalization leaves the feeling that it *expands* the realm of application and shows where the "natural" limits lie: e.g., the first principle of relativity generalizes the equality of inertial systems from mechanics to all of physics. In the same way other major scientific advances are characterized by the positing of an hypothesis that universalizes a limited situation: Newton's proposition of universal gravitation, Galileo's extension of terrestrial physics to celestial phenomena, Maxwell's generalization that abolished divisions between electric, magnetic, and optical phenomena, and others that can be readily supplied from such work as that of Helmholtz, Darwin, and Freud. An early philosophical warrant for this way of hypothesizing was given in Newton's Third Rule of Reasoning.

151. Einstein, *Remarks*, in Schilpp, *op. cit.*, p. 684.

152. Reichenbach, *The Philosophical Significance of the Theory of Relativity*, in Schilpp, *op. cit.*, p. 292; italics supplied. It is at this point that Reichenbach adds:

"The philosopher of science is not much interested in the thought processes which lead to scientific discoveries; he looks for a logical analysis of the completed theory, including the relationships establishing its validity. That is, *he is not interested in the context of discovery*, but in the context of justification."

To be sure, scientists on their own side sometimes appear to have mirror-symmetrical blind spots or lack in interest or preparation. But this need not end in a stalemate. On the contrary, one may hope that further investigations of the status and meaning of ad hocness can benefit from teamwork of the kind that has been so fruitful in scientific research generally—a collaboration among historians, philosophers, scientists, and other scholars who share the conviction that a problem is worth attacking and who are willing to pool their diverse strengths to work on it together.

153. Michelson, STUDIES IN OPTICS, p. 156.

154. Cf. Albert Einstein, *Time, Space and Gravitation* (1948), OUT OF MY LATER YEARS (London: Thames and Hudson, 1950), pp. 54–58. It was discussed in Holton, *Mach, Einstein, and the Search for Reality*, pp. 647, 667; pp. 230 and 252 in this volume.

155. Shankland, *Conversations with Albert Einstein*, pp. 53–54; italics added. An experiment along these lines was devised later and gave a null result, as Einstein had predicted. In the same interview Einstein deplored the current state of nuclear theory, but again added the same type of caution: "He felt that just the multiplication of facts and experimental data in nuclear physics would not clarify the situation or lead to a final correct theory. This is in marked contrast to the prevalent view that experimental facts will ultimately reveal regularity and thus give the hints that will lead to a theoretical solution. He disagreed completely with this view . . ." (p. 54).

156. Copies of correspondence in the Einstein Archive. Miss Dukas reports

that the famous remark "Raffiniert ist der Herr Gott, aber boshaft ist Er nicht" [God is subtle, but not malicious] was made by Einstein at a reception in 1921 after a lecture in Princeton when Einstein was asked for his views about Miller's experiment of 1921 that was reported to have yielded a positive result for the ether drift at higher altitudes.

157. R. S. Shankland, S. W. McCuskey, F. C. Leone, and Gustav Kuerti, *New Analysis of the Interferometer Observations of Dayton C. Miller,* REVIEWS OF MODERN PHYSICS, 27:167–178, 1955.

158. Shankland, *Conversations with Albert Einstein*, p. 52, footnote 20. There is a significant irony in this story. Shankland's group thought initially that the most likely solution to the puzzle of Miller's results lay elsewhere and invested three years' work in this way. Shankland writes, "it was not until early 1954, after the complete analysis of variance results were available that we were convinced that the periodic effects found by D. C. Miller were not due to statistical fluctuations or to his method of analysis. Only then did we plunge deeply into the study of the temperature effects to find the real cause of Miller's result" (p. 51, footnote 19).

A second irony is that another genial scientist, long before Einstein, had a similar intuitive response about this possible source of error in the Michelson-type experiment. In a recently discovered letter, Michelson, writing to Simon Newcomb, from his visit in Berlin on 22 November 1880, about his plans for his very first interferometer experiment, had reported the response of the august head of the laboratory, Hermann von Helmholtz: "I had quite a long conversation with Dr. Helmholtz concerning my proposed method for finding the motion of the earth relative to the ether, and he said he could see no objection to it, except the difficulty of keeping a constant temperature." The letter was first given in Swenson's thesis, *The Ethereal Aether*; see also footnote 11 in Shankland, *Michelson-Morley Experiment*, p. 19.

Yet another irony lies in Miller's own interpretation of Einstein's interest in and response to his work. In a letter from Miller to T. C. Mendenhall of 2 June 1921, Miller wrote, "Last week Prof. Einstein visited me and spent an hour and a half in talking over the ether-drift experiments. I found him exceedingly pleasant and not at all insistent on the theory of relativity, but apparently more interested in the results of the experiments than in anything else and quite willing to accept the results whether for or against the theory. At least he was sincere enough and cordial enough to leave this impression." (From a letter in the Archive of the Center of History and Philosophy of Physics, American Institute of Physics, New York. I thank Dr. Charles Weiner, its Director, for communicating the letter to me.) One plausible explanation of Miller's impression may be contained in Shankland's report on a question he asked Einstein about that visit: "I referred to Einstein's visit to D. C.

Miller at Case in 1921. . . . He told me that when he came to the United States that year he did not know a word of English. On the trip he picked up some by ear." Shankland, *Conversations with Albert Einstein*, p. 50.

159. Shankland, *Conversations with Albert Einstein*, p. 49.

160. *Conference on the Michelson-Morley Experiment*, p. 344.

161. Shankland, *Conversations with Albert Einstein*, p. 57.

162. *Ibid.*, p. 56.

163. The proceedings were published as *Professor Einstein at the California Institute of Technology*, SCIENCE, 73:375–381, 1931. All quotations below are from this source unless otherwise identified.

164. Jaffe, *op. cit.*, pp. 167–168.

165. *Ibid.*, p. 101.

166. PROCEEDINGS OF THE AMERICAN PHILOSOPHICAL SOCIETY, 93:544–545, 1949.

167. The English translations published by SCIENCE (*Professor Einstein at the California Institute of Technology*, p. 379) and PROCEEDINGS OF THE AMERICAN PHILOSOPHICAL SOCIETY (1949) differ only slightly. But for the sake of completeness I shall give here a more faithful translation of the first part of the German-language text:

"Dear Friends! From far away I have come to you, not to strangers but to men who for many years have been faithful comrades in my work. You, my honored Herr Michelson, began when I was only a small boy, not even a meter high. It was you who led the physicists into new paths, and through your marvelous experimental labors prepared for the development of the relativity theory. You uncovered a dangerous weakness in the ether theory of light as it then existed, and stimulated the thoughts of H. A. Lorentz and FitzGerald from which the special theory of relativity emerged. This latter, in turn, led the way to the general theory of relativity and to the theory of gravitation. Without your work this theory would today be scarcely more than an interesting speculation; your verifications furnished the real [or realistic] basis for this theory. Campbell's determination of the deflection of light in the case of light rays passing by the sun, St. John's determination of the red shift of spectral lines through the gravitational potential that exists at the sun's surface, . . . belong to the best supports of the General Relativity Theory."

168. As he wrote to his friend Conrad Habicht about the relativity paper: "The fourth work lies at hand in draft form, and is an electrodynamics of moving bodies, making use of a *modification* of the theory of space and time." In Carl Seelig, ALBERT EINSTEIN (Zurich: Europa Verlag, 1954), p. 89; see also p. 97; italics supplied.

169. Einstein, *Gedenkworte auf Albert A. Michelson,* ZEITSCHRIFT FÜR ANGEWANDTE CHEMIE, 44:685, 1931.

170. Bernard Jaffe, MEN OF SCIENCE IN AMERICA (New York: Simon & Schuster, 1944), p. 372. Also reprinted in Jaffe, MICHELSON AND THE SPEED OF LIGHT, pp. 100–101. The same implication without further illumination is found in the well-known and slightly breathless book by the psychologist Max Wertheimer, PRODUCTIVE THINKING. Wertheimer reports that from 1916 on he spent "hours and hours" with Einstein, "to hear from him the story of the dramatic developments which culminated in the theory of relativity" (p. 168). "When Einstein read about these crucial experiments made by physicists, and the finest ones made by Michelson, their results were no surprise to him, although very important and decisive. They seemed to confirm rather than to undermine his ideas" (p. 172).

171. N. L. Balazs, quoted in Michael Polanyi, PERSONAL KNOWLEDGE (Chicago: University of Chicago Press, 1958), p. 11.

172. *Ibid.,* p. 10. Polanyi goes on to use these statements to support his own conclusions: "The usual textbook account of relativity as a theoretical response to the Michelson-Morley experiment is an invention. It is the product of a philosophical prejudice. When Einstein discovered rationality in nature, unaided by any observation that had not been available for at least 50 years before, our positivistic textbooks promptly covered up the scandal by an appropriately embellished account of his discovery" (p. 11).

This remark led to remarkably vituperative attacks upon him from the extreme positivistic school. The ensuing debate was revealing in its own right, but its examination must be delayed to another occasion.

173. Translation in Einstein, IDEAS AND OPINIONS, pp. 73–76, under the title *H. A. Lorentz, Creator and Personality.*

174. As is not unusual, one cannot rely on published translations; we established the text by using the German original in the Einstein Archive:

"Das einzige Phänomen, dessen Erklärung auf diesem Wege nicht restlos, d. h. nicht ohne zusätzliche Annahmen, gelang, war das berühmte Michelson-Morley-Experiment. Dass dies Experiment zu der speziellen Relativitätstheorie hinführte, wäre ohne die Lokalisierung des elektromagnetischen Feldes im leeren Raume undenkbar gewesen. Der wesentliche Schritt war eben überhaupt die Zurückführung auf die Maxwell'schen Gleichungen im leeren Raume oder—wie man damals sagte—im Aether."

Two other translations that differ slightly from each other are given in Einstein, IDEAS AND OPINIONS, p. 75, and in the collection, H. A. LORENTZ: IMPRESSIONS OF HIS LIFE AND WORK, ed. G. L. de Haas-Lorentz (Amsterdam: North-Holland Publishing Co., 1957), p. 8.

175. See notes 29 and 30 above.

176. Copy in the Einstein Archive, Princeton.

177. Sir Edmund Whittaker, A History of the Theories of the Aether and Electricity, Volume II.

178. Copy in the Einstein Archive.

179. Shankland, *Conversations with Albert Einstein*, p. 50.

I wish to thank the Executor and Trustees of the Albert Einstein Estate, and particularly Miss Helen Dukas, for help and for permission to cite from the publications and documents of Einstein. Early versions of the paper were discussed in my History of Science Seminar, and a draft was presented as one of the Monday Lectures at The University of Chicago, November, 1967. I am also grateful for the opportunity which the Rockefeller Foundation and the Director of the Villa Serbelloni provided to write the final draft.

10 ON TRYING TO UNDERSTAND SCIENTIFIC GENIUS

Introduction

HISTORIANS of science return again and again to the epochal contributions of the Newtons and Niels Bohrs, the Darwins and Freuds. Although the history of science is not primarily the study of the work of "genius," historians cannot avoid encountering at every turn the primary or secondary effects of certain few extraordinary, transforming works. At the same time, men and women at that level of achievement are the most puzzling ones.

What is meant by genius in science? What are its characteristics? Can one understand it, or is that a contradiction in terms? I am not speaking merely of "creative" people, nor of men of "high attainment." I am aware of the large amount of literature on creativity, and of some fine studies of men of genius in the arts or in political affairs. But I do not find them very helpful for understanding the life or the work of a Fermi or an Einstein, and even less for discerning how his personality and his scientific achievements interact.

Einstein himself pointed to one difficulty with such a study: it may be hard to find commonalities among many cases from which to gain more understanding about a specific case. He wrote in an essay in 1918,

This article, in somewhat condensed form, was first published in THE AMERICAN SCHOLAR, Volume 41, Winter 1971–72, pp. 95–110.

which we shall have occasion to study further, that the small group of genial scientists who have found special favor with the Angel of the Lord "are mostly rather odd, uncommunicative, solitary fellows who despite these common characteristics, resemble one another really less than the host of the banished." Leopold Infeld, who worked for many years closely with Einstein, dismissed entirely the possibility of giving a definition of genius for, he wrote, "it is characterized just by the fact that it escapes classification."[1]

Yet I do not think the matter is altogether hopeless. On the contrary, it is precisely the attempt to seek some clues through the study of scientific publications, letters, and biographies of such a scientist that has given me the ambitious topic. But I should warn at the outset that we shall be left with many questions, some problems that look interesting, a few hypotheses, and no permanent answers. Moreover, I shall discuss here only one person, and what I have to say may or may not be applicable to other scientists.

Singularities

The first temptation is to proceed reductionistically, and to analyze the man of genius into externally visible, singular elements of his work and character. Those of us who have worked with such a person will have caught glimpses of such elements. The first is undoubtedly his insight into the phenomena of science in a way that amounts almost to a special perception of a kind that can hardly be communicated to others, or a tactile coexistence with natural phenomena: sometimes the mind seems to move into the problem of nature as if it were a hand slipping into a glove. Another element may be his clarity of thought as shown by the penetration of his questions, and by the simplicity and ingenuity of his *Gedankenexperimente*—experiments carried out in thought in just the idealized milieu that turns out to be needed.

Third, one may be startled by the intensity and wide scope of his alertness, for example, to small signals in the large "noise" of any experimental situation or of its description. One is likely also to be constantly impressed by his extraordinary energy and persistent dedication—in manipulation of equipment, in the making of apparatus or tools, in computing or writing. There is a marvelous overabundance evident in such a person, in a Kepler or a Gauss no less than in a Mozart. Connected with it is surely the ability to lend himself—no, to give his whole life over—to the development of a field or an area of thought, usually to the near exclusion of satisfactions or drives other persons find irresistible.

And lastly, one is likely to perceive an aura or atmosphere surrounding such a person's actions and expressions, which sets him apart in a way difficult to define. It is not merely the sometimes unreasonable degree of optimism about his own mission, the self-confidence and self-reliance that appear to others at times to be egocentric obstinacy. Rather, I speak of a basic feeling that such a man has, and that may be shared by those who know him well: that he is, in some sense, one of the "chosen" ones.

These and other characteristics may be more or less adequate earmarks of genius in a particular case. They do apply to Einstein, for example. But I have no illusion that such a list of singularities explains anything. Quite apart from the question of whether any reductionistic approach of this kind can succeed, each of the elements themselves, except possibly the last one, seems to be found also in second- and third-raters. Moreover, these elements do not appear to be either exhaustive or to have convincing and necessary connections, nor does each of them seem necessary.

If we now ascend to the next more serious level, one may well be expected to turn to the methods of psychoanalysis or psychohistory. Some successful and beautiful studies of men not in the sciences exist, although Frank Manuel has given us the one example of such a study of a physical sicentist, in A PORTRAIT OF NEWTON. At least for the particular case I wish to discuss here, psychobiographical analysis will, in my opinion, be most fruitful when used in conjunction with all the other tools of the historian of science, rather than being made the central method. While the personality and work of the genius appear to me to be qualitatively different from those of other scientists, I am not prepared, at least not yet, to think that our methodology and techniques of study must be significantly different from those used in more ordinary cases. Indeed, the most promising road seems to me this: first to identify some special puzzle or problem characterizing our understanding of the man of genius, and then to bring the whole range of the historian's professional tools to bear on this particular puzzle—with the hope that the special character of genius may be reflected in the special aspects of the solutions.

In my work on the historical origins of special relativity theory, I find in retrospect that the most haunting questions have not been those of the sort one usually encounters in the work of most other scientists—such as the trajectory of conceptual development, fascinating though this is. Rather, there is what first strikes one as a remarkable set of puzzling polarities, or, if you will, symmetries and asymmetries, in Einstein's style and life's work.

Let us make a brief list of some of these apparent polarities among which we shall presently choose one for detailed attention.

The folkloric image itself is that of the wisest of old men, who even looked as if he had witnessed the Creation itself; but at the same time, he seems also an almost childlike person. Einstein himself once said, in a remark that will take on some significance later, that he was brought to the formulation of relativity theory in good part because he he kept asking himself questions concerning space and time that only children wonder about.

Then there is his legendary, iron ability to concentrate, often for years, on a single basic problem in physics, regardless of contemporary schools and fashions. Similarly, there is his stubborn faithfulness to a clearly established personal identity, characterized by uncompromising rejection of every *Zwang* and external, arbitrary authority, in physics as well as in clothing or in the demands of everyday life. But opposite to this glorious obstinacy and solitary intransigence with which to search for the basic permanence and necessity behind nature's phenomena, there is also his ever-ready openness to deal after all with the "merely personal" from which he so longed to flee—to deal with the barrage of requests for help and personal involvements that appealed to his fundamental humanity and his vulnerability to pity.

Closely related is another opposition (to express it from the viewpoint of the external observer, in the only language easily available to us, but a language that may well mislead rather than reveal). Einstein is of course known as a grand public personage, radiant and lively, with profound wit and charisma. But from early childhood to his late years he was at the same time also characteristically a solitary person. Max Talmey, who observed him often between his eleventh and fifteenth years, wrote later that he had never seen him in the company of schoolmates or other boys of his age, but that he was usually aloof, absorbed in books and music. In 1936 Einstein wrote in a short "self-portrait": "I live in that solitude which is painful in youth, but delicious in the years of maturity." Einstein could oscillate between these states of the public and private person.[2] He once confessed: "My passionate sense of social justice and social responsibility has always contrasted oddly with my pronounced lack of need for direct contact with other human beings and human communities."

Then there is Einstein as the apostle of rationality, whose thought was characterized by an exemplary clarity of logical construction. On the other hand, there is the uncompromising belief in his own aesthetic sense in science, his warning not to look in vain for "logical bridges" from experience to theory, but to make, when necessary, the great "leap" to basic principles. As he wrote in a well-known passage, "To these elementary laws there leads no logical path, but only intuition, supported by being sympathetically in touch with experience [*Einfühlung in die Erfahrung*]."[3]

Then, on the one hand we have his well-known personal philosophy of liberal agnosticism, even a withering contempt for established religious authority of any sort; and on the other hand there is also a clear personal religiosity. As he says in one of his letters, "I am a deeply religious unbeliever."

Elsewhere[4] I have discussed yet another apparent conflict, that between Einstein as a scientific revolutionary and Einstein as a conservative who stressed the continuity of physics—as in his remark quoted by Carl Seelig: "With respect to the theory of relativity it is not at all a question of a revolutionary act, but of a natural development of a line which can be pursued through centuries."[5]

The Field and the Quantum

It is surely significant that these personal "odd contrasts" have their counterparts in polarities that run right through his scientific work. The most striking of these is the well-known dichotomy between Einstein's devotion to the thema of the *continuum*—expressed most eminently in the field concept—as the basis for fundamental, scientific explanation, and, on the other side, his role in developing quantum physics in which the key idea is *atomistic discreteness*. This merits some amplification.

His devotion to the continuum was not exceeded by that to any other thema, except possibly of symmetry and invariance (that is, of "relativity" itself). It held to the very end; a paper published the year before his death was *Algebraic Properties of the Field in the Relativistic Theory of the Asymmetric Field.* In his letters, even more strongly than in his articles, we find him incessantly defending the continuum against attacks on it by the quantum physicists. He once called the classical concept of the field the greatest contribution to the scientific spirit, and we must recall that in the first paragraph of his fundamental 1905 paper on relativity theory, Einstein motivates the whole discussion by describing the old, seemingly trivial, experiment of a current induced in a conductor that moves with respect to a magnetic field.

The field played a crucial role in his imagination even earlier. From his *Autobiographical Notes* and other testimony we know that his successful formulation of universal principles on which to reconstruct physics in 1905 depended on his fully understanding at last the solution to a paradox on which he had reflected for ten years, since the day it had occurred to him as a sixteen-year-old student at the Kanton Schule of Aarau in 1895–96.[6] We shall later find that it was probably not by accident that this key perception happened at Aarau.

He later described the paradox as it appeared to him in a vivid thought experiment: "If I pursue a beam of light with the velocity *c* (velocity of light in a vacuum), I should observe such a beam of light as a spatially oscillating electromagnetic field at rest. [For example, looking back along the beam over the space of one whole wavelength, one should see that the local magnitudes of the electric and magnetic field vectors increase point by point from, say, zero to full strength, and then decrease again to zero, one wavelength away.] However, there seems to be no such thing, whether on the basis of experience or according to Maxwell's equations." Only by postulating the principles of relativity was the surprising expectation shown to be in error and the physics of the field rescued from this absurdity. (One should note that, as Banesh Hoffmann has pointed out, it is hard to see why the situation described in the thought experiment should be considered surprising or impossible *if* one applies a Galilean instead of a Lorentz transformation to Maxwell's equations. Somehow, Einstein appears already to have tacitly been thinking that Maxwell's equation must remain unchanged in form for the observer moving along with the beam—thereby adopting a principle of relativity *ab initio.*)

But the field held Einstein enthralled even earlier, as shown in a youthful essay which J. Pelseneer discovered in Belgium not long ago.[7] Its title is nothing less than *On the Examination of the State of the Ether in the Magnetic Field (Über die Untersuchung des Aetherzustandes im magnetischen Felde.)* It is a suggestive piece, for it shows that Einstein had already encountered Hertz's work on the electromagnetic field, and that he was thinking up experiments to probe the state of the ether which, he said, "forms a magnetic field" around electric current. For this purpose he suggested sending a lightbeam into the magnetic field as a probe. Any effects on the measurable speed or wavelengths of such a beam would reveal the "elastic deformation" of the ether or field.

It would be an error to think of that essay in any way as a draft of ideas on which the later relativity theory was directly based, or even to regard it necessarily as his first scientific work. But what is most significant about it is the idea of the lightbeam as a probe of a field. From the contemplation of how to measure the wavelengths of such a beam, it would be only a small step to the recognition of the paradox Einstein discovered soon afterwards at the Aarau school.

We can go back even further when searching for the point where the thematic commitment to the continuum was formed. It is well known that, as a child of four or five, Einstein experienced what he called "a wonder" when his father showed him a simple magnetic pocket compass. It was an experience to which Einstein often referred. His friend Moszkowski reported him in 1922 to have said, "Young as I was, the remembrance of this occurrence never left me." His biographer Seelig wrote in 1954 that the compass "to this day is vividly engraved in his memory, because it practically bewitched him." Another (although less reliable) biographer reported that Einstein told him of that early part of his life: "The compass, and only the compass, remains in my memory to this day." In his autobiography, written at the age of sixty-seven, we read: "I can still remember—or at least I believe I can remember—that this experience made a deep and lasting impression on me. Something deeply hidden had to be behind things."

The scene is most suggestive. There is the mysterious invariance or constancy of the compass needle, ever returning to the same direction, despite the fact that the needle seems free from any action-by-contact of the kind that is usually unconsciously invoked to explain the behavior of material things; despite the vagaries of motion one may arbitrarily impose on the case of the compass from the outside; and regardless of personal will or external *Zwang* or chaos. If Einstein remembered it so well and referred to it so often, it may be because the episode is an allegory of the formation of the playground of his basic imagination.[8] For anyone interested in the genesis of scientific ideas or the motivation toward scientific study, these sketchy remarks will indicate that there are here problems that cry out to be worked on. These are, however, not our chief concerns here. What does matter is that this long loyalty to the explanatory power of the continuum was destined to be put to a severe test.

For Einstein, of course, was also the brilliant contributor to the physics based on the thema precisely the polar opposite to the continuum, namely, the discrete quantum—for example, in the conception that light energy is not continuously divisible, but proceeds in well-defined quanta or photons. By his own report Einstein came to quantum physics by studying what Planck's radiation formula may imply for the "structure of radiation and more generally . . . the electromagnetic foundations of physics." Einstein's first fundamental paper in quantum physics, entitled *On a Heuristic Point of View Concerning the Generation and Transformation of Light*, was finished at Bern in 1905, three days after he had turned twenty-six, and only about three months before the relativity paper.

The wonder is that it is so completely different (in all ways but one) from the relativity paper. That is, its object is close to what Einstein later called "mental gymnastics": to overcome a problem in physics at any cost, but without a basic reformulation. As its title says, it is dominated by a heuristic attitude. On the other hand, the relativity paper is Natural Philosophy in the deepest sense: rejecting everything arbitrary, even assumptions concerning the nature of matter, in order to find the nature of space and time that allows causal continuity and prepares for the great simplifications and unifications, first of all, that of the transformation properties of mechanics and electrodynamics.

But while the relativity theory became Einstein's own most enduring life preoccupation, he could not accept quantum physics seriously. To

Leopold Infeld, he said "I may have started it, but I always regarded these ideas as temporary. I never thought that others would take them so much more seriously than I did." Yet, he continued to make some of the most seminal contributions to it for a quarter of a century from 1905 on, with rarely a year going by in which he did not publish an article on this subject.

These, then, are some of the characteristics that seem to be polar opposites. It is, I think, significant that Einstein himself drew attention to the existence of such polar pairs in the work and personality of outstanding scientists—and precisely in the one essay in which he came closest to asking the question raised here: wherein lies scientific genius? He was, of course, not referring to himself. His essay, entitled *Motiv des Forschens*, was delivered in 1918 as an offering in honor of his early mentor, friend, and colleague, Max Planck. Because it illuminates strikingly the questions at hand (and, by the way, could still serve as a manifesto for many a scholar), a summary of Einstein's essay has been attached to this paper as an appendix. It will be noted that Einstein contrasts the "positive" and "negative" motives for doing research at that highest level, and the opposing demands of clarity and completeness, of logic and intuition, of private and public science.

If we now take it as granted that striking polarities exist, of both personality and work, in this one case,[9] we must ask next: are they important or accidental? Earlier, we disavowed the idea that a plausible list of individual characteristics adds up to an explanation or even a characterization of genial scientists. If such singularities were not significant, why should polarities be? What special abilities do they convey to their possessor? Does a man of genius bring to bear upon his work the harmonies and disharmonies, the strengths and conflicts within his person—and the pressures and conflicts of his environment? Regardless of how a man like Einstein came to have his particular characteristics (that may in any case not be a very interesting question in the present state of such research), we can ask whether there was some special way in which he *put to use* these dichotomies and conflicting polarities.

We are encouraged to expect a positive reply to such questions if we notice the existence and brilliant exploitation of polarities in Enistein's *individual contributions*. The most evident example is the presence and use of contrast in the original relativity theory paper itself: we find there both the positivism of the instrumentalist and operationist variety, which

Einstein uses in defining the concepts, and, on the other hand, the rational realism inherent in the a priori declaration of the two basic principles of relativity (moreover, two apparently contradictory ones, introduced seemingly in arrogant disregard both of current scientific sensibilities and of the contemporary demand to base them plausibly on scientific experiments). Einstein himself acknowledged this ambivalence. In response to the charge[10] "Einstein's position . . . contains features of rationalism and extreme empiricism . . .," he replied, "This remark is entirely correct. . . . A wavering between these extremes appears to me unavoidable."

A similarly creative use of apparent opposites can be found in Einstein's contribution to quantum physics, centering on the wave-particle duality. It really is the hallmark of Einstein's most famous contributions that he could deal with, use, illuminate, transform the existence of apparent contradictories or opposites, sometimes in concepts that had not yet been widely perceived to have polar character. One need only think of his bridging of mechanics and electrodynamics, energy and mass, space coordinates and time coordinates, inertial mass and gravitational mass.

The Significance of Asymmetry

We can now choose for detailed study a concrete example by means of which to find further hints on how personal characteristics interacted with scientific work. The example offers itself in fact at the very beginning of the basic 1905 paper on relativity theory. The title is *On the Electrodynamics of Moving Bodies*, and neither there nor later on is the phrase "relativity theory" used. None of Einstein's papers has this phrase in the title until 1911, long after others began to refer to his work in that way. Indeed, it is of the essence to know that for the first two years Einstein, in his letters, preferred to call his theory not "relativity theory," but exactly the opposite: *Invariantentheorie*. It is unfortunate that this splendid, accurate term did not come into current usage, for it might well have prevented the abuse of relativity theory in many fields.

In the very first sentence of the paper, there is a term that attracts our attention, especially now that we have become sensitized to it by the previous discussion on polarities: "It is known that Maxwell's electrodynamics—as usually understood at the present time—when applied to moving bodies, leads to asymmetries which do not appear to be inherent in the phenomena." It is not Maxwell's electrodynamics that is at fault; it is the way it is usually understood: a bold, not to say aggressive, statement from this relatively unknown young patent office employee. And this usual way of understanding has led to—what? An experimental puzzle? No. A theoretical impasse? No. It leads to *asymmetries* that do not appear to be inherent in the phenomena.

For example, you may think of inducing a current in a conductor—as Einstein immediately does in the style of a little thought-experiment. To calculate what current to expect when the conductor is moving with respect to a stationary magnet, you must use one kind of equation. When you calculate again on the assumption that now you will keep the conductor stationary and let the magnet move, you must use a different kind of equation—although the current actually produced is found to be identical in both cases, as has been known ever since Faraday first described the effect in 1831. The phenomenon is characterized by symmetry; but the machinery for calculation was characterized by polarity or asymmetry (until Einstein showed, later in the paper, how to "relativize" the problem so that the same equation may be used for both cases).

The importance of this passage, its historic veracity, and the fact that a very similar thought process led Einstein later to the *General* Theory of Relativity are brought out in a striking manner in parts of a hitherto unpublished manuscript in Einstein's handwriting, dating from about 1919 or shortly afterwards, now located in the Einstein Archives at the Princeton Institute for Advanced Study, and entitled (in translation from the German) *Fundamental Ideas and Methods of Relativity Theory, Presented in their Development.*

In the first nineteen pages of the manuscript, one finds largely an impersonal, pedagogic presentation of a familiar kind. But a personal account appears, in a rather surprising way, on pages 20 to 21 which introduce Part II, entitled *General Relativity Theory*:

> (15) *The fundamental idea of general relativity theory in its original form.* In the construction of special relativity theory, the following, [in the earlier part of this manuscript] not-yet-mentioned thought concerning the Faraday [experiment] on electromagnetic induction played for me a leading role.
>
> According to Faraday, during the relative motion of a magnet with

with respect to a conducting circuit, an electric current is induced in the latter. It is all the same whether the magnet is moved or the conductor; only the relative motion counts, according to the Maxwell-Lorentz theory. However, the theoretical interpretation of the phenomenon in these two cases is quite different. . . .

The thought that one is dealing here with two fundamentally different cases was for me unbearable [*war mir unerträglich*]. The difference between these two cases could be not a real difference but rather, in my conviction, only a difference in the choice of the reference point. Judged from the magnet, there were certainly no electric fields, [whereas] judged from the conducting circuit there certainly was one. The existence of an electric field was therefore a relative one, depending on the state of motion of the coordinate system being used, and a kind of objective reality could be granted only to the *electric and magnetic field together*, quite apart from the state of relative motion of the observer or the coordinate system. The phenomenon of the electromagnetic induction forced me to postulate the (special) relativity principle. [Footnote:] The difficulty that had to be overcome was in the constancy of the velocity of light in vacuum which I had first thought I would have to give up. Only after groping for years did I notice that the difficulty rests on the arbitrariness of the kinematical fundamental concepts [presumably such concepts as simultaneity].

When, in the year 1907, I was working on a summary essay concerning the special theory of relativity for the JAHRBUCH FÜR RADIOAKTIVITÄT UND ELEKTRONIK, I had to try to modify Newton's theory of gravitation in such a way that it would fit into the theory [of relativity]. Attempts in this direction showed the possibility of carrying out this enterprise, but they did not satisfy me because they had to be supported by hypotheses without physical basis. At that point, there came to me the happiest thought of my life, in the following form:

Just as is the case with the electric field produced by electromagnetic induction, the gravitational field has similarly only a relative existence. *For if one considers an observer in free fall, e.g. from the roof of a house, there exists for him during his fall no gravitational field—at least in his immediate vicinity.* (Italics in original.)

To return now to the clue that is offered by the use of the word "asymmetry" in Einstein's first relativity paper: at first glance it is surely curious that he used the term asymmetry for this apparent redundancy or lack of universality. Moreover, such terms as symmetry or asymmetry still referred at that time largely to aesthetic judgments, often thought to be the polar opposites of scientific judgments. In physics literature, symmetry arguments were quite uncommon (and even these are easier to find now in retrospect); the term itself was rarely used except in such branches as crystal physics. For example, the word symmetry is mentioned only casually in Mach's SCIENCE OF MECHANICS (although the concept is used implicitly in discussions involving the Principle of Sufficient Reason, in a form traced by Mach to Schopenhauer). As late as 1929, the eleventh edition of the large encyclopedia of physics of Müller-Pouillet indexes only a single noncasual use of the symmetry concept outside crystal physics—and that is, naturally, the entry for symmetrical tensors, and refers to Einstein's general relativity theory itself. There is no entry for symmetry or asymmetry in the *Sachregister* of the encyclopedic volume PHYSIK (E. Warburg, ed., B. G. Teubner, Leipzig, 1915). Nor for that matter is there any such entry in the eleventh edition of the ENCYCLOPAEDIA BRITANNICA (1910).

Only with the growth of the role of quantum mechanics, and more recently of elementary particle physics, has it become clear that the conservation laws of physics are closely connected with the concept of symmetry of space and time, as had been implicit in the Lagrangian and Hamiltonian methods of solving physical problems. And it is only with the help of hindsight that we have come to see, as Hermann Weyl[11] points out, that "the entire theory of relativity . . . is but another aspect of symmetry," in the sense that "the symmetry, relativity, or homogeneity of this four-dimensional medium [the space-time-continuum] was first correctly described by Einstein. . . . It is the inherent symmetry of the four-dimensional continuum of space and time that relativity deals with."

But all this was far in the future in 1905. In any case, Einstein's reference to asymmetry at the start of his paper (and other references to symmetry in the rest of the paper) was not a symmetry consideration of the same kind as those just mentioned. Few physicists, if any, can have thought in 1905 that there was something of fundamental importance in the asymmetry to which Einstein pointed.

And if one considers how many troubles there were in electrodynamics at the time, it must have seemed peculiar indeed to seek out this quasi-aesthetic discomfort, and to put it at the head. What Einstein's perception of asymmetry at this point *does* show us, however, is his remarkable and original sensitivity to polarities and symmetry properties of nature that later became recognized as important in relativity theory and in contemporary physics generally.

From everything we now know about Einstein, we have also been prepared to understand that his desire to remove an unnecessary asymmetry was not frivolous or accidental, but deep and important. At stake is nothing less than finding the most economical, simple, formal principles, the barest bones of nature's frame, cleansed of everything that is ad hoc, redundant, unnecessary.[12] To his assistant, Ernst Straus, Einstein said later: "What really interests me is whether God had any choice in the creation of the world." In fact, sensitivity to previously unperceived formal asymmetries or incongruities of a predominantly aesthetic nature (rather than, for example, a puzzle posed by unexplained experimental facts)—that is the way each of Einstein's three otherwise very different great papers of 1905 begin. In all these cases the asymmetries are removed by showing them to be unnecessary, the result of too specialized a point of view. Complexities that do not appear to be inherent in the phenomena should be cast out. Nature does not need them.

And Einstein does not need them. In his own personal life, the legendary simplicity of the man is an integral part of this reaching for the barest minimum on which the world rests. I will not need to recall here the many stories about this simplicity. Even people who knew nothing else about Einstein knew that he preferred the simplest possible clothing. We noted that he hated nothing more than artificial restraints of any kind. He once was asked why he persisted in using ordinary handsoap for shaving, instead of shaving cream, despite the fact that it was clearly less comfortable for him to shave that way. He said, in effect: "Two soaps? That is too complicated!"

Two processes to describe the same effect of the induced current? That is too complicated. Nature does not work this way, and Einstein does not work this way. The overlap between the two was once expressed by Einstein in a humble sentence: "I am a little piece of nature." We have here an important clue to our question of what may be meant by the easy term "genius": *there is a mutual mapping of the mind and life-style of this scientist, and of the laws of nature.*

At the same time, the desire to remove an unwarranted "asymmetry" contains a clue to a second connection between Einstein's work and his person. The area that seems worth exploring lies where two studies meet: one is the well-known use of symmetry and asymmetry arguments in mathematics, particularly geometry, and the other is the investigation of mathematical and other thought processes in children by Wolfgang Köhler, Max Wertheimer, and others of that school.

"What, Precisely, Is 'Thinking'?"

We know far too little about Einstein as a child, but he is commonly reported to have been withdrawn, slow to respond, quietly sitting by himself at an early age, playing by putting together shapes cut out with a jigsaw, erecting complicated constructions by means of a chest of toy building parts. Before he was ten, he was making, with infinite patience, fantastic card houses that had as many as fourteen floors. He is said to have been unable or unwilling to talk until the age of three. In an (unpublished) biography of Einstein, his sister Maja wrote in 1924: "His general development during his childhood years proceeded slowly, and spoken language came with such difficulty that those around him were afraid he would never learn to talk." Many pediatricians and psychologists might consider such evidence to indicate an almost backward child.

But it is coming to be more widely agreed that an apparent defect in a particular person may merely indicate an imbalance of our normal expectations. While it is patently absurd to think that a deficiency in one area "causes" or explains talent in another, a noted deficiency should at least alert us to look for a proficiency of a different kind in the exceptional person. The late use of language in childhood, the difficulty in learning foreign languages—one remembers that Einstein failed in foreign languages at the *Gymnasium* and again at the entrance examination in Zurich (one of the reasons for his having to go to the Kanton Schule in Aarau), that his vocabulary in English was fairly small—all this may indicate a polarization or displacement in some of the skill from the verbal to another area. That other, enhanced area is without doubt, in Einstein's case, an extraordinary kind of visual imagery that penetrates his very thought processes.

Although it seems to have been hardly noted so far, Einstein himself plainly signals this point early in his *Autobiographical Notes.* He asks, rather abruptly:

> What, precisely, is "thinking"? When at the reception of sense-impressions,

memory-pictures [*Erinnerungsbilder*] emerge, this is not yet "thinking." And when such pictures form series, each member of which calls forth another, this too is not yet "thinking." When, however, a certain picture [*Bild*] turns up in many such series, then—precisely through such return—it becomes an ordering element for such series, in that it connects series which in themselves are unconnected. Such an element becomes an instrument, a concept. . . . It is by no means necessary that a concept must be connected with a sensorily cognizable and reproducible sign (word). . . . All our thinking is of this nature of a free play [*eines freien Spiels*] with concepts. . . . For me it is not dubious that our thinking goes on for the most part without use of signs (words), and beyond that to a considerable degree unconsciously.

It is not accidental at all that this surprising passage comes just before Einstein tells of the two "wonders" experienced in childhood. One of these was the experience with the compass that we have mentioned. The other wonder, of a totally different nature, was the little book dealing with Euclidean plane geometry which, he recalled, was given to him at about the age of twelve. In this connection, Einstein described an early, successful use of his particular way of "thinking" in visual terms:

I remember that an uncle told me the Pythagorean theorem before the holy geometry booklet had come into my hands. After much effort I succeeded in "proving" this theorem on the basis of the similarity of triangles; in doing so it seemed to me "evident" that the relations of the sides of the right-angled triangle would have to be completely determined by one of the acute angles. Only something which did not in similar fashion seem to be "evident" appeared to me to be in need of any proof at all. Also, *the objects with which geometry deals seemed to be of no different type than the objects of sensory perception, "which can be seen and touched."* (Italics supplied)

The objects of the imagination were to him evidently persuasively real, visual materials, which he voluntarily and playfully could reproduce and combine, analogous perhaps to the play with shapes in a jigsaw puzzle. The key words are *Bild* and *Spiel*; and once alerted to them, one finds them with surprising frequency in Einstein's writings. Thus, responding to Jacques Hadamard, Einstein elaborates the point made above:[13]

The words or the language, as they are written or spoken, do not seem to play any role in my mechanism of thought. The psychical entities which seem to serve as elements in thought are certain signs and more or less clear images which can be "voluntarily" reproduced and combined. . . . But taken from a psychological viewpoint, this combinatory play seems to be the essential feature in productive thought—before there is any connection with logical construction in words or other kinds of signs which can be communicated to others. The above-mentioned elements are, in my case, of visual and some muscular type.[14] Conventional words or other signs have to be sought for laboriously only in a secondary stage, when the mentioned associative play is sufficiently established and can be reproduced at will.

Max Wertheimer, one of the founders of Gestalt psychology and a friend of Einstein, reports [15] that from 1916 on, in numerous discussions, he had questioned Einstein "in great detail about the concrete events in his thoughts" leading to the theory of relativity. Einstein told him, "These thoughts did not come in any verbal formulation. I very rarely think in words at all. A thought comes, and I may try to express it in words afterwards"; and later: "During all those years there was a feeling of direction, of going straight toward something concrete. It is, of course, very hard to express that feeling in words. . . . But I have it in a kind of survey, in a way visually."

In Einstein's published work, his visual imagination sometimes breaks through vividly. One thinks here, for example, of passages where he describes through experiments involving the picturesque tasks of coordinating watch readings, the arrival of light signals, the positions of locomotives and those of lightning bolts. Possibly Einstein's ability to deal with models and drawings in the patent office, and his own delight throughout his adult life with the workings of puzzle-toys, are additional clues of some significance. More important for our purposes, this ability of visualization is evident in the haunting thought experiment of the lightbeam, begun at Aarau, and even in the thought experiment proposed in his earlier essay of 1894–5, where he envisages probing the state of the ether in the vicinity of a current-carrying wire.

I have little doubt that the ability to make such clear visualizations of experimental situations was crucial in his task of penetrating to the relativity theory (for example, in the argument leading to the relativity of simultaneity). It is so to this day. As anyone knows who has tried to instruct students in relativity theory, the problem of getting a firm initial understanding of special relativity is not one of mathematics—at most, elementary calculus is required—but rather one of clear *Vorstellung*.[16] What helps a beginner most is precisely the ability to imagine vividly some thought experiments involving the perceptions and reports of two observers who are moving relative to each other. The style of thinking necessary at the outset is quite different from working, for example, through the machinery of a formalism such as characterizes the work of a Sommerfeld.

But in the long run, the strength of Einstein's visual imagination was not limited only to such uses. Rather, the hypothesis here proposed is that the special ability to think with the aid of a play with visual forms has deeper consequences. It animates the consideration of symmetries and a corresponding distaste for extraneous complexities from the beginning to the end—from the result embedded in his 1905 paper that the Lorentz transformation equations yield "contractions" and "time dilations" that are, for the first time, symmetrical for all inertial systems, to his long wrestling with the task of constructing a relativistic theory of gravitation, basing its field structure on a symmetric tensor g and the symmetric infinitesimal displacement tensor Γ, and finally to his self-imposed labors of finding a theory of the total field through the generalization of giving up the symmetry properties of the g and Γ fields. From the beginning to the end, Einstein's scientific thought was pervaded by questions of symmetry and the closely related concept of invariance.

But long before he wrote scientific papers, Einstein was Einstein—already at the age of three, playing silently, resisting verbal language, and refusing thereby to accept an externally imposed authority in names and rules by which many another child has had to "civilize" and give up his own curiosity and imaginative play. It is a world that by its very definition we hardly know how to describe. But it is the world in which, from all the evidence we have, the play with geometric and other visual images, and hence the perception of such transformation properties of forms as symmetry and asymmetry, appear to have been basic for the development of successful thought itself.

The ABC of Visual Understanding

The literature on the subject of the visual element of thought in children is small. This deprives us to some degree of one of the important strands in our net, and conversely alerts cognitive psychologists and educators to a promising research topic. But there was one pioneer in this field who, in an unexpected way, now comes into our story. He is Johann Heinrich Pestalozzi, the Swiss educational reformer. Born in 1746 in

Zurich, he was educated at the University of Zurich, and first tried farming in a small town in the Kanton of Aargau. During the French invasion of Switzerland in 1798, a number of children were left without parents or food at Lake Lucerne, and Pestalozzi devoted himself to their care and education during that winter. In 1801 he published his ideas on education in the book WIE GERTRUD IHRE KINDER LEHRT, and from then on he became a widely influential force in education. It is recorded that von Humboldt, Fichte, Mme. de Staël and Talleyrand visited him and his schools.

Basic to Pestalozzi's approach to education were the development of observation, the humanistic approach to each subject, the collaborative and sympathetic relation between teacher and student—and his view that "conceptual thinking is built on visual understanding (*Anschauung*)." His method was, for that reason, to put the ABC of visual understanding ahead of the ABC of letters.

An excerpt from his book will give some indication of the approach followed in his own schools and in those that were founded under his influence:

> I must point out that the ABC of visual understanding is the essential and the only true means of teaching how to judge the shape of all things correctly. Even so, this principle is totally neglected up to now, to the extent of being unknown; whereas hundreds of such means are available for the teaching of numbers and language. This lack of instructional means for the study of visual form should not be viewed as a mere gap in the education of human knowledge. It is a gap in the very foundation of all knowledge at a point to which the learning of numbers and language must be definitely subordinated. My ABC of visual understanding is designed to remedy this fundamental deficiency of instruction; it will insure the basis on which the other means of instruction must be founded.[17]

The "quintessence" of Pestalozzi's fundamental book WIE GERTRUD IHRE KINDER LEHRT was summarized by Albert Richter in these words[18]:

1. The foundation of instruction is the *Anschauung* (visual understanding).
2. Speech (words) must be connected with the *Anschauung*.
3. The time of learning is not the time of judgment, of critique.
4. In every subject, instruction shall begin with the simplest elements, and from that point on be brought forward stepwise in accord with the development of the child—that is to say, in psychological order (sequences).
5. One must rest at every point as long as is necessary for the particular material of instruction to become the pupil's free mental property.

6. Instruction has to follow the way of development, not the way of expostulation, memorization, information.

7. For the instructor, the individuality of his charge should be holy.

8. The main purpose of elementary instruction is not the acquisition of information and skills, but the development and strengthening of mental powers.

9. Ability should follow knowledge; then follows skill.

10. The interaction between instructor and pupil, and particularly school discipline, should be pervaded and guided by love.

11. Instruction should be subordinated to the purpose of education.

12. The foundation of moral-religious development of the child lies in the relationship between mother and child.[19]

One notes the modern and humane ring of many of these guidelines. Still, Pestalozzi himself is quoted as having made this final assessment: "When I look back and ask myself, what have I really achieved for the cause of human instruction? I find this: I have firmly anchored the first and fundamental axiom of instruction in the recognition of *Anschauung* as the absolute foundation of all knowledge."

Success in Aarau

How different a school founded on such principles must have been from that which Einstein fled when he left Munich as a boy of about fifteen—his regimented, militaristic school, so verbally oriented, and so deeply unappreciative of him as a student and as a person!

In the late summer of 1895, when Einstein arrived in Zurich to sit for the entrance examination to the Polytechnic Institute, he thought he had put all schools behind him. But then he failed the entrance examination, as we have noted. He had given up his native country and taken steps to renounce his citizenship, he had left his parents in Italy, he was a foreigner in Switzerland, he had failed to get into the Polytechnic Institute. The dislocation was complete.

There ensued a kind of moratorium for this boy who seemingly had failed in many ways (except, always, in mathematics and physics). The next test being scheduled a year later, he was advised by the director of the Polytechnic Institute to enroll in the Kanton Schule at Aarau, thirty miles northwest of Zurich, in the capital of the Kanton of Aargau. There is no doubt that this was a crucial turning point—for while he was at that particular school, everything somehow changed for him, from getting his first great *Gedankenexperiment* that led him to relativity, to finding true friends—indeed, one of them, the son of the teacher Jost Winteler in whose house Einstein was a happy and beloved boarder, eventually married Einstein's sister, Maja. Later, Einstein's thoughts often and gladly turned to this school. He corresponded, and occasionally met, with several of his classmates for years afterwards. He mentions the school, and his obtaining his *Abitur* there, in his autobiographical note of 1922 or '23, which he had to compose, as is the custom, on receiving the Nobel Prize, for official publication in *Les Prix Nobel*—even though, characteristically, Einstein's essay is otherwise embarrassingly short—only fourteen lines. Just a month before his death he remembered his school again in these words: "It made an unforgettable impression on me, thanks to its liberal spirit and the simple earnestness of the teachers who based themselves on no external authority."[20]

Some of his teachers, particularly the science instructor Fritz Mühlberg, seem to have interested and encouraged Einstein. There also appears to have been an easy atmosphere, relaxed, informal and democratic. In science instruction, too, the aims of general education and humanistic learning were kept foremost. The course prided itself not on memorization, but instead stressed the kind of work that would develop individual thinking. In addition to the more usual learning materials there were excursions, two kinds of drawing courses, work with specimens in the museum, and laboratory work. Maps and other visual materials seem to have been freely used.

These and other elements in the suggestive though fragmentary descriptions of the school by Einstein's most reliable biographers—Frank, Reiser and Seelig[21]—show that this school seems to have been characterized by many of the fundamental pedagogic guidelines laid down by Pestalozzi (even though in the course of time there almost inevitably will have been periods of rigidity in adhering to those principles, followed by periods of abreaction). This prepares us for the discovery that indeed the Kanton Schule of Aarau was first founded in 1802 by democratic patriots,[22] reportedly acting in the spirit of Pestalozzi—just a year after the publication of Pestalozzi's manifesto, and while Pestalozzi himself was running one of his schools at Burgdorf, less than fifty miles away. So it may not be an accident that Einstein became aware of the strength of his genial scientific imagination at this particular school; here, at last, he was in a place that did not squash, and may well have fostered, the particular style of thinking that was so congenial to him.

We began by noting some odd contrasts and polarities that appear in the work and life-style of Einstein. We have added others: early failure and success, handicap in one direction and extraordinary ability in another. In discussing the role of simplicity and symmetry, we found one clue to genius in the proposition that there is, at least in this case, a mutual mapping of the habits and life-style of the genial scientist and of nature's own laws. The investigation of the circumstances surrounding the *Gedankenexperiment* at Aarau, from which the relativity theory grew, supports a more generalized phrasing of our initial hpothesis: At least in this particular case, *there is a mutual mapping of the style of thinking and acting of the genial scientist on the one hand, and the chief unresolved problems of contemporary science on the other.*

Thus, what seemed to us at first to be puzzling internal polarities in Einstein may equally well be viewed as talents for dealing with the dichotomies that often have turned out to be at the base of the most unyielding problems of science. For example, we have noted earlier that epistemologically the 1905 relativity paper oscillates between the positivism of the Machist kind, needed for the definition of the concepts, and, on the other hand, a rational realism needed for the a priori declaration of the basic principles of relativity, and that Einstein confessed later to harboring both of these extremes in his own thoughts. But to this day, it is virtually inconceivable that he, or anyone else, could have attained a genuine formulation of relativity without just these two contrasting elements.

Much good and even excellent science can be done in a more monolithic ways, neglecting or avoiding any evidence of conceptual dichotomies. But it is not often stressed that such dichotomies are by no means unusual in science. They exist from its smallest observational protocol to the most over-arching theory. Ascending from the lowest level, we note first the antithesis between the obvious, observable, palpable, limited, material object such as a magnet needle, and, say, the field in which it is caught—tenuous, invisible, appearing usually rather mysterious to the beginning student, but commanding, and stretching into infinity. A further step up, and there are conceptually antithetical pairs such as matter and energy, space and time, the gravitational and electromagnetic field—even the theoretical and experimental activity, and the parsimony of a good theory versus the infinite number of actual cases it embraces.

Another step up, and we note the antitheses between the great themata—be it the continuum versus the discrete, or classically causal law versus statistical law, or the mechanistic versus the theistic world interpretation. Such thematic antitheses, expressed in the great current puzzles of science, have haunted such scientists as Newton, Bohr and Einstein, even when lesser scientists could afford the luxury of avoiding such confrontations, and doing more comfortable work on thematically unambiguous problems—an activity similar to that which Einstein once dismissed as seeking the thinnest part of a board in order to drill one's hole there. After all, genius discovers itself not in splendid solutions to little problems, but in the struggle with essentially eternal problems. And those, by their very nature, are apt to be problems arising from thematic conflicts.

It has been lately fashionable in some quarters to think that physical science normally progresses by moving on the whole fairly calmly in one direction and in one stream bed, and that such progress is interrupted only at certain periods of great upheaval in science.

But this can be true only in a limited sense. Not far below the surface, there have coexisted in science, in almost every period since Thales and Pythagoras, sets of two or more antithetical systems or attitudes, for example, one reductionist and the other holistic, or one mechanistic and the other vitalistic, or one positivistic and the other teleological. In addition, there has always existed another set of antitheses or polarities, even though, to be sure, one or the other was at a given time more prominent—namely, between the Galilean (or, more properly, Archimedean) attempt at precision and measurement that purged public, "objective" science of those qualitative elements that interfere with reaching reasonable "objective" agreement among fellow investigators, and, on the other hand, the intuitions, glimpses, daydreams, and a priori commitments that make up half the world of science in the form of a personal, private, "subjective" activity.

Science has always been propelled and buffeted by such contrary or antithetical forces. Like vessels with draught deep enough to catch more than merely the surface current, scientists of genius are those who are doomed, or privileged, to experience these deeper currents in their complexity. It is precisely their special sensitivity to contraries that has made it possible for them to do so, and it is an inner necessity that has made them demand nothing less for themselves.

This, it seems to me, is the direction in which to seek for answers to the question of why the Angel of the Lord chose them . . . or, at least, chose one of them.

Acknowledgment

Early drafts of this paper were discussed at the psychobiography seminar of Professor Erik Erikson at Stockbridge, Massachusetts; at Professor Dore Ashton's seminar at the Cooper Union School of Art and Architecture; and at a Van Leer Jerusalem Foundation Symposium held in honor of Professor S. Sambursky. I thank the participants, and particularly Professor Erikson, for illuminating discussions. I also wish to express my indebtedness to Miss Helen Dukas and to the Estate of Albert Einstein for permission to cite from Einstein's writings.

This work is part of a study sponsored by a grant from the National Science Foundation.

APPENDIX

The following is a summary of Einstein's essay *Motiv des Forschens*,[23] with translation of key passages (in quotes). In this remarkable address, given in 1918 in honor of Max Planck (and later mistranslated as *Principles of Research*), Einstein gave us what is perhaps the best autobiographical insight into his own view of the motivation for doing research in science. The analyses and suggestions given in the preceding pages may help in interpreting these charming but intensely charged passages, and in turn may receive illumination from them.

"The Temple of science is a multi-faceted building." In it, many engage in science out of joy in flexing their intellectual muscles, or for utilitarian ends. These are useful persons, to be sure, although external circumstances could easily have made them into engineers, officers, etc. If only such scientists existed, "the Temple would not have arisen."

"If there now came an Angel of the Lord to drive these persons out of the Temple," few scientists would be left in it. But one of them would be Planck, "and that is why we love him."

Now let us turn to those "who found favor with the Angel." They are mostly "rather odd, uncommunicative, solitary fellows, who despite these common characteristics resemble one another really less than the host of the banished."

"What led them into the Temple? The answer is not easy to give, and can certainly not apply uniformly. To begin with, I believe with Schopenhauer that one of the strongest motives that lead men to art and science is flight from the everyday life with its painful harshness and wretched dreariness, and from the fetters of one's own shifting desires. One who is more finely tempered is driven to escape from personal existence and to the world of objective observing [*Schauen*] and understanding. This motive can be compared with the longing that irresistibly pulls the town-dweller away from his noisy, cramped quarters and toward the silent, high mountains, where the eye ranges freely through the still, pure air and traces the calm contours that seem to be made for eternity."

"With this negative motive there goes a positive one. Man seeks to form for himself, in whatever manner is suitable for him, a simplified and lucid image of the world [*Bild der Welt*], and so to overcome the world of experience by striving to replace it to some extent by this image. This is what the painter does, and the poet, the speculative philosopher, the natural scientist, each in his own way. Into this image and its formation he places the center of gravity of his emotional life, in order to attain the peace and serenity that he cannot find within the narrow confines of swirling, personal experience."

The theoretical physicists' picture of the world [*Weltbild*] is one among all the possible pictures. It demands vigorous precision in the description of relationships. Therefore the physicist must content himself from the point of view of subject matter with "portraying the simplest occurrences which can be made accessible to our experience"; all more complex occurrences cannot be reconstructed with the necessary degree of subtle accuracy and logical perfection. "Supreme purity, clarity, and certainty, at the cost of completeness."

Once such a valid world image has been achieved, it turns out to apply after all to every natural phenomenon, including all its complexity, and in its completeness. From the general laws on which the structure of theoretical physics rests, "it should be possible to attain by pure deduction the description, that is to say, the theory of every natural process, including those of life, if such a process of deduction were not far beyond the capacity of human thinking. To these elementary laws there leads no logical path, but only intuition, supported by being sympathetically in touch with experience [*Einfühlung in die Erfahrung*]." It is true that this uncertain methodology may in principle give rise to many systems of theoretical physics with equal claim; but in fact it has turned out that at any time just one such system is generally accepted to be decidedly superior. "Though there is no logical bridge from experience to the basic principles of theory," in practice it is agreed that "the world of experience does define the theoretical system uniquely. . . . This is what Leibnitz termed happily 'pre-established harmony.' Physicists accuse many an epistomologist of not giving sufficient weight to this circumstance."

"The longing to behold that pre-established harmony is the source of the inexhaustible perseverance and patience with which Planck [and, we may add here, as throughout, Einstein] has given himself over to the most general problems of our science, not letting himself be diverted to more profitable and more easily attained ends. I have often heard colleagues trying to trace this attitude to extraordinary will power and discipline—in my opinion, wrongly. The state of feeling [*Gefühlszustand*] which makes one capable of such achievements is akin to that of the religious worshipper or of one who is in love; his daily striving arises from no deliberate decision or program, but out of immediate necessity. . . ."

NOTES

1. Leopold Infeld, ALBERT EINSTEIN, HIS WORK AND ITS INFLUENCE ON OUR WORLD, New York: Charles Scribner's Sons, 1950, p. 118.

2. Such an "oscillation" has been described, for example, by the perceptive biographer Antonina Vallentin in EINSTEIN: A BIOGRAPHY (London: Weidenfeld & Nicolson, 1954), p. 27.

3. In the lead essay in the collection SCIENCE ET SYNTHÈSE (Paris: Gallimard, 1967), p. 28, Ferdinand Gonseth writes on Einstein's methodological autonomy in his early work and concludes: "I believe that the seeds of all that followed were contained within his assumption of the freedom to discard a proof in favor of a conviction born of the pursuit of truth."

4. Gerald Holton, *Science and New Styles of Thought*, THE GRADUATE JOURNAL 7: 399–422, 1967, and reprinted here as Essay 3.

5. See also one of Einstein's earliest letters (1905), written to his friend Conrad Habicht and cited in Essay 5.

And to Maurice Solovine, Einstein wrote a year later, soon after his first papers on relativity were published and in the middle of its further development: "Nothing much here by way of science. Soon I shall enter the stationary and sterile stage of life in which one complains about the revolutionary orientation of youth."

6. As Einstein's biographer Philipp Frank pointed out, the paradox continued to preoccupy him ever more urgently throughout his whole period of study at the Polytechnicum in Zurich. "He studied all works of the great physicists to the purpose of finding whether they could contribute to the solution of this problem concerning the nature of light." Translated from Philipp Frank, EINSTEIN (Munich: Paul List Verlag, 1949), p. 40.

7. I am grateful to Professor Pelseneer for a copy of the six-page essay. The covering letter that Einstein had sent with the essay to his uncle, Caesar Koch, received later the added note in Einstein's own hand: "1894 or 95. A. Einstein

(date supplied 1950)." The work was probably composed while the fifteen- or sixteen-year-old boy was preparing his first attempt to enter the Polytechnic Institute in Zurich.

8. The significance and validity of childhood experiences of this sort among scientists and inventors has not be sufficiently studied, although interesting material exists. One recalls here Newton's reported construction of toys, and Ernst Mach's account of the lasting impression made on him by a visit, as a small child, to a wind-operated grain mill. Orville Wright, when asked when and how he and his brother, Wilbur, first became interested in the problem of flight, responded: "Our first interest began when we were children. Father brought home to us a small toy actuated by a rubber spring which would lift itself into the air. We built a number of copies of this toy, which flew successfully."

9. Shortly after a draft of this paper had been prepared, I came upon a congenial passage on the same subject, pointing to the existence of "certain tensions," in the foreword by Russell McCormmach, editor of HISTORICAL STUDIES IN THE PHYSICAL SCIENCES (Philadelphia: University of Pennsylvania Press, 1970), Volume II, pp. X–XI. In recent correspondence, Professor McCormmach kindly has drawn my attention also to C. P. Snow's profile of Einstein in A VARIETY OF MEN (New York: Charles Scribner's Sons, 1967), in which the existence of some paradoxes is noted. Needless to say, literary figures have also been subject to such analysis; one of the less familiar examples may be V. I. Lenin's scathing discussion of the "glaring contradictions in Tolstoy's works, views, doctrines," in *Leo Tolstoy as the Mirror of the Russian Revolution* (1908).

10. P. A. Schilpp, ed., ALBERT EINSTEIN: PHILOSOPHER-SCIENTIST (Evanston, Ill.: Library of Living Philosophers, Inc., 1949), pp. 679–680.

11. Hermann Weyl, SYMMETRY (Princeton, N.J.: Princeton University Press, 1952), pp. 17, 130, 132.

12. For a detailed discussion of Einstein's rejection of the ad hoc elements in physics, see Gerald Holton, *Einstein, Michelson, and the "Crucial" Experiment*, ISIS, 60, 2: 132–197, 1969, and reprinted here as Essay 9.

13. Published in Jacques Hadamard's book, THE PSYCHOLOGY OF INVENTION IN THE MATHEMATICAL FIELD (Princeton, N.J.: Princeton University Press, 1945), pp. 142–143.

14. Einstein might have added "auditory" to visual and muscular. See his remark to R. S. Shankland (AMERICAN JOURNAL OF PHYSICS, 31:50, 1963): "When I read, I hear the words. Writing is difficult, and I communicate this way very badly."

15. Max Wertheimer, in the chapter *The Thinking That Led to the Theory of Relativity*, in PRODUCTIVE THINKING (New York: Harper & Bros., 1945), p. 184. Wertheimer's own remarks on psychology are useful, but those on physics are far less soundly based.

16. This point has been additionally documented, most recently in the record of remarks by Einstein in 1938 and published in 1971: "I grasp things as quickly as I did when I was younger. My power, my particular ability, lies in *visualizing the effects, consequences, and possibilities*, and the bearings on present thought of the discoveries of others. I grasp things in a broad way easily. I cannot do mathematical calculation easily. I do them not willingly and not readily." (Italics supplied.)

17. Translation of a quotation as given in Rudolph Arnheim, VISUAL THINKING (Berkeley, Calif.: University of California Press, 1969), p. 299. Arnheim's book itself is very suggestive and should be consulted.

18. In the appendix to the fourth edition (Leipzig: 1880) of Pestalozzi's book, WIE GERTRUD IHRE KINDER LEHRT, pp. 184–185.

19. This attitude may well account in good part for the fact, as Richter reports (p. 184), that major opposition arose against Pestalozzi as a consequence of his remarks concerning religious education and that in particular this work was considered an "estrangement from Christ."

20. Cf. Carl Seelig, ed., HELLE ZEIT-DUNKLE ZEIT (Zurich: Europa Verlag, 1956), p. 9. [Roman Jakobson, in *Verbal Communication*, SCIENTIFIC AMERICAN, September 1972, draws attention to the inspirational role Jost Winteler, a founder of linguistics known for the introduction of such concepts as "configurational relativity" and symmetry properties of sound patterns, may have had on his young boarder.]

21. E.g., Carl Seelig, ALBERT EINSTEIN (Zurich: Europa Verlag, 1954), pp. 16 and 21–24. There also was a marked anticlerical tradition in the school at Aarau, as documented by some of Einstein's classmates.

22. Noted, for example, by Friedrich Herneck, ALBERT EINSTEIN (Berlin: Buchverlag der Morgen, 1967), p. 49.

23. Based on the essay published in English translation in IDEAS AND OPINIONS (New York: Crown Publishing Co., 1954), pp. 224–227.

III *On the Growth of Science*

11 THE DUALITY AND GROWTH OF PHYSICAL SCIENCE

IN THE explosive growth of physical science during the last fifty years, the old views on the manner in which science grows were among the earliest casualties. The ground is still shifting and the emergence of one dominating new view cannot as yet be expected. Whatever the compromises that may eventually be reached in a field now stretching from Whitehead to Carnap, one type of observation will probably receive a great deal more attention than it was accorded in many previous analyses. I am referring to the essential incongruities in science, which include the element of irrationality and contradiction in scientific discovery, the discrepancy between the precision of physical concepts and the flexibility of language, the conflict between the motivating drive and the rules of objectivity—in short, the whole complexity in the relations between the individual creative scientist on one hand and science as an institution on the other. After describing some of these problems in the following paragraphs, it may be possible to construct a model-mechanism for the growth of science which accounts for the presence and, indeed, the *necessity* of these incongruities.

This essay was published in slightly longer form in the AMERICAN SCIENTIST, Volume XLI, pp. 89–99, 1953, and is based, in part, on material in the author's book INTRODUCTION TO CONCEPTS AND THEORIES IN PHYSICAL SCIENCE, Cambridge, Addison-Wesley, 1952 (rev. ed., 1973).

The statement that there is no single scientific method has become a truism only rather recently. Until the latter part of the last century, scientists themselves seem to have thought quite generally that their discoveries were being made in some orderly, step-by-step fashion. In the PRINCIPIA, Newton formulated four "Rules of Philosophy" which he then used frequently in his proofs; and in the OPTICKS, he gave a concise description of how a scientist should go about his business. That passage is worthy of close reading, for it has ever since been at the heart of all such recipes for the pursuit of science:

> As in Mathematicks, so in Natural Philosophy, the Investigation of difficult Things by the Method of Analysis ought ever to precede the Method of Composition. This Analysis consists in making Experiments and Observations, and in drawing general Conclusions from them by Induction, and admitting of no Objections against the Conclusions, but such as are taken from Experiments, or other certain Truths. For Hypotheses are not to be regarded in experimental Philosophy. And although the arguing from Experiments and Observations by Induction be no Demonstration of general Conclusions; yet it is the best way of arguing which the Nature of Things admits of, and may be looked upon as so much the stronger, by how much the Induction is more general. And if no Exception occur from Phaenomena, the Conclusion may be pronounced generally. But if at any time afterwards any Exception shall occur from Experiments, it may then begin to be pronounced with such Exceptions as occur. By this way of Analysis we may proceed from Compounds to Ingredients, and from Motions to the Forces producing them; and in general, from Effects to their Causes, and from particular Causes to more general ones, till the Argument ends in the most general. This is the Method of Analysis: And the Synthesis consists in assuming the Causes discover'd, and establish'd as Principles, and by them explaining the Phaenomena proceeding from them, and proving the Explanations.

The faith in an essentially simple method of science re-expressed itself during the following two centuries in all fields of science, and even beyond in philosophy and the social studies. It seemed that Newton had found a key to the problem of intellectual advance, and that the most convincing proof was the gigantic achievement of the master himself.

And yet, on looking into the history of science, one is overwhelmed by evidences that all too often there is no regular procedure, no logical system of discovery, no simple, continuous development. The process of discovery has been as varied as the temperament of the scientists. Of course, individual research projects are as a rule unspectacular, within their small scope fairly routine and logically consistent; but precisely some of the most important contributions have initially depended on wrong conclusions drawn from erroneous hypotheses, misinterpretations of bad experiments, or chance discoveries. Sometimes a simple experiment yielded unexpected riches, whereas some most elaborately planned assaults missed the essential effect by a small margin. Great men at times had all the "significant facts" in their hands for an important finding, and yet drew trivial or wrong conclusions; others established correct schemes in the face of apparently contradictory evidence. Even the work of the great heroes, viewed in retrospect, sometimes seems to jump from error to error until the right answer is reached; indeed, this gift must be one of the deepest sources of greatness.

One concrete example of the irrational element in scientific discovery will suffice here: John Dalton's formulation of the atomic theory in the 1800's. Initially, Dalton had been interested in meteorology, and particularly in the problem of why the gaseous constituents of the atmosphere are so thoroughly mixed despite the differences in their specific gravities. In reading Newton's PRINCIPIA, Dalton had found the proof that *if* a gas consisted of particles repelling one another with a force proportional to the distance between them, then it would have to exhibit a reciprocal relationship between pressure and volume such as had been found by Boyle in actual experiments on existing gases. Dalton took this statement to be a *proof* of the proposition that real gases do indeed consist of particles endowed with the stated forces. Next he accepted the postulate, in natural accord with the contemporary caloric theory of heat, that each gas particle is surrounded by a sphere of caloric fluid, a fluid endowed with the quality of self-repulsion. Thirdly, he announced, partly on the basis of his own experiments, that the individual particles of one pure gas must differ in size from those of another gas. Finally, Dalton concluded that the thorough mixture of components in the earth's atmosphere was now explainable because mutually-repulsive contiguous particles of several different sizes would not be in equilibrium in strata. It is now well known that this work led Dalton to the epochal concepts of the chemical atom, atomic weight, the Law of Multiple Proportions, and so forth, but it is worthy of note that *each and every one of his steps as just given was factually wrong or logically inconsistent.* Newton's proof was a mathematical exercise, not applicable to real gases, and the contemporary theory of heat was soon to be overthrown, for the caloric fluid concepts contained inherent contradictions. Dalton's own experiments were frequently inadequate to support his conclusion, to say the least, and his final conclusion did not even follow from his own premises.[1]

Although it is of course only the master in his field who can turn to his full advantage the illogical and unexpected, the significant point is the growing recognition of the importance of these elements, e.g. in the findings of researchers on the psychology of invention that some of the outstanding theoreticians believe their conscious thought-processes to occur without the mediacy of communicable symbols or languages.[2] At the least it has become clearer in our day that the pursuit of science is itself not necessarily a science.

On the Nature of Concepts

If the first point be granted, consideration can be given to the second. It is an intentional juxtaposition: after the sometimes irrational or casual nature of discovery, the logical nature of physical concepts.

Three aspects of established physical concepts may be briefly noted. First, they are operationally definable in one form or another. Although this condition is in many ways restrictive, it is enormously successful in bringing disputing parties to a fairly precise point of agreement or disagreement. Therefore, the energies of investigators are now not regularly drained off by misunderstandings and long battles concerning definitions and rules of procedure. There are indeed large areas of possible disagreement among scientists, but such arguments can be resolved, often by having recourse to some series of measurements which both disputants, rightly or wrongly, acknowledge at the time to be decisive. Surely one of the impressive features of modern physical science is the rapidity of settling most major differences of opinion in the field.

Operationally definable concepts, however, do not by themselves guarantee us a science. A second attribute of physical concepts is their largely quantitative character. This is not only true of concepts like valence or mass, but also more generally. For example, *mechanical equilibrium* is not measured in units of 3, 4, or 5, but it does correspond to the state in which a body has *zero* acceleration. More generally, a concept like *electron,* in actual use, is primarily a summary term for a whole complex of measurables, namely 4.8×10^{-10} units of negative charge, 9.1×10^{-28} gram mass, and so forth.

There is a third characteristic of physical concepts, without which science would degenerate into a meaningless mass of data. Evidently the chaos of experience permits the formulation of an infinite number of concepts, all of them quantitative and meaningful in the operational sense; but it is its recurrence in a great many descriptions and laws, often in areas far removed from the context of initial formulation, that induces the selection of a particular concept. The meaning behind the statement that we "believe in the reality" of electrons is that the concept is at present needed so often and in so many ways—not only to explain cathode rays, the phenomenon that leads to the original formulation of the concept, but also for an understanding of thermionic and photoelectric phenomena, currents in solids and liquids, radioactivity, light emission, chemical bond and so on.

The observation that the established physical concepts are largely *operational, quantitative* and *diversely connected* summarizes the severe limitations on the scope of science. At the same time these three attributes combine into the most effective mechanism conceivable for assuring the rapid increase of the scientific enterprise, aiding in the unambiguous communication of problems, disagreements, and results, and knitting together the contributions of many men, separated though they may be by the barriers of disciplines, oceans, or centuries.

In this process of accounting for limitations and precision in the nature of concepts, a contradiction appears to have arisen. Science seems to depend on clear and prescribed types of concepts in one direction, and on a free license of creativity in the other. As H. D. Smyth has recently stated, "We have a paradox in the method of science. The research man may often think and work like an artist, but he has to talk like a bookkeeper, in terms of facts, figures, and logical sequence of thought."

This dilemma is resolved—and here is the second central point—*by distinguishing two very different activities, both denoted by the same word, "science"*: the first level of meaning refers to *private science* (let us term it S_1), the science-in-the-making, with its own vocabulary and modes of progress as suggested by the conditions of discovery. And the second level of meaning refers to *public science* (S_2), science-as-an-institution, textbook science, our inherited world of clear concepts and disciplined formulations. S_1 refers to the speculative, creative element, the continual flow of contributions by separate individuals, each working on his own task by his own, usually unexamined methods, motivated in his own way, and uninterested in attending to the long-range philosophical problems of science. S_2, in contrast, is science as the evolving compromise, as the growing network synthesized from these individual contributions by the general acceptance of those ideas which do indeed prove meaningful and useful to generations of scientists. The cold tables of physical and chemical constants, the bare equations in textbooks, form the hard core, the residue distilled from individual triumphs of insight, checked and cross-checked by the multiple testimony of general experience.

This distinction between the two meanings of the same word, which appears to be helpful in allaying some serious confusions in methodological discussions, is often not adequately made by those who analyze science. Scientists themselves are largely responsible for that condition; for in formalizing an individual contribution for publication, it is part of the game to cover up the transition from the private to the public stage, to make the results in retrospect appear neatly derived from clear fundamentals, until, as phrased in John Milton's lines, "so easy it seemed / Once found, which yet unfound most would have thought / Impossible!" Months of tortuous, wasteful effort may be hidden behind a few elegant paragraphs, with the sequence of presented development running directly opposite to the actual chronology, to the confusion of students and historians alike.

On Motivation

The better to illuminate the duality of science by a specific example, a brief consideration of the complex problem of the *motivation* of scientists is now in order.

When Whitehead said, "Science can find no individual enjoyment in Nature; science can find no aim in Nature; science can find no creativity in Nature," he may have referred to the stable aspect of science, but surely not to the transient one; that is, to S_2, not S_1. Codified science is successful exactly in so far as it is indeed indifferent to human enjoyment and concerns; but the very opposite is true of the individual investigator. His creativity depends on a complete interpenetration of his person and his work. His dedication may spring from some nonrational, mystical or religious conviction which was often acknowledged in other years when scientists were freer with their personal secrets. For example, Galileo, a pious man, looked upon the laws of nature as proofs of the Deity equal to that of the Scriptures, and Newton revealed in a letter: "When I wrote my treatise [PRINCIPIA] about our system, I had an eye upon such principles as might work with considering men for the belief of a Deity; and nothing can rejoice me more than to find it useful for that purpose."

Although obvious cases like Kepler, Newton, and J. R. Mayer are rare, the nonrational substructure and the motivating forces manifest themselves in various other ways. Consider that ancient trend in science, the preoccupation and satisfaction with laws involving integral numbers. At the very outset of Galileo's historic work on the law of free fall, we find him drawing prominent attention to this fact.

Throughout the development of physics and chemistry and to the present day similar references are encountered. The work of Bode, Döbereiner, Balmer, and many of their respective contemporaries sometimes bordered on sheer play with numbers. A faint trace of this attitude re-emerges today in the use of "magic numbers" in nuclear theory—a term perhaps more suggestive than is generally assumed.

Of the multitude of supporting examples the following extract is sufficiently representative. Wolfgang Pauli, on accepting the Nobel Prize, 1945, was speaking of his great teacher's early work on line spectra:

> Sommerfeld however preferred . . . a direct interpretation, as independent of models as possible, of the law of spectra in terms of integral numbers, following as Kepler once did in his investigation of the planetary system, an inner feeling for harmony . . . . The series of whole numbers 2, 8, 18, 32 . . . giving the length of the periods on the natural system of chemical elements, was zealously discussed in Munich, including the remark of the Swedish physicist Rydberg that these numbers are of the simple form $2n^2$, if n takes on an integral value. Sommerfeld tried especially to connect the number 8 with the corners of a cube.

Apart from the preoccupation with theistic or numerical propositions, there are several other types of driving forces in scientific work. One of these is exemplified by Count Rumford's explanation why he could not accept the caloric theory after "the fact of the transmission of heat through the Torricellian vacuum was established beyond any doubt": "I must freely confess that, however much I might desire it, I never could reconcile myself to it [the caloric theory], because I cannot by any means imagine how heat can be communicated in two ways entirely different from each other."

It was a wrong reason for rejecting a fallacious theory. But this inability to allow two different mechanisms for conduction and radiation illustrates the common and powerful drive toward a holistic world picture. New conceptual schemes tend to be simple and unitary even though reason may tell us that theories have generally become more correct by becoming more complex.

Finally one may illustrate the powerful hold of a fourth "nonscientific" preconception with a quotation from Max Planck, who revealed in his autobiographical notes that it appeared to him that the independence of black body radiation from the physical and chemical nature of the emitter "represents something absolute, and since I had always regarded the search for the Absolute as the loftiest goal of all scientific activity, I eagerly set to work." It might by said that the foundations of quantum theory were laid under metaphysical auspices.

The avowed motivation of many other scientists appears equally startling. Surely it is precisely because the drive toward discovery is in a sense irrational that it is so powerful, even under the most adverse conditions. The very progress of science has often depended on what might have been noted at the time as the dogged obstinacy of its devotees, and what is now renamed their inspired tenacity. Names like Faraday and Joule come to mind instantly in this connection.

Inevitably there arises a serious question in such a recital: do not such personal and varied involvements endanger the very business of science, the search for "objective truths"? The resolution of this paradox is illuminating. Consider, as the specific example, the case of Eddington, who had such strong convictions that he could write a "Defense of Mysticism." Whether or not his persuasion was "meaningless" from the point of view of public science (S_2) it may have been the mainspring of his devoted search for truth. What is here important is that in his voluminous and distinguished scientific writing no overt sign of his personal mysticism can be found, nothing that might not equally well have been written by a gifted investigator with exactly the opposite metaphysical orientation, or even by one belonging to the "hard-boiled" school, quite indifferent to such questions.

This then is the third main point. In modern science personal persuasions no longer intrude explicitly into the published work, *not because they do not exist*—on the contrary, they are as essential in S_1 as they ever were—*but because, in S_2, they are now generally neglected.* Few scientists can forget the story of Julius Robert Mayer who came close to complete oblivion when he attemped to support his work almost entirely with that preposterous mixture of bad logic and bad metaphysics which had in fact led him to his conclusions. If he has become immortal, it is thanks to the fact that his inspired idea of the existence of a general Law of Conservation of Energy found champions who gave it theoretical extension and experimental confirmation.

The fruitlessness of metaphysical discussions in the sciences has finally brought about a curious case of atrophy; the personal metaphysical tenets of scientists, although sometimes very strong, are in a free society generally so varied, so vague, in fact technically so inept that in a sense they cancel out. We may go one step further, and state that even though a science (S_1) without a metaphysical substructure has never been possible, it may be argued that our science (S_2) will be healthy only so long as our scientists formally remain poor metaphysicians. In only those places where one codified, generally accepted set of dogmas exists—as in some of the old scholastic universities or in modern totalitarian states —can an extraneous and ultimately detrimental idea in a scientific publication still survive the scrutiny of colleagues.

On the Growth of Science

As contrasted with his largely unconscious motivations and procedures, the intellectual discipline imposed upon the physical scientist is now quite as rigorously defined as the conventionalized form for research papers. This superposition of discipline and convention upon the results of free creation represents a dualism in the work of the scientist which runs parallel to the dualism in the nature of science itself. I believe that the argument may be extended to explain the present form of the scientific activity, and perhaps even its spectacularly successful growth. Starting with the key recognition of a distinction between the two meanings of science, the difference between S_1 and S_2 is comparable to the double interpretation given to animal species—either as a group of diverse struggling individuals or as an abstraction catalogued and described in zoology texts at the present stage of evolution.

I find it helpful to regard science as a growing organism analogous to a biological species, for the growth in both cases depends to a large extent on the operation of four quite analogous mechanisms. First, there is a mechanism of continuity; both a species and a science can persist only if there is some stable method for transmitting the structure in a definite way from generation to generation. In biology the principle of continuity is found in the processes of heredity based on the highly specific nature of the genes, and in science continuity is identifiable with the specific operational and quantitative nature of important concepts. Without this measure of unambiguous continuity, scientists could not communicate their work to one another and to their pupils. It is not merely a truism that the convenience and freedom of communication is vital to the existence and growth of knowledge. The zealous secrecy with which the alchemists guarded their results doomed their efforts to collective stagnation and was one of the reasons for the delay in the rise of modern chemistry. Again today the growing imposition of secrecy in basic research (when it is done under military sponsorship) is raising fears that the arteries of science are being dangerously constricted.

Modifying the first mechanism is a second, the mechanism of mutation, the constant opportunity for individual variations. In the case of biological species, of course, the process of mutation is made possible by various chemical and physical influences on the genes, and on chromosome partition and recombination; in science, mutations are assured by its essential freedom and by the boundless curiosity of the human mind. Scientists have always vigorously defended the right to pursue any promising lead, to publish and to exchange scientific information freely. More than most groups they know that truths can be found only on the free market place of ideas. And this accentuates again the vital importance of academic institutions and foundations which enable men to follow the unpredictable paths of knowledge without constraint or interference.

A third mechanism in the growth of science is multiplicity of effort. To assure continuity and growth despite the low rate of occurrence of really great modifications, and in the absence of some obvious single master plan by which to proceed, science and the species alike must rely on a large number of individual attempts, from which may ultimately emerge those few types that are indeed useful. The uncountable fossils of bygone members of the species, like the uncountable pages of the individual scientific researches of the past, are mute testimonies of the wastefulness of the process of evolution. It would seem that apart from a favorable environment in which to grow, knowledge, like many zoological types, must be present in amounts larger than some threshold value; but given that condition, a free science, ever productive of its own tools for further advancement, tends to become a self-perpetuating or even an explosive chain reaction.

Finally, a selection mechanism is at work in the growth of science whereby certain of the seemingly innumerable contributions and unpredictable mutations are incorporated into the continuous stream of science—a conflict among ideas not greatly different from the fight for existence in living nature which permits a species to adapt itself to a changing environment. The survival of a variant under the most diverse and adverse conditions is mirrored in science by the survival of those discoveries and concepts that find usefulness in the greatest variety of further applications—of those conceptual schemes that withstand the constant check against experience.

The operation of this selection mechanism appears most forcefully in the historical controversies attached to our basic theories, that is, to the theories that contained the new, often implausible assumptions. As Whitehead reminds us, all truly great ideas seemed somewhat absurd when first proposed. Indeed, they may be called great ideas precisely because it took the unusual mind to break through the pattern of contemporary thought, to discern the truth behind its mask of a grotesque or trite disguise, to dare propose the unbelievable or question the obvious. Almost every great innovator, from Copernicus to Niels Bohr, has had to meet initially the skepticism or active opposition of his colleagues.

In the light of those triumphs, is the scientific fraternity not unreasonably conservative? Not at all. Conflict is a fundamental necessity in the evolution of ideas. We must not permit our judgment to be distorted by references to stories with famously happy endings. After all, Galileo might have been wrong; with Newton yet to come, his insistence and insight were far less convincing and important then than they appear now. The more urgent fact is that "revolutionary" ideas arise only very rarely compared with the large number of workable or even great ideas conceived within the *traditional* setting, so that the individual scientist is wisely predisposed to favor the type of advance which he knows and believes in from personal experience. He, rather like Galileo's opponents, quite rightly must defend his fundamental concept of nature against large-scale destruction, particularly at the earlier stages when the innovators cannot present very many results and confirmations of their new ideas.

Sometimes, indeed, the discoverers themselves are so strongly committed to established ideas, so startled by the implications of their new results, that they predict the storm of condemnation, or even hesitate to draw the final conclusions. The most famous recent instance was Hahn and Strassmann's statement on the brink of their experimental discovery that neutron bombardment of uranium results in nuclear fission: "We cannot yet bring ourselves to make this leap, in contradiction to all previous lessons of nuclear physics."

On returning now to our argument from the evolutionary processes governing the growth of science, the significance of the bitter struggles that new theories may create within science can be appreciated more fully. The fitness of new knowledge, like the fitness of a species, is most convincingly demonstrated and most advantageously molded by vigorous contest. The situation outside the sciences is not so very different; predominant religious and social concepts have not developed quickly or conquered quietly. In order to change the direction of development in a field of learning, people's minds must be changed. Even in science this is a slow process, sometimes an impossible one. Max Planck said, with perhaps only a little too much bitterness about his own early struggles: "An important scientific innovation rarely makes its way by gradually winning over and converting its opponents: it rarely happens that Saul becomes Paul. What does happen is that its opponents gradually die out, and that the growing generation is familiarized with the ideas from the beginning." And as if to prove his point, it was Planck himself who, five years after Einstein's first publication on the photon theory of light, angrily commented that all the fruits of Maxwell's great work would be lost by accepting a quantization of energy in the wavefront "for the sake of a few still rather dubious speculations."

These are examples of the conflicts which may find a central place in a modern understanding of the structure and development of science. Although our particular evolutionary view is based on a heuristic model which calls for considerable expansion, it does gather up a satisfyingly large number of problems, and provides yet another basis for the great moral which the progress of science teaches its students: faith in the marvelous ability to arrive eventually at truths by the free and vigorous exchange of intelligence.

NOTES

1. Fuller details are well developed in Leonard K. Nash, THE ATOMIC MOLECULAR THEORY, Cambridge: Harvard University Press, 1950. Other examples of this sort will be found in the other Case Studies which the Harvard University Press has been publishing under the editorship of Mr. Conant. See also Gerald Holton, INTRODUCTION TO CONCEPTS AND THEORIES IN PHYSICAL SCIENCE, (Cambridge: Addison-Wesley, 1952), Chapters 17–19.

2. See, for example, W. I. B. Beveridge, THE ART OF SCIENTIFIC INVESTIGATION, New York: W. W. Norton & Co., 1950.

12 MODELS FOR UNDERSTANDING THE GROWTH OF RESEARCH

12 MODELS FOR UNDERSTANDING THE GROWTH OF RESEARCH

THE FACT that an important social invention has occurred, one that is destined to transform a part of society, sometimes goes unrecognized for a surprisingly long time. A case of this sort was the nineteenth-century development of science as a profession. Another case exists at present. It is to be found in the particular way by which scientists have come to organize and coordinate their individual research pursuits into a fast-growing commonwealth of learning. The new pattern for doing basic research in science is worth studying for its intrinsic merits. This essay hopes to spell out what it now means to be active in basic scientific work.

There are more important reasons still for sketching here the operation of this new commonwealth of learning. One reason is the fact that this pattern carries specific lessons for the conduct and organization of effective scholarly work in any field, no matter how different or remote from science it may be and must remain; one such lesson should be in the definition of a scale for measuring the adequacy of support. The second reason is, conversely, the realization that scientific work may best be understood as one of the products of the general intellectual

This is a somewhat condensed and revised version of the essay originally published in DAEDALUS, Spring 1962, and included in EXCELLENCE AND LEADERSHIP IN A DEMOCRACY, Stephen R. Graubard and Gerald Holton, eds. (New York: Columbia University Press, 1962), pp. 94–131.

metabolism of society, and hence that in the long run the growth of science depends critically on the growth of all fields of scholarship.

The Stimuli for Growth

As a profession, science has been remarkably little studied, except for a handful of books and reports that seem to be covered over by the growing flood of changing statistical data. I shall choose basic research in physics as carried on today in the United States to characterize some common features of all the sciences. The choice is quite appropriate from several points of view. For example, the number of its academic practitioners has not grown at an inordinate rate compared with other fields of study. In the year 1914 there were only 23 doctorates awarded in physics in the United States out of a total of 505 for all fields, 244 of which were in science.[1] At that time, the Ph.D. degree granted in physics amounted to 4.6 percent of all Ph.D. degrees for the year, or 9.4 percent of those in all the sciences. Remarkably enough, a recent survey shows that half a century later we have virtually the same proportions. The 484 Ph.D. degrees in physics accounted for 5.1 percent of all Ph.D. degrees, and for 9 percent of all those in science.[2]

The great rise of research output in physics in the last half century did not entail a corresponding loss of numbers in other areas. Indeed, in a sense, there has been a relative decrease in the number of basic-research physicists, since now one-half of all new Ph.D.'s in physics are heading for governmental or industrial research and administration employment for which there was no equivalent in previous years and where a much smaller fraction of men are doing basic research than are in academic employment.[3] Thus the remaining group of Ph.D.'s in physics is of intermediate size—that is, it is comparable to the graduating Ph.D. classes in history, political science, mathematics, religion, or English literature.[4] Those who do not stay in the academic life serve, of course, to link physics to applied science, as is also the case in the larger fields of chemistry and biology.

Physics is a good profession to choose for this analysis because there exists an immense variety in the group, from the man in the small college who, with two or three colleagues, does all the work of the department and still finds time to think about new physics, to the man whose full time is spent in the laboratory of a large research institute. Any two physics-research projects picked at random are likely to have less in common with each other than does the statistical average of all physics research compared with the statistical average of almost any other experimental science. And yet there is in this group, taken as a whole, a strong sense of cohesiveness and professional loyalty. Despite the variety, despite the specialization that makes it difficult to follow what is being done in the laboratory next door, despite the important differences between basic and applied, large and small, or experimental and theoretical physics, its practitioners still clearly conceive of themselves as doing in different ways work in one identifiable field. There are no large cleavages and disputes between sizable factions representing fundamentally different styles.

If we first focus specifically on the professional life of a representative physicist, it is essential to remember at the outset the brevity of time in which basic research on a significant scale has been done in this country, or, for that matter, anywhere. The word "scientist" itself did not enter the English language until 1840. Until about the turn of the century, the pattern was that of work done by isolated men. Experimental research was often financed with one's own funds. Even in a relatively large department, advanced students were rare. Thus, Harvard University, one of the earliest in the United States to grant Ph.D. degrees in physics, had a total of six theses before 1900, and thereafter an average of about two per year until World War I. During that war, it has been reported, "there was no classification of physicists. When the armed forces felt the need of a physicist (which was only occasionally), he was hired as a chemist."[5] Having an adequate laboratory space of one's own in most universities was an unfulfilled wish even for the outstanding experimentalist—though this applied more to Europe than to the United States. For example, in 1902, at the peak of their research, four years after their discovery of polonium and radium and after many years of pleading for more space in which to do their extensive chemical and physical work, the Curies still had only their old wooden shed at Rue Lhomond and two small rooms at Rue Cuvier. On being proposed for the *Légion d'Honneur*, Pierre Curie wrote to Paul Appell: "Please be so kind as to thank the Minister, and inform him that I do not feel the slightest need of being decorated, but that I am in the greatest need of a laboratory."

The growth of science between the wars needs little discussion. The driving force was in part the needs of an increasingly sophisticated, technologically oriented, competitive economy, and in part the sheer excitement induced in more and more students (drawn from a widening base in the population) by beautiful ideas ever more rapidly revealed, such as the quantum theory and early nuclear physics. But the rate at which exciting ideas are generated is correlated with the ability of a field of study to "take off" into self-amplifying growth.

To the economic and intellectual stimuli of earlier days the Second World War added the new stimulus of the threat of nuclear power in the hands of the Germans, who had been the foremost nation in scientific achievement. Einstein's letter of 1939 to President Roosevelt dates the moment after which the scale of research support changed in a surprisingly short time by more than an order of magnitude. Since 1940, Federal funds for science alone have grown over one hundred-fold.[6]

What mattered here, however, was really not so much the hot war and the cold war, for wars by themselves had not in the past unambiguously promoted the growth of science. Rather, it was a development unprecedented in recorded history: the demonstration that a chain of operations, starting in a scientific laboratory, can result in an event of the scale and suddenness of a mythological occurrence. The widespread fascination and preoccupation with science—in itself an essential element in its continued growth—find here their explanation at the elemental level.

In our society there had always been a preoccupation with the scientific hero who comes back with a major revelation after having wrestled with his angel in self-imposed isolation (i.e., Newton, Röntgen) or in relative obscurity (i.e., Curie, Einstein). Now, a whole secret army of scientists, quartered in secret cities, was suddenly revealed to have found a way of reproducing at will the Biblical destruction of cities and of anticipating the apocalyptic end of man that has always haunted his thoughts. That one August day in 1945 changed the imagination of mankind as a whole—and with it, as one of the by-products, the amount of support of scientific work, including accelerators, field stations, observatories, and other temples.

To a physicist, nothing is so revealing as relating qualitative changes to quantitative changes. Man can cope surprisingly well with large rates of change in his environment without himself changing significantly. His psyche can take in its stride rapid rearrangements in the mode of life—collectively, for example, those owing to a large increment in the life span, and, individually, those owing to great deterioration of health.

Precipitous changes of condition over fairly short periods are well-tolerated. But the traumatic experience of *one* brief, cataclysmic event on a given day can reverberate in the spirit for as long as the individual exists, perhaps as long as the race exists. Hiroshima, the flight of Sputnik and of man in space—these were such mythopoeic events. Every child will know hereafter that "science" prepared these happenings. This knowledge is now embedded in dreams no less than in waking thoughts; and just as a society cannot do what its members do not dream of, it cannot cease doing that which is part of its dreams. This, more than any other reason, is the barrier that will prevent scientific work from retreating to the relative obscurity of earlier days.

Who Are the Scientists? A Representative Case

The element of discontinuity in the general experience of our time merely reinforces the discontinuities in the experiences of contemporary science. The rate at which events happen is again the important variable. For, when a field changes more and more rapidly, it reaches at some point a critical rate of activity beyond which one has to learn by oneself, not merely the important new ideas, but even the basic elements of one's daily work. This is now true of many parts of physics and of some other fields of science, not only for the most productive and ingenious persons, but for anyone who wishes to continue contributing. The recent past, the work of one or two generations ago, is not a guide to the future, but is prehistory.

Thus the representative physicist is far more his own constantly changing creation than ordinary persons have ever been. His sense of balance and direction cannot come from the traditional past. It has to come from a natural sure-footedness of his own—and from the organism of contemporary science of which he strongly feels himself a part. None of the novels or the representations in the mass media which I have seen have portrayed him with success, perhaps because they missed the fact that this is the component that really counts.

Though I am referring to statistical data, the man I have selected to typify my comments is not a statistical average but rather a summary of traits, each of which is well-represented in the profession and all of which, taken together, will be generally agreed to among physicists as representing a worthy and plausible specimen. I go into some detail, partly because not only the novelists but even the anthropologists have so far failed to penetrate this part of the forest to provide a good description of the new tribe. But I also want to make a basic point about the humane qualities of training and professional life. First of all, I note that our man, like the majority of his colleagues, is young, perhaps thirty-five years old, or just three years short of the median age of fulltime-employed scientists in the United States.[7] Even so, he has already had nine years of professional experience and increasingly creative work, having finished his thesis at the age of twenty-six, after a graduate study period of about four and a half years.[8] In completing this work, like 25 percent of all physics graduate students in the country,[9] he was supported by fellowships. During his last two years he worked on a research assistantship, helping an experimental group in the construction of a new type of beta-ray spectrometer and submitting as his thesis early measurements he made with it in connection with this work.[10] Thus, like the majority of graduate students in physics, his education was financed from the outside and proceeded without significant delays.

After graduation, he hoped to obtain a postdoctoral fellowship—perhaps the best way for the really good scholar to consolidate his grasp of his material and to map out a field for himself before plunging into the routine of professional life. But there are not yet enough such programs, and he did not receive an award. He found a position at a middle-sized university. In selecting academic life—the only aspect of the profession to be treated here—he has become one of approximately 8,000 physicists in colleges and universities, as against twice as many working in industry and half as many in the government.[11]

He knows from folklore that in 1945 there were only a few really good departments of physics in the United States; but now there are some thirty universities with research programs lively enough to yield between five and forty-three theses each year,[12] and there are many more good small departments. The availability of funds has helped to spread excellence in basic research widely and rapidly. He would have an even larger choice in liberal arts colleges, but he has become rather used to cooperative experimental research of the size and with the tools that are usually associated with the larger universities. Significantly enough, there are twice as many physical scientists on the faculties of universities as in liberal arts colleges; they form a larger fraction of the total faculty on campus; and—most important for this particular experimentalist—whereas the average liberal arts college employs one non-faculty professional staff person in physical science for every ten physical science faculty members, at the average university the proportion is better than one to one.[13] This implies much better backing from technical personnel in universities, particularly for those inclined to do large-scale experimentation; however, the ratio in some colleges will probably soon improve, when the regional joint facilities now being developed among colleges in several areas are completed.

Our physicist has to his credit a number of publications—several short papers and one long review paper. He is considered a productive person, interested in one of the main excitements (which to him has recently become an experiment in the field of high-energy physics), and, to some degree (less than one would perhaps like) in his undergraduate students. These he meets relatively rarely by the standards of his predecessors. One formal lecture course, more rarely two, is a typical class schedule for a physics professor at a major university; it allows sufficient time for work, for contact with graduate students, and for the long seminars with colleagues in which one carries on one's continuing self-education. Summers are given by members of his small group to research on the same contract with the government agency that sponsors the project. During these months there may be some extra salary for faculty and assistants. When necessary, there are trips to one of the seven major laboratories sponsored by the Atomic Energy Commission but administered for unclassified academic research by a regional group of universities.[14]

These circumstances, to repeat, are not typical of all scientists, but representative of a type of new scientist now often encountered. What is emerging is the picture of a research-minded scholar who lives in a world that has arranged fairly adequate support to help him carry through his ideas whenever such help is possible. This help shows up in a number of other important (or even quite trivial) ways. For example, postdoctoral fellowships bring good research talent at no extra cost to the project, for a year or two at a time. Or when an important article in a Russian-language journal appears, it will be found in one of the translation journals of the American Institute of Physics.

An insight into the sources from which basic-research sponsorship usually comes and the places where the work is done may be obtained by a quick count of the acknowledgments cited in the program abstracts for a recent meeting of the American Physical Society.[15] Of the 480 papers contributed, 18 percent are from colleges and universities without indication of foundation or government support; 43 percent acknowledge such support (from the National Science Foundation, the Atomic Energy Commission, the United States Air Force, the Office of Naval Research, the National Aeronautics and Space Administration, or others) ; 21 percent are papers on basic research done in and largely financed by industry; 16 percent were done in government (including national) laboratories by persons employed there; and the remaining 2 percent include sponsorship from private foundations such as Sloan and Ford.

Our man's federally funded contract, therefore, is financed quite typically; it is not a large contract, and of course no part of the work is hampered by restrictions on publication nor, indeed, does it have any directly foreseeable applications to defense activities.[16] The amount of the grant available to our man and to a senior colleague and collaborator who is acting as "principal investigator" is perhaps $46,000 for a two-year period. About half this sum is for the purchase and construction of equipment; the rest is largely for services, including graduate student research assistants. Though the original request for funds was considerably cut, there is enough to pay for some work by the machine shop, electronics technicians, secretarial help, the draftsman's office or the photographer, and for publication and reprint charges. The contract support, therefore, is adequate.

Our physicist is better off than a considerable number of other academic physicists in less convenient circumstances. Many, in smaller colleges particularly, are hard-pressed. And, on the other hand, this man is perhaps not differently situated from many an equally talented and productive young man or woman in fields outside the sciences. Nonetheless, it is clear by the standards of the recent past in physics itself that here is a new type of scholar. Indeed, he and each of many colleagues like him has available for life the security, means, and freedom to do research that Alfred Nobel hoped to give by his prize to the few outstanding persons in the field. Most significantly, our new scientist is new in that he does not regard himself as especially privileged. To be sure, there is much anxiety at each application for new or renewed funding for research support, and there may be periods of real scarcity. But on the whole the facilities for doing creative work are being accepted and used by him without selfconsciousness.

This is the point. For whatever reasons, right or wrong, that society has chosen to make this possible, the circumstances exist for getting scholarly work done by more people than might otherwise do it, and for providing humane conditions of training for the oncoming generation.

There are at once a number of urgent objections, of course. One might say that it is not difficult to construct utopias for any field, given enough money. On the one hand, the money involved is not so large that a major country cannot easily afford it, and the amount small on any scale except that of depression-reared experience or the starvation-oriented practices in all too many other equally worthy fields of scholarship;[17] on the other hand, this is not a paper utopia, but a working system for employing people's minds and hands in the time-honored mission of adding to the sum of the known.

Alternatively, the opposite objection may be heard: that really good ideas do not flourish without an element of personal hardship. But, despite the support intended by well-known stories (true, false, and sentimentalized), the evidence now is altogether the other way round. The once-in-a-generation ideas may still, as always, come from the most unexpected places; yet, throughout history, transforming ideas, as well as great ideas only one magnitude less high, have not appeared in science at a rate equal to a fraction of the present rate. The sacrifice implied by the sum of thousands upon thousands of wretched student and research years under inadequate conditions in the past can surely be no source of satisfaction, even if the additional expenditures had not, after all, shown a better yield in science. I suspect that another Marie Curie, a Kepler, even a Roger Bacon, would not be damaged by more help, or by the availability of cooperative research facilities for those inclined to use them.

I will refrain from elaborating on the point that the new scientist now seems to have at least as much time and energy for other socially valuable activities as previous generations did. Because in themselves they are either not new or not intrinsically unavoidable parts of the present pattern of science, I shall equally refrain from elaborating here some of the persistent and well-known complaints raised by scientists themselves: the volume of material to digest, the imbalance between different special fields, the encroachment of bureaucracy and of military technology, the need for keeping some "big science" efforts big and in the news by artificial means, the usual difficulty of finding persons of real judgment to be advisers or to help run the scientific organizations themselves, and the poverty of many teaching efforts.[18]

I shall also neglect here the occasional pirate who is drawn to the scientific field, as in earlier times a man of talent with a like soul would have found scope for his aspirations in the service of a queen or a Boniface III. The obligations and opportunities of power and all it entails now lie on many of the most outstanding scientists, and abuse is exceedingly rare.[19]

There remains a third major objection. Has this useful and often pleasant arrangement not been bought at too high a price? It is popularly suspected that somewhere in the background there is a group of high military officers whose interest and decision ultimately control, from year to year, whether or not academic research shall flourish, just as the Renaissance patron determined whether the studio would continue or not. It is not, after all, only the intrinsic merit of the subject that now gives it vigorous life, but also the weapons-aspect of its occasional by-product, vigorously exploited by applied scientists and engineers in industry and government.

This is of course frightening and confusing ground. In part these widely held conceptions are not true, or at least no longer true. The influence of government (particularly that of the military branch) on science has not been without an effect in the opposite direction. As some scientists have become increasingly effective and trusted in their roles as advisors, a noticeable educative influence has made itself felt in Washington. Indeed, it is nowadays more typical for scientific advisors to try to turn off what appear to be hastily conceived projects initiated by the Pentagon.

And yet, the deeper intent of the objection cannot be either disproved, or evaded, or sustained. It is at the same time bitterly true and false, as would be a refusal to sanction the rising standard of living in our present, artificially inflated economy. The problem posed is at bottom the same for the academic scientists as it is for anyone from grammar school teacher to legislator who participates in the life of a nation which is so closely geared to an arms race. (One suspects that if tomorrow it were discovered how to destroy multitudes by reciting poems, the physicists would have to move into the garrets, and poets would be enticed into the laboratory space.) But while the hope of gaining indirectly military benefits from basic science motivates the Federal agencies that support physics, the large majority of academic scientists themselves have clearly declared again and again their eagerness to work toward a peaceful resolution of the crisis that is to a degree responsible for the high level of their support. In fact, it is largely from the work of such scientists that one may hope for the development of ideas, understanding, and techniques that will help in achieving what mankind never before took to be a serious task, the control of armaments and of international aggression.

Requirements for Growth: Mobility, Organization, Leapfrogging

While it would not be either possible or necessary in this context to describe in detail the research project that engages our physicist's attention, let us turn from his personal background to the general rules of action of the profession. We leave him as he is contemplating a possible modification in the use of a liquid-hydrogen bubble chamber, a device for making apparent the passage of elementary particles such as those generated in accelerators. The triggering event for this thought was a brief article, the heading of which is duplicated in Figure 1.

It will be instructive to study this figure with care. It contains a great deal of information about the metabolism of a lively field of scholarship, denoted even in the very name of the journal. The PHYSICAL REVIEW is perhaps *the* definitive physics journal in America, though it is only one of the many good journals in which basic research in physics is published. In 1958, the sheer bulk (7,700 pages in that year), the continuing rate of expansion, and the delay between the receipt and publication of articles made it necessary to detach from the PHYSICAL REVIEW the "Letters to the Editor," in which brief communications are made. This resulted in the separate, quickly printed, semimonthly publication, PHYSICAL REVIEW LETTERS. The article indicated in Figure 1 came out a month after its receipt; under the older system it might have taken twice as long.

Why is this speed so important? One explanation could be that this profession is made up of fiercely competitive people. It is true that egos are strong and competition naturally present. But in the United States, at least, it proceeds in a low key; personal relationships, though perhaps lacking some color and warmth, are almost invariably friendly.

There are three explanations for this fact. First, the authority of scien-

VOLUME 7, NUMBER 6 PHYSICAL REVIEW LETTERS SEPTEMBER 15, 1961

HELICITY OF THE PROTON FROM Λ DECAY*

J. Leitner, L. Gray, E. Harth, S. Lichtman, and J. Westgard
Syracuse University, Syracuse, New York

M. Block, B. Brucker, A. Engler, R. Gessaroli, A. Kovacs, T. Kikuchi, and C. Meltzer
Duke University, Durham, North Carolina

H. O. Cohn and W. Bugg
Oak Ridge National Laboratory, Oak Ridge, Tennessee

A. Pevsner, P. Schlein, and M. Meer
Johns Hopkins University, Baltimore, Maryland

and

N. T. Grinellini, L. Lendinara, L. Monari, and G. Puppi
University of Bologna, Bologna, Italy
(Received August 16, 1961)

*This research is supported in part by the Office of Naval Research, U. S. Atomic Energy Commission, Office of Scientific Research, and the National Science Foundation.

Figure 1. The heading of a short announcement of results in PHYSICAL REVIEW LETTERS, 7:264, 1961.

tific argument does not lie in personal persuasiveness or in personal position but is independently available to anyone. Second, there is the general loyalty to the common enterprise, mentioned previously. And most importantly, scientists as a group seem to be self-selected by a mechanism that opposes aggressive competition. Anne Roe, in summarizing her long studies in this field, reports in an essay, *The Psychology of Scientists:*

> Their interpersonal relations are generally of low intensity. They are reported to be ungregarious, not talkative—this does not apply to social scientists—and rather asocial. There is an apparent tendency to femininity in highly original men, and to masculinity in highly original women, but this may be a cultural interpretation of the generally increased sensitivity of the men and the intellectual capacity and interests of the women. They dislike interpersonal controversy in any form and are especially sensitive to interpersonal aggression.[20]

Thus the theory of aggressive competition is not likely to be correct in explaining the speed often felt to be necessary. Rather, one must look to other causes. I will select two quite obvious ones, which seem to me among the most important. One is the intense interest in what has been found. The other is the natural desire not to be scooped by other groups known to be interested in the same topic. And here it is important to note a major cause for this possibility—the fact that research is usually carried out in the open. It would be most unusual for a typical academic physicist not to instruct any visitor who shares his interests on the detailed current status of his research, even if, and precisely because, this same visitor is working on the same "hot" lead. This principle of openness is one of the basic aspects of the scientific ethos.

We now read the names of the authors given in Figure 1, and are perhaps surprised by their number. To be sure, a commoner number of collaborators would be two, three, or four, although ten percent of the authors of the other papers in the same issue of the journal are sole authors. Yet it is neither the longest list of authors to be found, nor is it unrepresentative. Here let me signal three points. One is the cooperation in research that is implied within each group, as well as among widely dispersed groups; another is the distribution in this country (and indeed internationally) of the cooperating enterprises (some long established, others not known as little as twenty years ago to have had strong research interests in physics); the third is the authors' remarkably heterogeneous backgrounds that are implied. The list of names makes the point more bluntly than could any comment of mine.

This last point is perhaps the most important of these factors in explaining the growth of science in our time. Nowhere else can one find a better *experimental* verification of the general worth of the democratic doctrine, which is often uttered but rarely tested seriously. Social and geographic mobility in a field of work, as in society itself, is the essential prerequisite for a full exploitation of individual talent. The success of contemporary science all over the world despite the great variety of social and political settings is merely a striking case study of this proposition.[21]

The gathering of talent brings not merely rewards proportionate to the amount of talent but also rewards that are, at least in the early stages of a new field, nonlinear and disproportionate. In other words, the contributions of n really good persons working in related areas of the same field are likely to be larger (or better) than n times the contribution of any one of them alone in the field. This is true of a group as well as of individuals who do not work in physical proximity to one another.

With respect to the former, the particular way group work or cooperative research functions was long ago discovered and exploited by industrial laboratories and by medical researchers. Although some group research existed as far back as the seventeenth century, and beginnings of cooperative research even on something like the present scale of groups had been made, notably in the Cavendish Laboratory and E. O. Lawrence's laboratory at Berkeley, physicists did not really understand its full merits until the creation of the World War II laboratories (the Manhattan District, the Massachusetts Institute of Technology Radiation Laboratory, the Harvard Radar Countermeasures Laboratory, and others). Not only did they learn what it means to do science when the rest of society is really backing science (a lesson not forgotten); more particularly, they discovered how to work together in groups, despite the fact that a member may be neither particularly inclined to gregariousness nor even informed in detail on the subject of his neighbor's specialization.

What took place here was analogous to impedance matching, the method by which an electronics engineer mediates between the different components of a larger system. That is, special coupling elements are introduced between any two separately designed components, and these allow current impulses or other message units to pass smoothly from one to the other. Similarly, in these quickly assembled groups of physicists, chemists, mathematicians, and engineers, it was found that the individual members could learn enough of some one field to provide impedance matching to one or a few other members of the group. They could thus communicate and cooperate with one another somewhat on the model of a string of different circuit elements connected in one plane, each element being well enough matched to its immediate neighbors to permit the system to act harmoniously. While an applied organic chemist, say, and a pure mathematician, by themselves, may not understand each other or find anything of common interest, the addition of several physicists and engineers to this group increases the effectiveness of both chemist and mathematician, *if* each scientist is sufficiently interested in learning something new.

That this system worked was a real discovery, for the individual recruits had come largely without any experience in group research. And while during the war the system of cooperative research was tried out successfully on applied, or "mission-directed,"[22] research on a large scale, it was continued after the war in many places in basic science, at first on a much smaller scale—and it was still found to work to great advantage.

Another and even more important effect of group work on the growth of a field exists among eager groups in the same field who are, however, not side by side but located at some distance from one another. One research team will be busy elaborating and implementing an idea—usually that of one member of the group, as was the case with each of the early accelerators—and then will work to exploit it fully. This is likely to take from two to five years. In the meantime, another group can look, so to speak, over the heads of the first, who are bent to their task, and see beyond them an opportunity for its own activity. Building on what is already known from the yet incompletely exploited work of the first group, the second hurdles the first and establishes itself in new territory. Progress in physics is made not only by marching, but even better by leapfrogging.

We can turn for a specific illustration to accelerators, not because they are glamorous or unique, but because quantitative data are easy to find there. Ernest Rutherford suggested in 1927 that the nucleus should be explored by bombarding it with artificially accelerated particles, because the natural projectiles available from radioactive sources are neither continuously controllable in speed nor of high enough energy. This gave rise at the Cavendish Laboratory in the early 1930's to the design and construction by J. D. Cockcroft and E. T. S. Walton of an accelerator for protons. Its first successful operation is represented by a black circle near the left edge of Figure 2.[23] Improvements since then have increased the top operating energy, e.g., in the proton linac, from the original one million electron volts (1 Mev) to about 60 Mev (note the nonlinear, i.e., logarithmic scale on the ordinate). But in the meantime, a profusion of new machines of quite different types have made their appearance, one after another. The cyclotron of E. O. Lawrence and M. S. Livingston (1932) was a radically different machine, and it immediately rose to higher operating energies; but this curve later flattened out (owing to the impossibility of a fixed-frequency resonance accelerator of this type to impart effectively more energy to particles when these have already achieved a significant relativistic mass increase).

The electrostatic generator, initiated by Van de Graaff at the Massachusetts Institute of Technology, entered the situation at about this time, with less energy but with useful advantages in other ways. It differed from its two main predecessors qualitatively (i.e., in the funda-

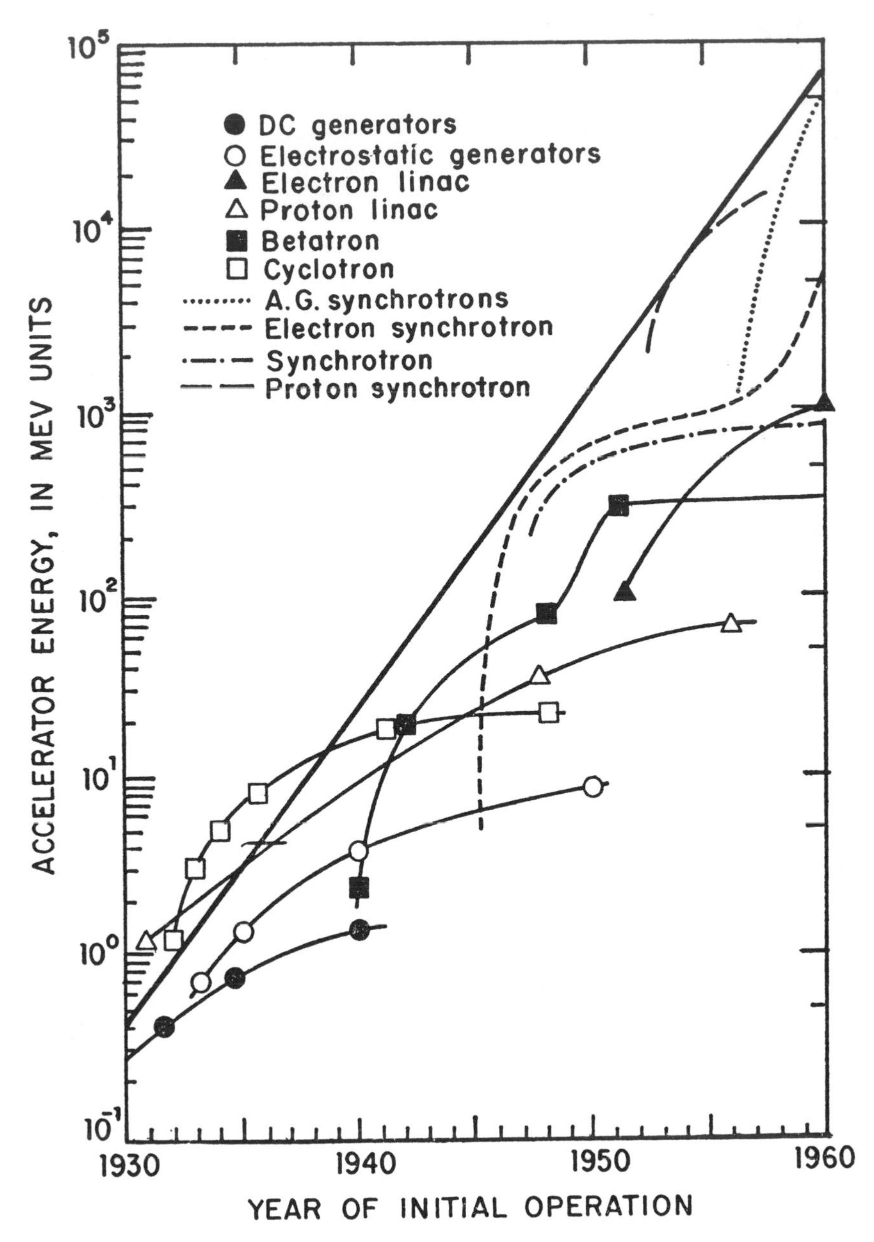

Figure 2. The increase of operating energy in particle accelerators. (Courtesy of M. S. Livingston.)

mental method of achieving the accelerating voltage), as indeed these differed from each other. In 1940 the betatron—again a fundamentally different machine—started with a design by D. W. Kerst at the University of Illinois, and then entered regions of higher and higher energies, where new phenomena could be expected to occur. New machines are continuing to come from different groups and widely dispersed laboratories; the leapfrogging process is clearly at work and opens up more and more spectacular fields for basic research.

One cannot help noticing an unexpected but crucial result in Figure 2. The heavy straight line (which would be an exponentially upturned curve if it were on an ordinary plot instead of on the semilog coordinates) of course indicates roughly the approximate maximum accelerator energy available to physicists in any year. This line shows that the top energy increased on the average by a factor of about ten every five years—for example, from about 500 Mev in 1948 to about 5,000 Mev (i.e., 5 Bev) in 1953. At this rate, the 33,000 Mev Alternating Gradient Synchrotron at the Brookhaven National Laboratory, first operated on 29 July 1960, was ready none too soon.

This ten-fold (i.e., order-of-magnitude) increase in energy every five years entails a corresponding opening up of interesting results and new fields of work, each of which will keep research projects going for a long time. The multiplication of fields and results constitutes a graphic example of what is meant by an increase in scientific activity in one area. This, too, is a particular and peculiar pattern of physical science—although, of course, the time for a doubling of range or scale is not so short in most other areas of physics.[24] The driving force here is in large part a simple and general psychological one: particularly when the more onerous material constraints on the realization of an ingenious new idea are removed, the really original person is not likely to be interested in spending his creative energy on something that produces much less than a three-fold, five-fold, or preferably an order-of-magnitude change. This has always been true, even when the financial considerations prohibit the realization of the idea, or when costs are inherently no great factor. A five- to ten-fold increase in accuracy of measurement or of prediction; an extension of the accessible pressure range from 2,000 atmospheres to 10,000, then to 50,000, then to above 200,000; an eight-fold increase in the volume of space seen by a new telescope—these are obviously interesting and worthy goals. On the other hand, to increase the precision or range in an area by, say 30 percent is good, but is not likely to generate special enthusiasm in an individual or a particular group.

The natural pace, therefore, is that of doubling (or more), and of doing so rapidly. As in developments in the military missile field, the urge is strong to design an accelerator which will be beyond the one now being readied for its first tests. Leapfrogging has become somersaulting. But not all physics is accelerator-bound, just as not all science is physics, and so a balance is preserved in the large.

These considerations apply directly only to experimental physics, and even then only to those research projects that go after an extension of knowledge that can be associated with an increase of some numerical index such as range or accuracy. It therefore does not refer to such experimentation as the investigation of G. P. Thomson, which was intended to confirm whether or not an electron beam exhibits wave properties, and it also does not refer to much theoretical work. Models to deal with these cases are nevertheless possible—for example, by using as a quantifier the criterion of the inclusion in one framework of previously unrelated elements, and the production of new, unrelatable entities—and such models produce the same general conclusions concerning the increase of pace.

Diffusion Speed and Critical Rates

Nothing is more striking in a high-metabolism field such as physics or experimental biology than the usefulness of the present. For example, M. M. Kessler[25] has found that 82 percent of the references cited in research papers published in the PHYSICAL REVIEW during the last few years are references to other recent articles in scientific journals. Half of these articles cited are less than three years old! Reference to the more distant past decreases quite sharply; only 20 percent of all references are seven years old or more.

After journal citations, the next most frequent references (about 8 percent) are to private communications, unpublished or to be published; if the latter, they are usually in preprint form, the old standard method of communicating in a specialty field, a method which has now grown markedly. References in PHYSICAL REVIEW articles to books turn out to rank only third, or 6½ percent (the remaining 3½ percent of references being to industrial reports, theses, etc.). Even these books seem increasingly often to be edited volumes of various articles. The net effect, then, is that of the diffusion and use of information at high speed.[26]

There are other ways in which scientific information diffuses and is used. Nothing, surely, is a more viscous medium for diffusion than the educational system of college and high school. How do the advances of science fare there? We know that the situation is not yet satisfactory, and we can understand the difficulty that must arise whenever the diffusion time is radically different from the natural pace of research. An example is the treatment of special relativity theory in a long-established senior-level physics text, such as F. K. Richtmyer's INTRODUCTION TO MODERN PHYSICS. In the first edition (1928) the theory of relativity occupied about a page. Six years later came the second edition, with twelve pages on this topic, gathered in an appendix. The third edition, eight years later, had a separate, regular, thirty-page chapter in the text. And in subsequent editions of this outstanding text the material has properly spread throughout the book so that it is meaningless to make an estimate of the actual space given it. But then, little had been added to special relativity theory as a separate research topic since long before 1928.

Alternatively, by making a cut through the educational system another way, one can follow the progress of ideas as they move from the research desk down to the schoolroom. The emanation electroscope was a device invented at the turn of the century to measure the rate at which a gas such as thorium loses its radioactivity. For a number of years it seems to have been used only in the research laboratory. It came into use in instructing graduate students in the mid-1930's, and in college courses by 1949. For the last few years a cheap commercial model has existed and is beginning to be introduced into high school courses. In a sense, this is a victory for good practice; but it also summarizes the sad state of scientific education to note that in the research laboratory itself the emanation electroscope has long since been moved from the desk to the attic. The high rate of turnover of ideas in science presents almost insoluble problems for a conventional educational system in which information about the events at the top are propagated slowly and without a short-circuiting of any of the intermediate elements below.

In order to have a better model of the process by which knowledge in

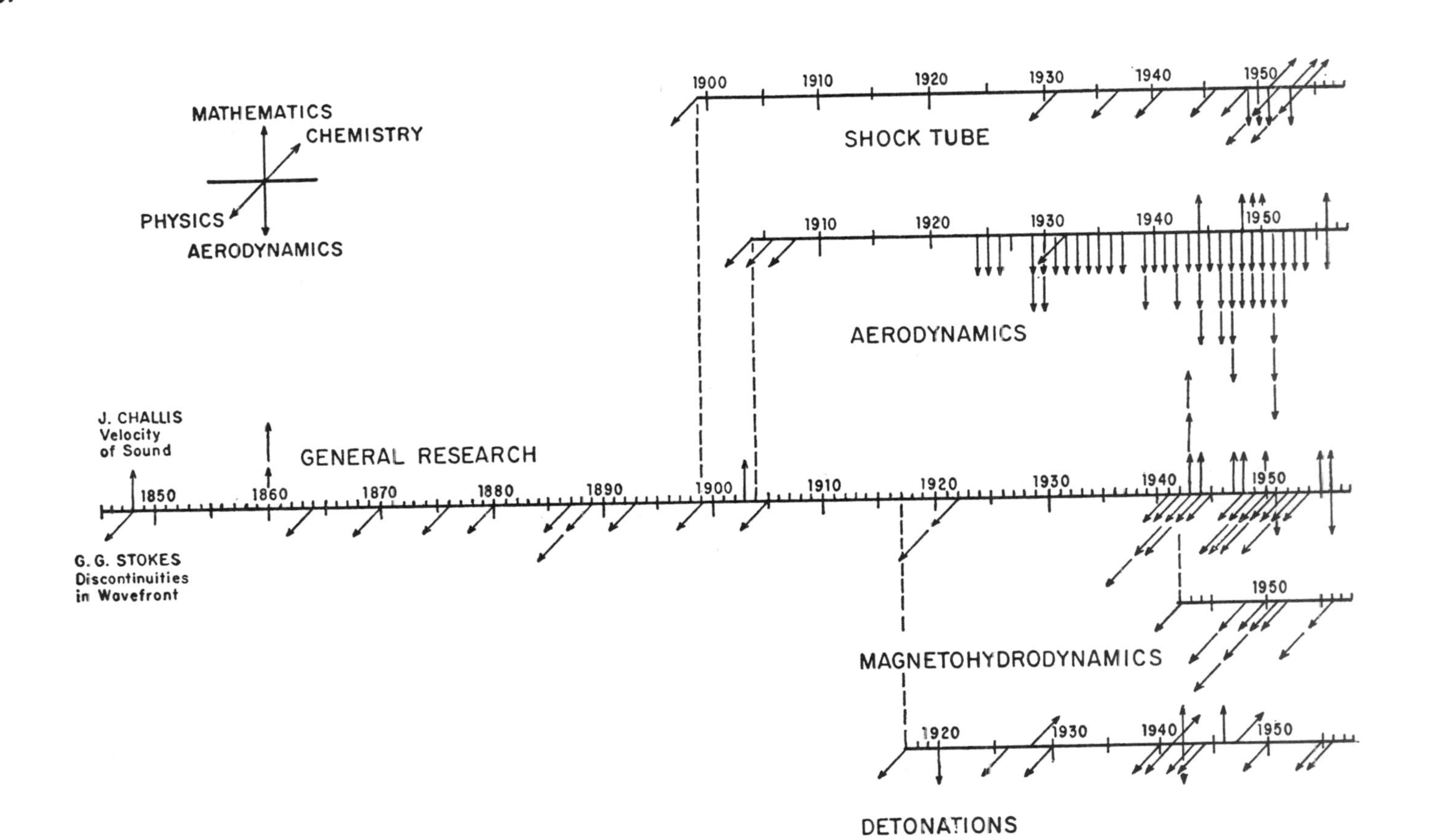
MATHEMATICS
CHEMISTRY
PHYSICS
AERODYNAMICS
SHOCK TUBE
AERODYNAMICS
J. CHALLIS
Velocity
of Sound
GENERAL RESEARCH
G. G. STOKES
Discontinuities
in Wavefront
MAGNETOHYDRODYNAMICS
DETONATIONS
1850
1860
1870
1880
1890
1900
1910
1920
1930
1940
1950

a research field advances, we must think about the rate of diffusion along yet another dimension. In all fields of scholarship, the inputs for a lively research topic are not restricted to a narrow set of specialties, but can come from the most varied directions. In physical science it is easy to document this process of the diffusion of knowledge from many sides, over a period of time, into one research area—on the part of individuals, and quite independently of the effectiveness of groups dealt with earlier. Figure 3 is a schematic design intended to give, in rough approximation, both a feeling for what may be meant by the "growth of a field" and an overview of the cumulative effects of contributions from various scientific specialties.[27]

The field chosen is that of shock waves. It is a "classical" research subject that originated in 1848 when the British mathematician and physicist G. G. Stokes and the astronomer and mathematician J. Challis communicated their struggles with solutions of the equation of motion in a gas as developed by Poisson in 1808. Stokes was led to propose, on theoretical grounds, that a steep gradient in velocity and density should exist in the gas if a large disturbance were propagated in it. Both their contributions are represented by the two arrows at the far left, the directions of the arrows indicating the specialty fields involved.

The successive events are similarly indicated. For example, further basic work in the mathematics of wave propagation by Riemann and by Earnshaw follows in 1860, and other arrows placed on the "General Research" line refer to contributions in mathematics by men such as Hadamard (1903), Chandrasekhar (1943), and Kantrowitz (1951), or in physics by Mach (1876, 1887, 1889), Bethe (1942), von Neumann (1943), and Truesdell (1951). New specialty fields branch off as shown from time to time, some having pronounced technological orientation; but it is illustrative of the difficulties of clear separation that a branch such as magnetohydrodynamics (where the initial arrow indicates the work by Alfvén in 1942) now plays a fundamental part in both basic and applied fusion research. The increasing activity is evident throughout. As these lines go forward, one may well expect further branchings

Figure 3. A representation of the development of basic research and of some applications. Each arrow represents a major contribution. Its direction indicates the specialty field involved (see coordinate system at the left); for example, an arrow rising perpendicularly from the time axis represents mathematics.

at the growing edge from any of the five present lines, and fundamental contributions along any of the four dimensions. It is becoming more and more evident that departmental barriers are going to be difficult to defend.

Another illustrative interpretation of cumulative growth is obtained by following, on a shorter time scale than Figure 3, the effect and interrelationship of a few particularly creative and stimulating persons within a field. Figure 4 represents the results of a recent study,[33] tracing in general terms the rise of the fields of molecular beams, magnetic resonance, and related work in pure physics. In particular, it is focused on one part of the extensive achievement of I. I. Rabi, both in developing the original molecular beam techniques, and in selecting and stimulating a group of productive associates or students.[28]

This description is analogous to making a large magnification of a small part of the previous figure to determine its "fine structure." After working with Otto Stern in Hamburg, Rabi in 1929 effected a branching-off from previous lines of research (analogous therefore to Alfvén's arrow for 1942 at the head of the magnetohydrodynamics line in Figure 3, or the arrow on the aerodynamics line for Prandtl in 1904). It can be seen that soon after, both in independent laboratories as well as in those of Rabi and his associates, the applicability of the early techniques, and the originating of new questions now suggesting themselves in neighboring parts of the same fields, provoked a rapid branching into several new directions. The excitement of this field as a whole and its fruitfulness are attested by the large rate of inflow of new persons, including many outstanding experimental and theoretical physicists.

The course of the future is clearly going to be a continuing multiplication on the same general pattern. And although the growth is more eye-catching at the end portion of each branch, there is still a fruitful harvest in many of the lower boxes in Figure 4. Thus, molecular beams themselves remain important in current research. Finally, the connections with the technological exploitations of these advances have not been represented; but one should be aware that such connections almost invariably exist, and in this case they could be shown at several points (for example, maser, atomic clock).

A Simple Model for the Growth of Research in Science

We may now correlate the descriptive details in a simple qualitative model of the growth process of scientific research. It is too ambitious to expect such a model to tell us "how science works," but it should help us to understand its more bewildering and spectacular aspects.

A hypothetical construction should start with a "zeroeth-order" approximation; that is, we know it to be inadequate from the beginning, but we also know how to improve it to attain a first-order approximation and, if possible, higher-order approximations later. Such a start is provided by Newton's analogy of having been on the shore of the known, "while the great ocean of truth lay all undiscovered before me." Scientists do indeed seem generally to think about basic research in terms of some such picture. They often have described it as if it were a voyage of discovery launched on uncharted waters in the hope of reaching a new shore, or at least an island. To be sure, neither research nor a sea voyage is undertaken without some theory that serves as a rough chart. Yet such vague terms are used, even when the promise of end results would strengthen the cause of the hopeful explorer. Thus during a Congressional inquiry to ascertain the large financial needs for future accelerator constructions, the scientists—quite properly—gave Congress no more definite commitment of returns on the considerable investment it was asked to undertake than this:[29]

> It is, therefore, likely that the next decade will see the discovery of unexpected phenomena as well as the development of hitherto unknown techniques of particle detection and identification, and new means of particle acceleration and containment. Since it is impossible to predict the nature of these developments, it is very difficult to take their effect into account in any ten-year cost preview.

Taking the analogy of the voyage of exploration as sufficiently suggestive for the moment, we see that on the average a single searcher will expect the number of new islands he discovers to increase with time, perhaps more or less linearly. The same will be true if his is not the only ship that has started out, and if we assume the expeditions to be still few and not yet in contact with one another so as to affect the individual search patterns.

Hence the number of unknown islands yet to be found in a finite ocean (that is, the number of interesting ideas—not "the facts"—supposed to be still undiscovered in this pool) will be expected to drop off in time, somewhat as line I in Figure 5(a) does. In developing a model

Figure 4. Connections among the contributions in an expanding part of basic physics.

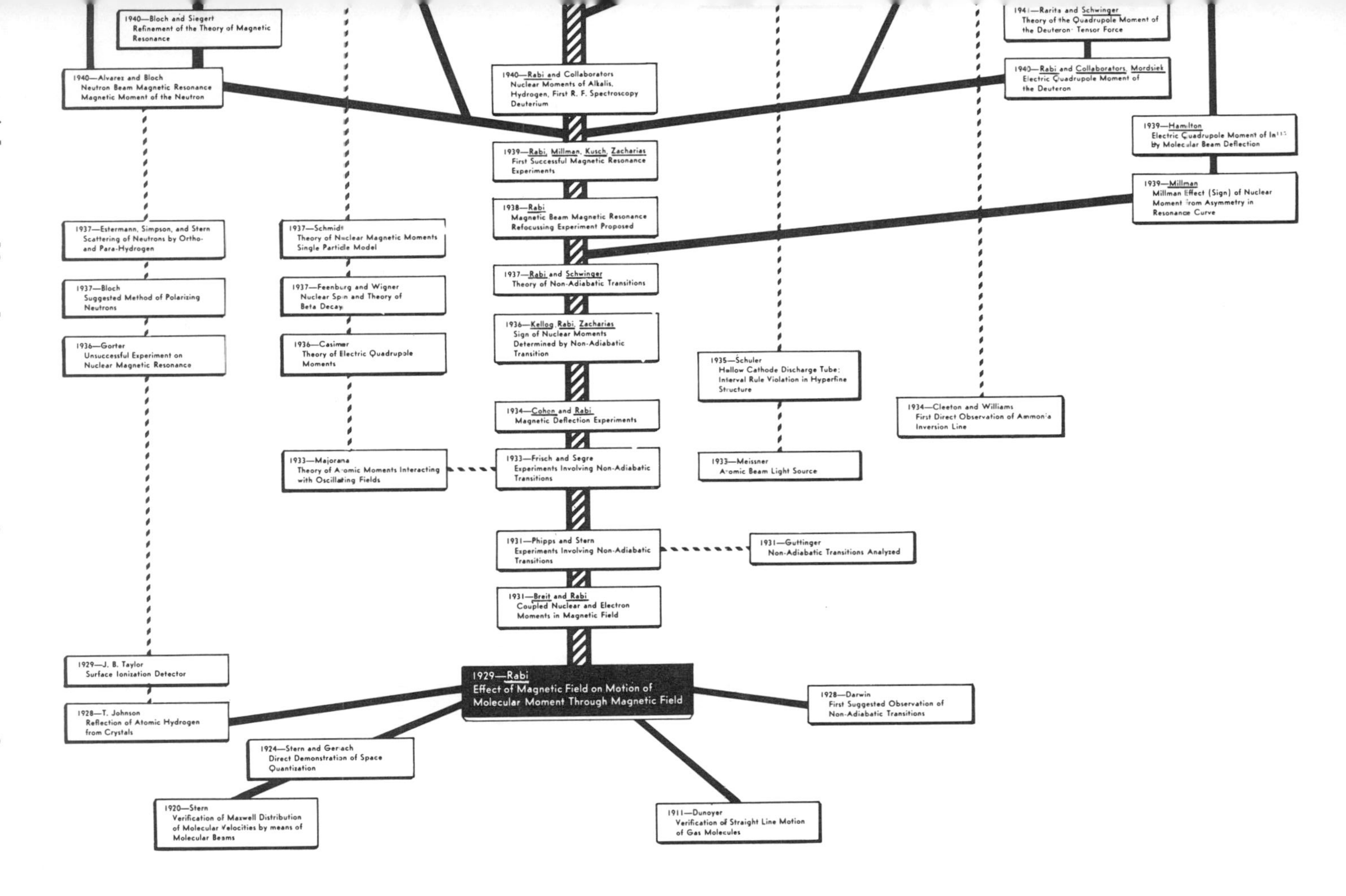

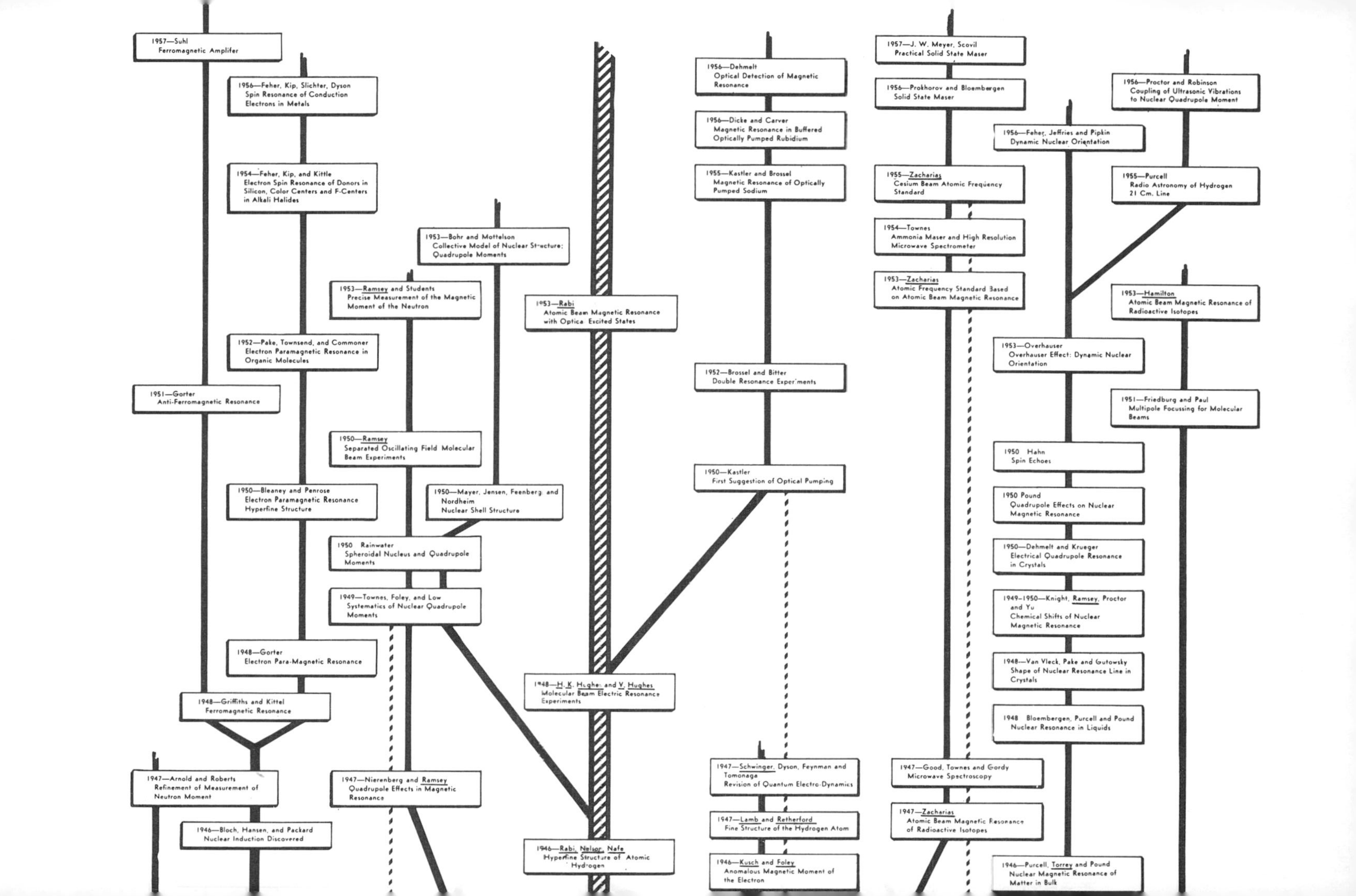

1957—Suhl Ferromagnetic Amplifer
1956—Feher, Kip, Slichter, Dyson Spin Resonance of Conduction Electrons in Metals
1954—Feher, Kip, and Kittle Electron Spin Resonance of Donors in Silicon, Color Centers and F-Centers in Alkali Halides
1952—Pake, Townsend, and Commoner Electron Paramagnetic Resonance in Organic Molecules
1951—Gorter Anti-Ferromagnetic Resonance
1950—Bleaney and Penrose Electron Paramagnetic Resonance Hyperfine Structure
1948—Gorter Electron Para-Magnetic Resonance
1948—Griffiths and Kittel Ferromagnetic Resonance
1947—Arnold and Roberts Refinement of Measurement of Neutron Moment
1946—Bloch, Hansen, and Packard Nuclear Induction Discovered
1953—Bohr and Mottelson Collective Model of Nuclear Structure; Quadrupole Moments
1953—Ramsey and Students Precise Measurement of the Magnetic Moment of the Neutron
1950—Ramsey Separated Oscillating Field Molecular Beam Experiments
1950—Mayer, Jensen, Feenberg and Nordheim Nuclear Shell Structure
1950 Rainwater Spheroidal Nucleus and Quadrupole Moments
1949—Townes, Foley, and Low Systematics of Nuclear Quadrupole Moments
1947—Nierenberg and Ramsey Quadrupole Effects in Magnetic Resonance
1953—Rabi Atomic Beam Magnetic Resonance with Optical Excited States
1948—H. K. Hughes and V. Hughes Molecular Beam Electric Resonance Experiments
1946—Rabi, Nelson, Nafe Hyperfine Structure of Atomic Hydrogen
1956—Dehmelt Optical Detection of Magnetic Resonance
1956—Dicke and Carver Magnetic Resonance in Buffered Optically Pumped Rubidium
1955—Kastler and Brossel Magnetic Resonance of Optically Pumped Sodium
1952—Brossel and Bitter Double Resonance Experiments
1950—Kastler First Suggestion of Optical Pumping
1947—Schwinger, Dyson, Feynman and Tomonaga Revision of Quantum Electro-Dynamics
1947—Lamb and Retherford Fine Structure of the Hydrogen Atom
1946—Kusch and Foley Anomalous Magnetic Moment of the Electron
1957—J. W. Meyer, Scovil Practical Solid State Maser
1956—Prokhorov and Bloembergen Solid State Maser
1955—Zacharias Cesium Beam Atomic Frequency Standard
1954—Townes Ammonia Maser and High Resolution Microwave Spectrometer
1953—Zacharias Atomic Frequency Standard Based on Atomic Beam Magnetic Resonance
1947—Good, Townes and Gordy Microwave Spectroscopy
1947—Zacharias Atomic Beam Magnetic Resonance of Radioactive Isotopes
1956—Proctor and Robinson Coupling of Ultrasonic Vibrations to Nuclear Quadrupole Moment
1956—Feher, Jeffries and Pipkin Dynamic Nuclear Orientation
1955—Purcell Radio Astronomy of Hydrogen 21 Cm. Line
1953—Hamilton Atomic Beam Magnetic Resonance of Radioactive Isotopes
1953—Overhauser Overhauser Effect: Dynamic Nuclear Orientation
1951—Friedburg and Paul Multipole Focussing for Molecular Beams
1950 Hahn Spin Echoes
1950 Pound Quadrupole Effects on Nuclear Magnetic Resonance
1950—Dehmelt and Krueger Electrical Quadrupole Resonance in Crystals
1949–1950—Knight, Ramsey, Proctor and Yu Chemical Shifts of Nuclear Magnetic Resonance
1948—Van Vleck, Pake and Gutowsky Shape of Nuclear Resonance Line in Crystals
1948 Bloembergen, Purcell and Pound Nuclear Resonance in Liquids
1946—Purcell, Torrey and Pound Nuclear Magnetic Resonance of Matter in Bulk

for discovery, we shall now build a series of simple graphs on Figure 5(a) to summarize in an easily perceived form some qualitative trends.

But if Figure 5(a) itself were a proper model for discovery, science, like geographical exploration or gold-mining, would sooner or later be self-terminating. In fact, the end should come sooner rather than later, because the news of discoveries in a fruitful ocean spreads interest in them. New explorers will rush in, as shown by the early part of curve P in Figure 5(b). This influx by itself will assure that the quantity of ignorance remaining decreases with time in a manner shown not by curve I but by curve I′ in Figure 5(c); that is, it will drop more nearly exponentially than linearly. If one also takes into account the fact that communication among the searchers shown on curve P improves the effectiveness of each one's search (a main function of communication, after all), then the middle portion of curve I′ should really drop off even more steeply, causing I′ to have the shape of an inverted sigma; and this is precisely what the partial graphs detailed in Figure 2 indicated. In either case, however, the specified field will in time become less attractive, and the number of investigators will be decreasing somewhat as shown. Curve P thus indicates directly the size of the profession at any time, and indirectly—by the steepness of the slope of P—the intensity of interest or attractiveness of the field with respect to net recruitment (the inflow minus the outflow of people).

We shall soon have to add some mechanism to explain why science as a whole increases in interest and scope instead of deteriorating, as in Figure 5. Nevertheless, we already recognize that for some specific and limited fields of science this model is useful. Thus in 1820 Oersted's discovery of the magnetic field around wires that carry direct current, and

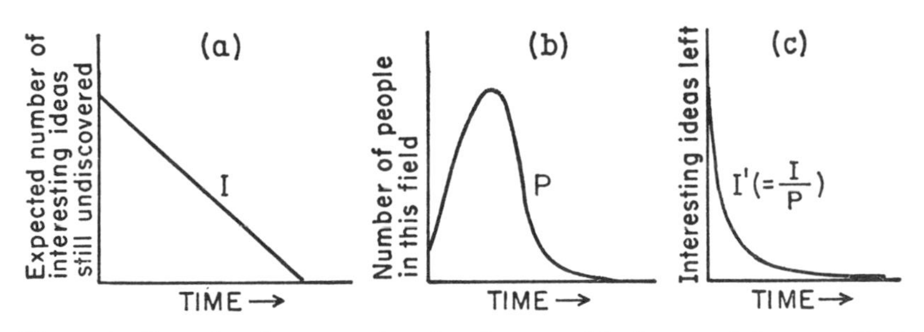

Figure 5. Zeroeth-order approximation for a model of research in a specified area.

the theoretical treatments of the effect by Biot, Savart, and Ampère in the same year, sparked a rapidly rising number of investigations of that effect; but it was not long before interest decreased, and by the time of Maxwell's treatise (1873) no further fundamental contributions from this direction were being obtained or even sought.

In fact, the same statement now applies (even in a good program) to virtually every topic presented in depth to physics students throughout their undergraduate training, and to a number of their typical graduate courses—except for students' own thesis fields. So, while Figure 5 may also be applicable to other areas of scholarship, the impressively different feature in physical science is that the time span for curve I′ has become quite short when compared with the time span of an active researcher's professional life, and frequently even when compared with the new recruits' period of training.

Figure 6 shows again in curve I′ the decrease of ignorance, together with a time scale (T, 2T, 3T, etc.) along with abscissa, drawn in such a way that the amount I′ has dropped roughly to half the initial value when period T has elapsed, to one-quarter after total time 2T, to one-eighth after 3T, etc. T is thus the "half life" of the suspected pool of

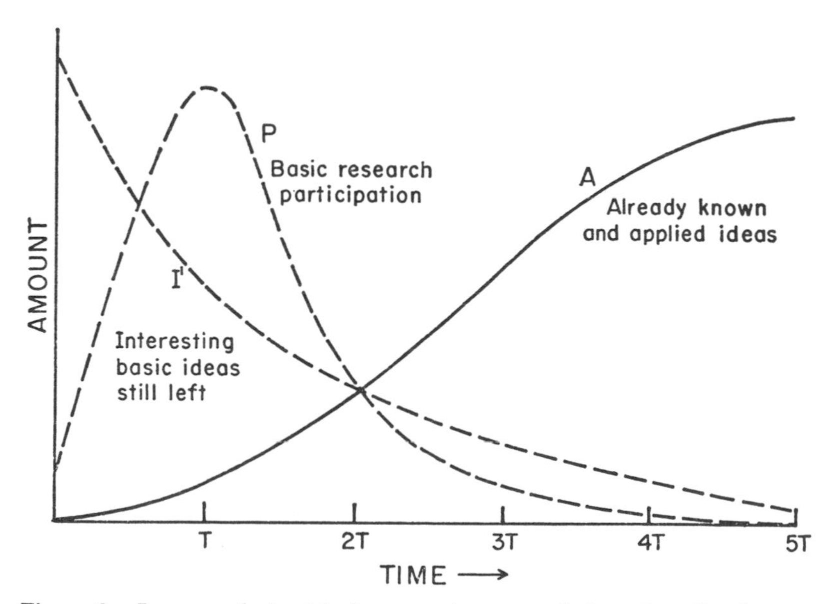

Figure 6. Inverse relationship between the accumulation of application and the interest in a basic-research field.

interesting basic ideas. The statements of the last paragraph imply that T is now short, perhaps between five and fifteen years for a specific, lively field on the frontiers of physical science.[36] This is also in accord with the data cited earlier, which showed that in reports of new research the references to published work fall overwhelmingly within the most recent years.

While it is not intended here to give an accurate idea of the absolute scales, the relative positions, or the detailed shape of the curves, P has been placed so as to indicate that the number of active researchers will reach a maximum when a large part of the presumed total of interesting ideas has already been discovered. This suspicion and the sense of dwindling time also contribute to the evident pressure and the fast pace. It appears to me that a critical slope for I′ exists. When the rate of decrease indicated by I′ for the specific research field is not so large (that is, when T is of the order of the productive life span of individuals, or longer), the profession organizes its work, its training methods, and its recruitment quite differently than if the value for T is only a few years. There are recent examples, as the case of oceanography, of a science passing from the first phase into the second, taking on many of the sociological characteristics of physics as a profession.

By means of Figure 6 we can briefly consider the application of new findings in basic research, as indicated in curve A. Such applications include use in other fields (for example, radioisotopes in medicine), and use for applied research and development. Curve A is meant particularly for the last of these, for example in the development of an industrial product. Clearly, a curve P′ that would be similar to P could be drawn to show how the number of people engaged in applied research is likely to grow and ultimately to diminish, for it is their work which A traces out.

Such a curve P′ would have the same general shape as P, but it would be displaced to the right of curve P. For it is clear that the longer one waits before beginning to apply fundamental ideas, the more nearly one's work will seem to be based on complete knowledge. Today, however, curve A does not wait to rise until I′ has reached very small values. We can readily understand this in terms of three factors: the competitive pressures within an industry, the natural curiosity of talented people, and the needs of basic research itself—which, in experimental physics at least, is now closely linked with the availability of engineering developments of basic discoveries. A curve P′ for applied research participation will therefore overlap curve P for basic-research participation, and indeed these two populations will often draw on the same sources. For example, Kessler[30] reminds us that articles in the PROCEEDINGS OF THE INSTITUTE OF RADIO ENGINEERS refer with considerable consistency to the publication of basic research in physics; in a relatively new applied field, such as transistors, reference to articles in the PHYSICAL REVIEW occur not much less frequently than citations of PHYSICAL REVIEW articles do in basic-research journals. In the past much blood has been shed over distinctions between pure and applied research. It may be fruitful to assume that a critical difference lies in the relative positions on the time axis of curve P showing the basic-research population and a corresponding curve P′ that could be drawn for the applied research population. The fruitful interaction of basic and applied science will be indicated by the overlap of these two populations, in time as well as in the sources from which they draw their material.

A First-order Approximation

We are now ready to attempt a first-order approximation to improve our model for the progress of scientific research. For this purpose we examine Figure 7(a), where curve D is simply the mirror image of I′, plotted in the same plane. That is, whereas I′ presented the decrease of ignorance, D presents the increase of total basic "discoveries" made in the finite pool of interesting ideas. The beginning of curve D indicates necessarily the occasion that launched the expeditions in this field, say the discovery in 1934 of artificial radioactivity by the Joliot-Curies while they were studying the effect of alpha particles from polonium on the nuclei of light elements.

Up to this point their research had followed an older line, originating in Rutherford's observation in 1919 of the transmutation of nitrogen nuclei during alpha-particle bombardment. The new Joliot-Curie observation, however, inaugurated a brilliant new branch of discovery. We suddenly see that the previous model (Figure 5) was fatally incomplete because it postulated an *exhaustible* fund of ideas, a limited ocean with a definite number of islands. On further exploration, we now note that an island may turn out to be a peninsula connected to a larger land mass. Thus in 1895 Röntgen seemed to have exhausted all the major aspects of X-rays, but in 1912 the discovery of X-ray diffraction in crys-

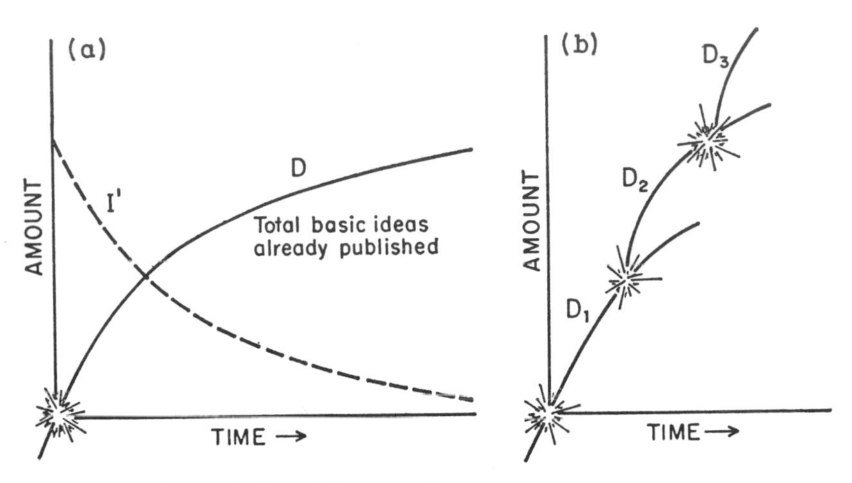

Figure 7. The escalation of discovery lines.

tals by von Laue, Friedrich, and Knipping transformed two separate fields, those of X-rays and of crystallography. Moseley in 1913 made another qualitative change by showing where to look for the explanation of X-ray spectra in terms of atomic structure, and so forth. Similarly, the Joliot-Curie findings gave rise to work that had one branching point with Fermi, another with Hahn and Strassmann. Each major line of research given by line D in Figure 7(a) is really a part of a series D_1, D_2, D_3, etc., as in Figure 7(b). Thus the growth of scientific research proceeds by the *escalation* of knowledge—or perhaps rather of new areas of ignorance—instead of by mere accumulation.

By means of this mechanism, we can understand at the same time the pace, the proliferation, and the processes of diffusion and branching shown in Figures 3 and 4. When an important insight (including a "chance" discovery) causes a new branch line D_2 to rise, fruitful research usually continues on the older line D_1. But many of the most original people will transfer to line D_2, and there put to work whatever is applicable from their experience along D_1. (Perhaps now the most important thing to know is when to drop D_1 and go on to D_2.) If the early part of D_2 rises steeply—because there is now a new area of ignorance that can be filled in at a rapid initial rate—then D_2 will appear as an exciting field, and will be very attractive to researchers. Many will switch to D_2; but the largest source of ready manpower is the new recruits to the field. Hence the lively sciences have a constant need to "grow," or at least to enlarge the profession. (This may well run eventually into difficulties as the limitation of available talent—or of the willingness of society to provide the needed funds—sets an upper boundary to possible growth.[31])

The newest recruits, therefore, are likely to be serving their apprenticeship at the newest and most rapidly growing edges. This is an excellent experience for them. But rapidity of growth depends on the inflow of research talent, and at the same time it is also *defined* by the output achieved. Thus there appears a danger of self-amplifying fashions: from a long-range point of view, too many people may be crowding into some fields and leaving others undermanned. One partial remedy has been for the less fashionable fields to set up their own professional specialty organizations and their own training and recruitment programs—a process which, once initiated, further polarizes the narrowing subsections within science as a whole.

With the concept of escalation in mind, let us finally re-examine Figure 2, where we found a leapfrogging process to ever higher accelerator energies. The two concepts are intimately related; Figure 2 indicates the application of the escalation process of Figure 7(b) to a narrow and particularly vigorous specialty, that of accelerator design.[32] The same analysis may be applied directly to other experimental fields which do not have such strong increases in the value of an easily identifiable variable. But it should again be stressed that advances in most theoretical aspects of science, and in not a few experimental ones, do resist quantification, and that then no analogue to Figure 2 can be readily drawn. Nor should this be unexpected. In the end, what any advance must be judged by is not some quantifiable improvement in a specialty, but the qualitative increase in the depth of understanding it contributes to a wide field. For this reason, our model for the growth of science must, and should, remain qualitative.

We may now summarize. We have described what is considered an adequate system designed to support the pursuit of interesting ideas that add to man's basic knowledge—a system that aids researchers to do this sooner rather than later, and with work and luck to make a large difference to the state of knowledge of their field. We have noted that —at least as long as reasonable funds are made available—scientific work can so arrange itself as to maximize the effectiveness of collaborators, of encouragement by one's fellow-men, and of the stimulus of new results, all of which keep morale high. We have noted also the open invitation to talent, no matter what real or imagined barriers to it may exist elsewhere; the aid given from student days forward for continued education; the predominantly youthful character of the profession, with the inflow of bright young people that is steadily growing; and the sense of building on the contribution of others.

This, I believe, would be a fair description of the major features of most basic-research sciences as professions in the United States today, whatever their faults may be in detail. But it is to be noted that none of these traits is inherently and necessarily restricted to the profession of science. (Indeed, until recently, perhaps the majority of these traits did not describe any science well.) The description in the paragraph above might well apply to almost any field of scholarship, as it now does to some of them. In this sense I regard this contemplation of the physical sciences as useful, not because their methods are to be imitated, but because they have achieved a state of operation that need have nothing to do with science as science, but only with academic science as a profession in which the achievement of excellence at every level of performance is the overriding criterion for the way the profession organizes its work.

One must of course distinguish between what is unique to science and what is not. It is not the point to say that historians must be mature men before they can be historians, or that the Romance languages did not help build bombs and have no need of cyclotrons, and that the Navy is not waiting for break-throughs in theology. It is also not to the point to say that science is unique in its attention to quantifiable knowledge, in its need for cumulative growth, or in its luck or its ability to survive periods of acceleration in growth. Certainly, the clamor for more money and more manpower for its own sake is always wrong, even in science, and it can be fatal outside science. Perhaps, indeed, we need no increase in the rate of scholarly production of studies in Byzantine art or even in the history of biology. *But even in science,* the quantitative aspects of "growth" are merely indices of deepening understanding. Therefore, the question now must be: allowing for differences between the needs of special fields of scholarship, what can we do to help each of the particular fields realize its full measure of excellence?

NOTES

1. Raymond T. Birge, *Physics and Physicists of the Past 50 Years,* PHYSICS TODAY, 9: No. 5, 20–28, 1956.

2. STATISTICAL ABSTRACT OF THE UNITED STATES (Washington, D. C.: U. S. Department of Commerce, 1961), pp. 130, 359. [More recent data shows that the same trend continues. The same applies to virtually all major points made in what follows. Ed.]

3. Byron E. Cohn, ed., REPORT OF THE CONFERENCE ON CURRICULA FOR UNDERGRADUATE MAJORS IN PHYSICS (Denver: University of Denver, 1961), p. 12.

4. STATISTICAL ABSTRACT OF THE UNITED STATES, *op. cit.*, p. 130.

5. Birge, *op. cit.*, p. 23.

6. Great care must be taken not to use any easily counted measure (money, persons, pages of articles, energy of accelerated particles) to stand for increases in what really "counts," namely, in the qualitative understanding and the qualitative rate of increase of that understanding. The numbers are useful to a degree, but the effects of numerical increases in hands, minds, and tools for science are highly nonlinear.

7. SCIENTIFIC MANPOWER BULLETIN No. 12, Washington, D.C.: National Science Foundation Publication 60–78, 1960.

8. INTERIM REPORT OF THE AMERICAN INSTITUTE OF PHYSICS SURVEY OF GRADUATE STUDENTS (mimeographed). With fellowship or scholarship aid, the median time now taken for the doctoral degree in physics is 4.31 years; without such aid, 4.84 years.

9. *Ibid.*

10. INTERIM REPORT OF THE AMERICAN INSTITUTE OF PHYSICS SURVEY OF GRADUATE STUDENTS, *op. cit.* More than one-third of all graduate students in experimental or theoretical physics held research assistantships in 1959–1960, and 31 percent were holding teaching appointments while studying. This, with fellowships and scholarships, means that only a relatively small fraction was not helped one way or the other, though 30 percent reported that inadequate finances were still a retarding factor in their graduate work.

11. THE LONG RANGE DEMAND FOR SCIENTIFIC AND TECHNICAL PERSONNEL (Washington, D. C.: National Science Foundation Publication 61–65, 1961), pp. 42, 45. All too often the sciences are discussed as though they were all physics. To obtain perspective, the total number of United States physicists of all kinds (under 30,000) should be measured against the total number of professional scientists and engineers in industry alone: without counting sci-

entists in government or physicians who do medical research, the number is 850,000 (SCIENTIFIC AND TECHNICAL PERSONNEL IN INDUSTRY, Washington, D.C.: National Science Foundation Publication, 1961.) If government scientists, research-minded physicians, and science teachers are added, the total is 1,400,000 (INVESTING IN SCIENTIFIC PROGRESS [Washington, D.C.: National Science Foundation, 1961], p. 18). It had been estimated that only 27,000 in this large group are basic-research scientists, and that 15,000 of the latter are particularly active—the "real" scientists, as it were. (Naval Research Advisory Committee, BASIC RESEARCH IN THE NAVY. 1959, Volume 1, p. 29.)

12. C. B. Lindquist, *Physics Degrees during the 1950's*, PHYSICS TODAY, 15:19–21, 1962.

13. SCIENTIFIC MANPOWER BULLETIN No. 13 (Washington, D.C.: National Science Foundation Publication 61–38, 1961), pp. 3, 6.

14. *Background Information on the High Energy Physics Program and the Proposed Stanford Linear Electron Accelerator Project: Report of the Joint Committee on Atomic Energy*, CONGRESSIONAL RECORD, 87th Congress, 1st Session, 1961, p. 38. One example is the Brookhaven National Laboratory, where approximately half the operating time of the principal accelerators is reserved for the resident staff, and the rest is for visiting groups from universities and other domestic and foreign institutions.

15. *Programme*, BULLETIN OF THE AMERICAN PHYSICAL SOCIETY (Series II), 7:7–93, 1962.

16. BASIC RESEARCH IN THE NAVY, *op. cit.*, p. 53. Basic-research sponsorship by the Navy, Army, Air Force, Atomic Energy Commission, and other branches of the government (and in other countries by their equivalents) is generally justified in such terms as these: the project is one "with which the Navy should be in communication lest a breakthrough of vital importance occurs. A classic example of the latter was early Navy work in nuclear physics which ultimately permitted more rapid utilization of nuclear power for ship propulsion. It is not possible to define firm boundaries as to Navy interest because of the unpredictability of basic research results and the complex interrelationships between fields of science."

17. No fact of science has ever been as difficult to verify as the figure given out as basic-research expenditure. For example, the budget submitted by the President on 19 January 1962 contained $12.4 billion for "Research and Development" (including that for the Department of Defense and Space Research and Technology). Of this sum $1.6 billion was said to be for "basic" research and training, including the programs of the National Institutes of Health, the National Science Foundation, and Agricultural Research, as well as large sums for the Atomic Energy Commission, Space, and unspecified items for the Department of Defense. Since in the past years the total sum spent for basic

research from all sources has been about twice what the Federal government supplied, one might arrive at a total bill of from $2.5 to $3 billion for basic research in all sciences for the fiscal year 1962, or about half of one percent of the Gross National Product. However, a more likely figure for 1962, particularly if use is made of a stricter interpretation of "basic research," is half this sum (or an average of about $8 per person living in the United States) for all basic scientific research in physics, metallurgy, experimental psychology, biology, etc. [Ten years later, this last figure has become even smaller.]

18. For brilliant discussions of some of these and related points, see the essays by Merle A. Tuve, *Basic Research in Private Research Institutes*, SYMPOSIUM ON BASIC RESEARCH, ed. Dael Wolfle (Washington, D.C.: American Association for the Advancement of Science, 1959), pp. 169–184; and by A. M. Weinberg, *Impact of Large-Scale Science on the United States*, SCIENCE, 134: 161–164, 1961.

19. A thorough and sympathetic study of the situation is in CONFLICT OF INTEREST AND FEDERAL SERVICE (Cambridge: for the Association of the Bar of the City of New York by the Harvard University Press, 1960), Chapter 7.

20. Anne Roe, *The Psychology of Scientists*, SCIENCE, 134:458, 1961.

21. Any relaxing of social, economic, or other barriers which prevent talent from finding its proper scope is to be encouraged. Physicists would do well to ponder whether the amazingly low number of women in physics (2½ percent) in the United States is not indicative of such barriers, particularly in view of the larger fraction typical of other technically advanced countries. Disturbing difficulties of another kind are discussed in Russell Middleton, *Racial Problems and the Recruitment of Academic Staff at Southern Colleges and Universities*, AMERICAN SOCIOLOGICAL REVIEW, 26:960, 1961. On the other hand, the obvious distribution of the authors' names in Figure 1 sets a certain norm for any field. The standard of social mobility implied by this case has very little to do with respect to science per se, but everything with the seriousness of one's interest in the excellence of scholarship.

22. This is the place to mention (without entering into it) the debate on the difficult problem of distinguishing among basic research, applied research, development, technology, quality control, and technical services. These form a continuous spectrum, and precise definitions do not survive the test of using them and talking about them. Suffice it to say that different panels of physicists and engineers working together usually manage to discriminate between these activities on the basis of brief descriptions. For a discussion, see Dael Wolfle, *The Support of Basic Research: Summary of the Symposium*, in Wolfle, *op. cit.*, pp. 249–280.

23. From M. S. Livingston, THE DEVELOPMENT OF HIGH ENERGY ACCELERATORS, New York: Dover Publishing Company, p. 3 (in press [published

1966]); reproduced by permission. A similar chart is in M. S. Livingston and J. P. Blewett, PARTICLE ACCELERATORS (New York: McGraw-Hill, 1962), p. 6.

24. Exponential increases in range or accuracy have long been a part of scientific advance, but the doubling rate was smaller. Thus between 1600 and 1930, approximately, the accuracies of measuring time and astronomical angular distance each increased fairly consistently at an average doubling time of about 20 years. For data, see H. T. Pledge, SCIENCE SINCE 1500 (New York: Harper & Brothers, 1959), pp. 70, 291.

25. M. M. Kessler, in PROCEEDINGS OF THE WESTERN JOINT COMPUTER CONFERENCE, 247–267, 1961, and private communications.

26. Not surprisingly, the speed of advance implies a degree of waste and a number of simultaneous efforts along virtually identical lines. I have discussed elsewhere other reasons for the necessity of some wastefulness and for synchronicity in scientific work; for example, in *On the Duality and Growth of Physical Science*, AMERICAN SCIENTIST, 41:89–99, 1953 and reprinted here as Essay 11. Nothing here should be taken as a defense of much that is merely expensive, large-scale gadgetry, but which passes for science under such labels as "Space."

27. Based on data presented in BASIC RESEARCH IN THE NAVY, *op. cit.*, Volume 1. I thank Dr. Bruce S. Old for arranging the release of the material for use here.

28. It should be understood that this chart does not pretend to an exhaustive description of all work in this field, and in particular does not indicate any work by these persons in other fields.

29. *A Ten-year Preview of High Energy Physics in the United States.* Detailed Backup for Report of Ad Hoc Panel of the President's Science Advisory Committee and the General Advisory Committee to the A. E. C., Dec. 12, 1960, in *Background Information on the High Energy Physics Program and the Proposed Stanford Linear Electron Accelerator Project, op. cit.*, p. 24.

30. Needless to say, one might cite a number of interesting research fields in physics in which the time scale is longer.

31. See Dael Wolfle, AMERICA'S RESOURCES OF SPECIALIZED TALENT (New York: Harper & Brothers, 1954), p. 192.

32. The model here proposed may be elaborated so as to deal with other features of scientific growth, for example, the manner in which work along lines D_2 and D_3 reflects on continued progress along D_1. Thus, after the early falling-off of contributions along the original lines of electrodynamics, interest was revived first by Maxwell's and Hertz's work in electromagnetic waves, then later by Lorentz's and Einstein's, and most recently by plasma physics.

13 THE CHANGING ALLEGORY OF MOTION

I

IN HIS influential book, THE ORIGINS OF MODERN SCIENCE, 1300–1800, Herbert Butterfield indicated the role which Motion has played in the history of thought: "Of all the intellectual hurdles which the human mind has been faced with and has overcome in the last fifteen hundred years, the one which seems to me to have been the most amazing in character and the most stupendous in the scope of its consequences is the one relating to the problem of motion. . . ."[1]

Herbert Butterfield does not use superlatives lightly, and yet he has chosen to underemphasize the matter on three counts. First, the problem of motion in science and philosophy has an even longer history. The scholastic adage *ignorato motu ignoratur natura* (who knows not motion, knows not nature) had earlier been also a theme, in its changing contexts, in Eleatic, Atomistic, Platonic, and Aristotelian writings. In the last, particularly, we can still see the concept of motion in a rich primal context, one in which movement as we now understand it plays a subordinate role. Aristotle's definition of motion preserves a generality which we can reconstruct only with great effort: *Motus est actus entis in potentia secondum quod in potentia est* (motion is the actuality of that which *is* potentially, viewed from the standpoint of potential being). Here motion denotes any transition from potentiality to actuality,

This essay in somewhat longer form was originally published in SCIENTIA, Volume LVII, pp. 1–10, 1963.

whether this change be generation or corruption of a substantial form, whether it be alteration in quality or in quantity, or whether it refers to occupation of a different place (local motion in the narrow sense).

This view, as E. J. Dijksterhuis has correctly said in THE MECHANIZATION OF THE WORLD PICTURE, introduces emphatically the proposition "that the subject matter of science is change,"[2] but this view also emphatically denies that the study of change is solely the prerogative of science. Aristotle's concept of motion is, I would suggest, a great allegory, in which local motion is only one of many attributes. As it has turned out, local motion took on a predominant position in science because to some extent the other kinds of motion could be reduced to it. For example, qualitative change in physics or chemistry or biology—whether it is a change of phase from gas to liquid or solid, or a change from one chemical substance to another, or the development of organization in a cell, or the decay of an elementary particle into different products—is understood by first invoking the relative motion or rearrangement of constituent parts. In the Aristotelian context, this would not be the centrally important aspect of qualitative change, but in the context of present knowledge, the motion of particles and the propagation of energy (e.g., in fields) are central tools of explanation. Science has progressively unmasked movement and change to find local motion behind them.

Moreover, this process of unmasking continues. In addition to its greater antiquity, the problem of motion is having a longer life. The problem of motion was not settled by Galileo and Newton, but seems to confront us in science anew with every great advance. It was in this century that Einstein showed the speed of light to be the maximum speed in free space for any physical object. Beyond that speed, all others are inherently impossible. Niels Bohr then showed that periodic motions are quantized. That is, an electron cannot orbit at any arbitrary distance from the nucleus, but is constrained to certain "allowed" orbits whose radii are related to one another as the ratios of whole numbers (such as 1 : 4 : 9 . . .).

Then quantum mechanics of the 1920's brought out the statistical nature of the motion of subatomic particles and of light energy. No longer can one think of the point-by-point progress of an electron in its orbit, or of a photon passing through an opening in an opaque obstacle and going on in a straight line to a screen beyond it. Given the state of motion of electrons or photons in their specified environment at one instant, one can assign the *probability* of having electrons or photons (why say "the same ones"?) materialize or register at another location and at a later instant.

The simultaneous determination of the position and the velocity of a moving particle had always been tacitly considered to be possible with arbitrary precision; but now such measurement was revealed to be afflicted with an inherent, coupled indeterminacy. As one attempts to make one of these more precise, the other necessarily becomes less precise. And recently, yet another restriction on conceivable motions appeared which rocked physics to its very foundations. In the radioactive decay of some nuclei, it was shown that the motion of the emitted β-ray is not equally probable in both the up and down directions, for example, but that instead one direction is preferred.

On the human scale of motion, the Galileo-Newton discussion still suffices. But on the atomic and subatomic levels—the levels on which the behavior of macroscopic objects finds its explanation in scientific terms—the history of recent science has been a history of accumulating restrictions with respect to motion. An infinite variety of motions which are imaginable, and therefore may have been thought to be possible, have turned out to be impossible. In every other respect we have seen a proliferation—of types of forces, of kinds of particles—but with respect to motion there has been a reduction, a structuring, a progressive taming. With each step, the Aristotelian allegory of motion as all-pervading, undelineated change has been more and more stripped down.

II

There is another sense in which the remark on motion cited at the beginning is too restrictive. It is the faint implication that the problem of motion is to be dealt with mainly by the scientist. A large collection of articles on motion assembled by the distinguished designer and artist Gyorgy Kepes, reemphasizes the fact that the problem of motion is as central and variegated in art as it is in science.[3] More to the point for our purpose here, the influence of the arts on scientific sensibilities, while not as easy to exhibit by example as is the influence in the reverse direction, is just as real. Erwin Panofsky's classic study, GALILEO AS A CRITIC OF THE ARTS, made this case as carefully as we are likely to see it made.[4]

It will be recalled that Panofsky asked why Galileo had failed to use or even refer to Kepler's laws of planetary motion in his long and strenuous fight on behalf of the Copernican system of planetary astronomy. This failure, which hobbled Galileo's case severely by our present standards, had always deeply puzzled historians of science, for Kepler's contribution was then (and is still) by far the most natural and convincing door to Copernicanism. Galileo's insistence on superposed circular motions was both more clumsy and less accurate. Panofsky proposed that Galileo's decision rested on aesthetic grounds. He could not bring himself to accept Kepler's elliptical planetary orbits, for the ellipse, the wretchedly distorted version of the godlike circle, was the very signature of the Manneristic style of art that Galileo so despised.

Surprisingly, it would seem from this example that the effects of the arts on science may not be salutary. If Galileo had not been so well trained and genuinely involved in the visual arts, he might have been able to make a better case for heliocentric astronomy. But the curious thing is that ultimately he was right. The analysis of planetary motion by elliptical orbits is repugnant not only to classicistic aesthetics, but also to the unsophisticated hand calculator and the impartial computing machine. Periodic motions are dealt with most simply by regarding them as the result of a suitable number of superposed circular (harmonic) motions of different amplitude, frequency, and phase. In the end, we now use Kepler's diagrams to imagine planetary motion, but we adopt Galileo's commitment to the circle to understand the motion computationally.

It seems to me that this case illustrates a general principle. When a creative person in one field responds to another field—a Galileo transferring aesthetic criteria from the arts to astronomical work, or conversely, a Cubist perceiving in relativity theory directives for painting—the result *on the surface of it* is apt to be disappointing. Even within a given field, be it physics or painting, analogy is more often dangerous than fruitful; and between unlike fields the transfers that invite themselves are often so superficial that they amount to little more than puns. But this need not be true on another level of transfer—a level whose existence may be *proved* only in the light of later advances, but which had been accessible to the perception of the genial Creator. Thus it was nearly 200 years after Galileo's decision to stick to circles that Fourier discovered the fact that any function of a variable can be expanded in a series of sines of multiples of the variable, thereby enabling us to subject Keplerian motion to Galilean analysis.

I do not wish to propose a concrete mechanism of prescience to explain the ultimate rightness of Galileo's decision (though "rightness" in a sense quite different from that which he could have specifically known). Rather, I am proposing that we deal with these matters by considering that such men, while forced to phrase their criteria and decisions in a language appropriate to the contemporary state of knowledge (including their own), may nevertheless also understand the problems they are wrestling with in a more general context—possibly on a nonverbal level, as Einstein suggested in Jacques Hadamard's THE PSYCHOLOGY OF INVENTION IN THE MATHEMATICAL FIELD.[5] Pretelescopic observational astronomy, seventeenth-century aesthetic theory, and harmonic analysis appear to be, if one examined the textbooks in these fields, separate and unrelated fields of study. What the delayed triumph of Galileo illustrates, however, is that we may also consider them to be different specialized views of a more general allegory, one of which a Galileo sees more than he can readily describe.

III

I do not want to throw doubt on the possibility of *direct* routes of valid transfer from science to art or to history or to any humanistic study, or conversely, from humanistic study to scientific inspiration. Thus Sigmund Freud, in the very first letter in the new collection by Ernst L. Freud, wrote to Emil Flüss (16 June 1873) describing one particularly successful—and significant—part of the General Examination (*Matura*) he had just passed before entering the University: "The Greek essay succeeded better, 'praiseworthy,' the only one,— a thirty-three-verse passage from King Oedipus; I had read this passage also on my own and made no secret of it."[6]

And yet, more important than the direct connections that may be occasionally surmised are the indirect connections, *the sharing of a common allegory.* The practicing artist who is intrigued with science may conceivably find something directly meaningful in the study of the working physicist's differential equations governing the motion of larger bodies, or the operator relations governing the behavior of electrons. But it seems to me much more important that he should understand that these equations are painfully wrought attempts to fashion out of the general concept of motion a limited concept that accentuates some features of it and suppresses others. By paying close attention to vestigial clues, and even better by studying the history of the formation of scientific concepts, he will discover those places where scientific study is (or used to be) connected to the same general allegory which nourishes the artist's own specialized conceptions.

He may also discover closely related difficulties plaguing science and art. For example, differential equations, the usual tool for describing motion in physics, accentuate continuity and are generally helpless in the face of discontinuity. Yet measurement, the process that gives meaning to the equations, is inherently a discontinuity-producing process. On the subatomic level, the measurement of the velocity of an object changes its velocity. On the scale of larger objects, the measurement of their speed generally necessitates doing something that is equivalent to taking snapshots, frozen, cartoon-strip slices of experience framed in hard, black outlines against the flowing background of continuing action.

The representation of motion in its full allegorical sense is therefore as impossible in science as it is in painting, or, for that matter, in a literary work. The need to use civilized mathematical functions is incongruous with respect to the discontinuous nature of self-conscious experience; and conversely, the need to take data during experimentation is incongruous with respect to the continuous nature of the processes which are being investigated. Such difficulties are, of course, now of no interest or concern to most scientists. Science may be defined as the study of areas in which such questions have become meaningless. Nevertheless, they are the navel-like scars that show where modern science had to separate itself from earlier natural philosophy.

This process of separation from the generalized meaning of motion during the rise of modern scientific conceptions can be spoken of as a *deallegorization of motion.* The most recent stage has been mentioned earlier: the discovery that whole ranges of thinkable motions are physically impossible or improbable. Before that, in inverse order, came the relativization of motion, the introduction of virtual motions into science, and, in the period from the thirteenth to the seventeenth century, the quantification of motion.

Each of these stages represented a stripping-away of anthropomorphic and other subjective associations from the definition of motion. But the last-mentioned was the major turning point, not only for the definition of motion, but for the discovery of the very possibility of science as we now understand it. Hence, it is particularly significant to study this development if one wishes to retrace the steps to the common allegory.

IV

Until Newton's time the question of *causation* of motion (dynamics) had two answers. On the one hand, bodies were thought to be subject to propulsion by virtue of some inherent or innate property, be it the attribute of gravity owing to the preponderance of Earth essence, or the action of an *anima* or *virtus*, or of an impetus, or (as still in Newton) a *vis* residing in the moving object. On the other hand, bodies were thought to be also subject to external influences, whether these be the hierarchical structure of the cosmos which guided the body to its proper level, or the tangential pressure and lateral attraction exerted by a line of force that emanates from another object, or the action of the surrounding medium, including external particles during collision (as in the chance collisions of atoms in Democritus, or the vortices of Descartes), or other influences propagated through an ether, a field, or at a distance.

Before these dynamical conceptions of motion as responses to internal or external afflictions could give way to the modern conception of motion as a *relationship of coupled objects,* two developments had to take place in kinematics. The description of motion (without regard to what may have caused it) was shown to be expressible first in terms of geometrical figures, and then in terms of numerical statements. As A. C. Crombie has shown in ROBERT GROSSETESTE AND THE ORIGINS OF EXPERIMENTAL SCIENCE,[7] in the first half of the fourteenth century two methods were in use for expressing functional relationships: one the "word-algebra" of Thomas Bradwardine in Oxford, the other the use of graphs identified with Nicole Oresme at Paris. "These fourteenth-century writers were still primarily concerned with the question how, in principle, to express change of any kind, whether in quantity or in quality, in terms of mathematics, and any treatment they gave of particular optical or dynamical problems was in most cases simply to illustrate a point of method. Yet they succeeded in taking the first step toward the creation out of the statically conceived Greek mathematics [where the prototype problems were Euclid's propositions and Archimedes' lever in a state of equilibrium] of the algebra and geometry of change that were to transform science in the seventeenth century."[8]

Tentatively and innocently, the mathematization of the allegory of motion had begun. Oresme drew figures, e. g., triangles and even three-dimensional objects, to represent the "quantity of a quality" (e.g., how much speed a body has), with the horizontal dimension standing for *extensio,* the height indicating *intensio.*[9] This was not yet a graph. The interest was in the whole figure, which was to reveal the "properties intrinsic to the quality" depicted.

One result, presented by Oresme in DE CONFIGURATIONE QUALITATE, became particularly important. It was a geometric proof of a proposition earlier deduced on other grounds at Merton College sometime before 1335. This was the rule that "a uniformly accelerated or retarded movement is equivalent, so far as the space traversed in a given time is concerned, to a uniform movement of which the velocity is equal throughout to the instantaneous velocity possessed by the uniformly accelerated or retarded movement at the middle instant of time."[10] The proof depends on the fact that (i) if a uniformly changing quality such as decreasing speed may be represented by a triangle ABC, and (ii) if a rectangle ABGF, representing a constant motion, is drawn so that its height AF or BG is just half as large as the long vertical side AC of the triangle, then (iii) the area of the triangle and that of the rectangle—representing in each case the total effect of the motion, e. g., the displacement achieved—by simple geometry are equal for both. By inspection it is evident, however, that if DE represents in any way the "degree of intensity" of the changing motion at its midpoint, then one may see that this should be the magnitude of the constant motion which in the end will achieve the same result as the changing motion. (See Figure)

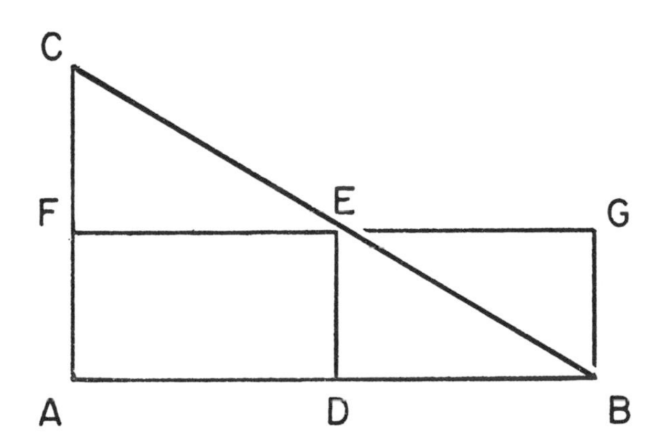

In more modern form such triangular and rectangular figures—now considered essentially as graphs of speed versus time—come to the surface again in Galileo's DISCOURSES ON TWO NEW SCIENCES. The Mean-Speed Rule is an essential part of his discussion of uniformly accelerated motion, the foundation of his whole kinematics, and hence of the proof of his neo-Platonic epistemology.

For, as Alexandre Koyré has said,[11] Galileo's work was an experimental proof of Platonism as a methodology of science. The scholastics had always been able to point to the two main failures of Platonism: on the one hand there was no good theory of terrestrial motion—even Archimedes, the greatest Platonist of antiquity, had only given a science of statics—and on the other hand there was no successful mathematization of *quality*. Now, in his work on the mechanics of falling bodies, Galileo had met both challenges.

As to the first of these: in the Third Day of Galileo's DISCOURSES of 1638, Theorem I, Proposition 1 is in fact the Mean-Speed Rule—replete with diagrammatic proof closely analogous to Oresme's of three centuries before—and with its aid, Galileo draws the famous, ecstatic conclusion, "So far as I know, no one has yet pointed out that the distances traversed, during equal intervals of time, by a body falling from rest, stand to one another in the same ratio as the odd numbers beginning with unity."

We now would put it in the equivalent form that the distance covered in free fall is proportional to the square of the elapsed time. But Galileo has his eye on *numbers,* 1, 3, 5, 7 . . . . Geometrical figure and number: these are the language of nature, as the neo-Platonists had always hoped to show. Here at last in the problem of falling motion, was the proof that triangles and other geometrical figures, and single whole numbers, are indeed the signs with which the Book of Nature has been written. As Koyré said, "La decouverte galiléenne transforme l'échec du platonisme en victoire. Sa science est une revanche de Platon."[12]

What of the second challenge? The mathematization of quality had proved possible for such qualities as motion and size, but not for others, such as taste, the sensation of heat, color (though most of these subsequently were indeed also found to have quantifiable aspects). Galileo's decision was simple: to banish the unquantifiable qualities from science—or more properly, to withdraw the attention of science from the realm of unquantifiables. As he wrote in IL SAGGIATORE, "I think that these tastes, odors, colors, etc., on the side of the object in which they seem to exist, are nothing else than mere names, [and] hold their residence solely in the sensitive body; so that if the animal were removed, every such quality would be abolished and annihilated. Nevertheless, as soon as we have imposed names on them, particular and different from those of the other primary and real accidents, we induce ourselves to believe that they also exist just as truly and really as the latter."

The divisions of concepts according to primary (quantifiable) and secondary qualities, and the abandonment by science of the latter, was, of course, a crucial reduction of the allegory of motion from its original sense of change and movement of every kind. It has frequently been observed that this division, and the consequent mathematical interpretation of nature, has had incalculable consequences on modern thought. This was, E. A. Burtt said, "the first stage in the reading of man quite out of the real and primary realm. . . . Man begins to appear for the first time in the history of thought as an irrelevant spectator and insignificant effect of the great mathematical system which is the substance of reality."[13]

It was unavoidable that we should finally come to this issue, which lies at the bottom of most discussions on the relation between art and science. I have elsewhere argued at length that the implication in statements such as that cited above is unwarranted and erroneous.[14] Far from making man an irrelevant spectator, the insights granted to those who take the trouble to learn the language of science have demonstrated a previously unsuspected capacity of man's mind. The de-emphasis of secondary qualities in science was not a wanton act of dehumanization, but rather a strategic decision to reach a worthy human goal, that of understanding nature (including, ultimately, man's nature) in a new way.

The difficulty has perhaps been not that this new way was too hard, but that it turned out to be all too easy. Once the scientists of the seventeenth century had found the key to this particular gate, the road that opened beyond led more and more speedily and deeply into remote and fascinating territory, further and further from the original ground of understanding the world in terms of multiform, undifferentiated, anthropomorphic allegories.

One cannot help regarding this as important progress; and one also cannot help noticing the loss. Only a few at any time seem to be able to move with some assurance on both sides of the gate equally. Galileo himself, of whose aesthetic criteria for planetary orbits we spoke earlier, was an example; there are not many others. The rest of us are far down on some road to de-allegorization—each on his own. For just as there are primal allegories other than that of motion (for example, those of space, time, matter, organism, life, death) from which science has progressively moved, so also are there in each case other roads than those of science, leading away from the primal allegories. Perhaps I may be allowed to suggest that even art, and particularly art, has in this respect suffered the same fate as science. And the treatment of motion in modern visual art may well prove the point. To my eye, at any rate, the distance between primal motion as expressed in ritual dancing or Navaho sandpainting on the one hand, and motion expressed in cinematography, mobiles, and action painting on the other, is as large as the distance between primal motion in Greek nature-philosophy, and motion as indicated on a tachometer or an oscilloscope screen.

For the process of reducing, transforming, and abstracting the allegory is the hallmark, not just of science, but of our whole developing culture itself.

NOTES

1. Herbert Butterfield, THE ORIGINS OF MODERN SCIENCE, 1300–1800 (New York: Macmillan Co., 1958), p. 3.

2. E. J. Dijksterhuis, THE MECHANIZATION OF THE WORLD PICTURE, trans. C. Dikshoorn, Oxford: Clarendon Press, 1961.

3. This article was included under the title *Science and the Deallegorization of Motion*, THE NATURE AND ART OF VISION, ed. Gyorgy Kepes (New York: George Braziller, 1965), pp. 24–31. This is one of the series VISION + VALUE edited by Mr. Kepes.

4. Erwin Panofsky, GALILEO AS A CRITIC OF THE ARTS, The Hague: Martinus Nijhoff, 1954.

5. Albert Einstein, *Testimonial from Professor Einstein*, Appendix II to THE PSYCHOLOGY OF INVENTION IN THE MATHEMATICAL FIELD, ed. Jacques Hadamard (Princeton, N. J.: Princeton University Press, 1945), pp. 142–143.

6. Sigmund Freud, BRIEFE 1873–1939 (Frankfurt am Main: S. Fischer Verlag, 1960), p. 6.

7. A. C. Crombie, ROBERT GROSSETESTE AND THE ORIGINS OF EXPERIMENTAL SCIENCE, Oxford: Clarendon Press, 1953.

8. *Ibid.,* p. 178.

9. A. C. Crombie, MEDIEVAL AND EARLY MODERN SCIENCE (New York: Doubleday & Co., 1959), Volume II, Chapter 1.

10. *Ibid.,* p. 93.

11. See particularly Alexandre Koyré, ÉTUDES GALILÉENNES (Paris: Hermann, 1939), Volumes II and III.

12. *Ibid.*, Volume III, p. 280.

13. Edwin A. Burtt, THE METAPHYSICAL FOUNDATIONS OF MODERN PHYSICAL SCIENCE, 2nd ed. (London: Routledge & Kegan Paul, 1932), p. 80.

14. Gerald Holton, *Modern Science and the Intellectual Tradition,* THE INTELLECTUALS, ed. George Bernard de Huszar (Glencoe, Illinois: Free Press, 1960), pp. 180–191; also in SCIENCE, 131:1187–1193, 1960; and here as Essay 14.

IV *On Education*

14 MODERN SCIENCE AND THE INTELLECTUAL TRADITION

When future generations look back to our day, they will envy us for having lived at a time of brilliant achievement in many fields, and not least in science and technology. We are at the threshold of basic knowledge concerning the origins of life, the chemical elements, and the galaxies. We are near an understanding of the fundamental constituents of matter, of the process by which the brain works, and of the factors governing behavior. We have launched the physical exploration of space and have begun to see how to conquer hunger and disease on a large scale. Scientific thought appears to be applicable to an ever wider range of studies. With current technical ingenuity one could in principle hope to implement most of the utopian dreams of the past.

Hand in hand with the quality of excitement in scientific work today goes an astonishing quantity. The world-wide output is vast. And the amount of scientific work being done is increasing at a rapid rate, doubling approximately every 10 or 20 years. Every phase of daily and

This is a slightly abbreviated form of an essay published in Science, Volume 131, pp. 1187–1193, 1960, and reprinted in The Graduate Journal, Volume IV, pp. 48–63, 1961. A shorter form was published in the anthology, The Intellectuals, ed. George B. de Huszar, Glencoe, Ill.: Free Press, 1960.

national life is being penetrated by some aspect of this exponentially growing activity.

It is appropriate, therefore, that searching questions are now being asked about the function and place of this lusty giant. Just as a man's vigorously pink complexion may alert the trained eye to a grave disease of the circulatory system, so too may the spectacular success and growth of science and technology turn out, on more thorough study, to mask a deep affliction of our culture. And indeed, anyone committed to the view that science should be a basic part of our intellectual tradition will soon find grounds for concern.

Some of the major symptoms of the relatively narrow place science, as properly understood, really occupies in the total picture are quantitative. For example, while the total annual expenditure for scientific research and development in this country is at a high level, basic research—the main roots of the tree that furnishes scientific knowledge and the fruits of technology—has a share of only a few percent.[2] Correspondingly, of the large number of trained scientists and engineers, only about 2 percent have been estimated to be responsible for the major part of the creative work being done in basic research. A nationwide survey found that nearly 40 percent of the men and women who had attended college in the United States confessed that they had taken not a single course in the physical and biological sciences. Similarly, in contrast to the overwhelming amount of, and concern with, science and technology today, the mass media pay only negligible attention to their substance; the newspapers have been found to give less than 5 percent of their (nonadvertising) space to factual presentations of science, technology, and medicine; and television stations, only about 0.3 percent of their time. In short, all our voracious consumption of technological devices, all our talk about the threats or beauties of science, and all our money spent on engineering development should not draw attention from the fact that the pursuit of scientific knowledge itself is not a strong component of the operative system of general values.

The Atomization of Loyalties

In the qualitative sense, and particularly among intellectuals, the symptoms are no better. One hears talk of the hope that the forces of science may be tamed and harnessed to the general advance of ideas, that the much deplored gap between scientists and humanists may be bridged. But the truth is that both the hopes and the bridges are illusory. The separation—which I shall examine further—between the work of the scientist on the one hand and that of the intellectual outside science on the other is steadily increasing, and the genuine acceptance of science as a valid part of culture is becoming less rather than more likely.

Moreover, there appears at present to be no force in our cultural dynamics strong enough to change this trend. This is due mainly to the atrophy of two mechanisms by which the schism was averted in the past. First, the common core of their early education and the wide range of their interests was apt to bring scholars and scientists together at some level where there could be mutual communication on the subjects of their individual competence; and second, the concepts and attitudes of contemporary science were made a part of the general humanistic concerns of the time. In this way a reasonable equilibrium of compatible interpretations was felt to exist, during the last century, between the concepts and problems of science on the one hand and of intelligent common sense on the other; this was also true with respect to the scientific and the nonscientific aspects of the training of intellectuals. Specialists, of course, have always complained of being inadequately appreciated; what is more, they are usually right. But although there were some large blind spots and some bitter quarrels, the two sides were not, as they are now in danger of coming to be, separated by a gulf of ignorance and indifference.

It is of course not my purpose here to urge better science education at the expense of humanistic and social studies. On the contrary, the latter do not fare much better than science does, and the shabby effort devoted to science is merely the symptom of a more extensive sickness of our educational systems. Nor do I want to place all blame on educators and publicists. Too many scientists have forgotten that especially at a time of rapid expansion of knowledge they have an extra obligation and opportunity with respect to the wider public and that some of the foremost research men, including Newton and Einstein, took great pains to write expositions of the essence of their discoveries in a form intended to be accessible to the nonscientist. And in the humanities, too many contributors and interpreters seem to scoff at Shelley's contention in his *Defence of Poetry* that one of the artist's tasks is to "absorb the new knowledge of the sciences and assimilate it to human needs, color it with human passions, transform it into the blood and bone of human nature."

It is through the accumulation of such neglects, just as much as through deterioration in the quantity and quality of instruction given our future intellectual leaders, that the acceptance of science as a meaningful component of our culture has come to be questioned. Again, this process is to a large extent merely one aspect of the increasing atomization of loyalties within the intelligentsia. The writer, the scholar, the scientist, the engineer, the teacher, the lawyer, the politician, the physician—each now regards himself first of all as a member of a separate, special group of fellow professionals to which he gives almost all his allegiance and energy; only very rarely does the professional feel a sense of responsibility toward, or of belonging to, a larger intellectual community. This loss of cohesion is perhaps the most relevant symptom of the disease of our culture, for it points directly to one of its specific causes. As in other cases of this sort, this is a failure of image.

Pure Thought and Practical Power

Each person's image of the role of science may differ in detail from that of the next, but all public images are in the main based on one or more of seven positions. The first of these goes back to Plato and portrays science as an activity with double benefits: science as pure thought helps the mind find truth, and science as power provides tools for effective action. In Book Seven of the REPUBLIC, Socrates tells Glaucon why the young rulers in the Ideal State should study mathematics: "This, then, is knowledge of the kind we are seeking, having a double use, military and philosophical; for the man of war must learn the art of number, or he will not know how to array his troops; and the philosopher also, because he has to rise out of the sea of change and lay hold of true being. . . . This will be the easiest way for the soul to pass from becoming to truth and being."

The main flaw in this image is that it omits a third vital aspect. Science has always had also a mythopoeic function—that is, it generates an important part of our symbolic vocabulary and provides some of the metaphysical bases and philosophical orientations of our ideology. As a consequence the methods of argument of science, its conceptions and its models, have permeated first the intellectual life of the time, then the tenets and usages of everyday life. All philosophies share with science the need to work with concepts such as space, time, quantity, matter, order, law, causality, verification, reality. Our language of ideas, for example, owes a great debt to statics, hydraulics, and the model of the solar system. These have furnished powerful analogies in many fields of study. Guiding ideas—such as conditions of equilibrium, centrifugal and centripetal forces, conservation laws, feedback, invariance, complementarity—enrich the general storehouse of imaginative tools of thought.

A sound image of science must embrace each of the three functions. However, usually only one of the three is recognized. For example, folklore often depicts the life of the scientist either as isolated from life and from the beneficent action[1] or, at the other extreme, as dedicated to technological improvements.

Iconoclasm

A second image of long standing is that of the scientist as iconoclast. Indeed, almost every major scientific advance has been interpreted—either triumphantly or with apprehension—as a blow against religion. To some extent science was pushed into this position by the ancient tendency to prove the existence of God by pointing to problems which science could not solve at the time. Newton thought that the regularities and stability of the solar system proved it "could only proceed from the counsel and dominion of an intelligent and powerful Being," and the same attitude governed thought concerning the earth's formation before the theory of geological evolution, concerning the descent of man before the theory of biological evolution, and concerning the origin of our galaxy before modern cosmology. The advance of knowledge therefore made inevitable an apparent conflict between science and religion. It is now clear how large a price had to be paid for a misunderstanding of both science and religion: to base religious beliefs on an estimate of what science cannot do is as foolhardy as it is blasphemous.

The iconoclastic image of science has, however, other components not ascribable to a misconception of its functions. For example, Arnold Toynbee charges science and technology with usurping the place of Christianity as the main source of our new symbols. Neo-orthodox theologians call science the "self-estrangement" of man because it carries him with idolatrous zeal along a dimension where no ultimate—that is, religious—concerns prevail. It is evident that these views fail to recognize the multitude of divergent influences that shape a culture, or a person. And on the other hand there is, of course, a group of scientists, though not a large one, which really does regard science as largely an iconoclastic activity. Ideologically they are descendants of Lucretius, who wrote on the first pages of DE RERUM NATURA, "The terror and darkness of mind must be dispelled not by the rays of the sun and glittering shafts of day, but by the aspect and the law of nature; whose first principle we shall begin by thus stating, nothing is ever gotten out of nothing by divine power."

Ethical Perversion

The third image of science is that of a force which can invade, possess, pervert, and destroy man. The current stereotype of the soulless, evil scientist is the psychopathic investigator in science fiction, or the nuclear destroyer—immoral if he develops the weapons he is asked to produce, traitorous if he refuses. According to this view, scientific morality is inherently negative. It causes the arts to languish, it blights culture, and when applied to human affairs, it leads to regimentation and to the impoverishment of life. Science is the serpent seducing us into eating the fruits of the tree of knowledge—thereby dooming us.

The fear behind this attitude is genuine but not confined to science; it is directed against all thinkers and innovators. Society has always found it hard to deal with creativity, innovation, and new knowledge. And since science assures a particularly rapid, and therefore particularly disturbing, turnover of ideas, it remains a prime target of suspicion.

Factors peculiar to our time intensify this suspicion. The discoveries of "pure" science often lend themselves readily to widespread exploitation through technology. The products of technology—whether they are better vaccines or better weapons—have the characteristics of frequently being very effective, easily made in large quantities, easily distributed, and very appealing. Thus we are in an inescapable dilemma—irresistibly tempted to reach for the fruits of science, yet, deep inside, aware that our metabolism may not be able to cope with this ever-increasing appetite.

Probably the dilemma can no longer be resolved, and this increases the anxiety and confusion concerning science. A current symptom is the popular identification of science with the technology of superweapons. The bomb is taking the place of the microscope; Wernher von

Braun, the place of Einstein, as symbols for modern science and scientists. The efforts to convince people that science itself can give man only knowledge about himself and his environment, and occasionally a choice of action, have been largely unavailing. The scientist *as scientist* can take little credit or responsibility either for facts he discovers—for he did not create them—or for the uses others make of his discoveries, for he generally is neither permitted nor specially fitted to make these decisions. They are controlled by considerations of ethics, economics, or politics and therefore are shaped by the values and historical circumstances of the whole society. It is, however, also appropriate to say here that there has been only a moderate success in persuading the average scientist of the proposition that the privilege of freely pursuing a field of knowledge having large-scale secondary effects imposes on him, in his capacity as citizen, a proportionately larger burden of civic responsibility.

There are other evidences of the widespread notion that science itself cannot contribute positively to culture. Toynbee, for example, gives a list of "creative individuals," from Xenophon to Hindenburg and from Dante to Lenin, but does not include a single scientist. I cannot forego the remark that there is a significant equivalent on the level of casual conversation. For when the man in the street—or many an intellectual—hears that you are a physicist or mathematician, he will usually remark with a frank smile, "Oh, I never could understand that subject"; while intending this as a curious compliment, he betrays his intellectual dissociation from scientific fields. It is not fashionable to confess to a lack of acquaintance with the latest ephemera in literature or the arts, but one may even exhibit a touch of pride in professing ignorance of the structure of the universe or one's own body, of the behavior of matter or one's own mind.

The Sorcerer's Apprentice

The last two views held that man is inherently good and science evil. The next image is based on the opposite assumption—that man cannot be trusted with scientific and technical knowledge. He has survived only because he lacked sufficiently destructive weapons; now he can immolate his world. Science, indirectly responsible for this new power, is here considered ethically neutral. But man, like the sorcerer's apprentice, can neither understand this tool nor control it. Unavoidably he will bring upon himself catastrophe, partly through his natural sinfulness, and partly through his lust for power, of which the pursuit of knowledge is a manifestation. It was in this mood that Pliny deplored the development of projectiles of iron for purposes of war: "This last I regard as the most criminal artifice that has been devised by the human mind; for, as if to bring death upon man with still greater rapidity, we have given wings to iron and taught it to fly. Let us, therefore, acquit Nature of a charge that belongs to man himself."

When a science is viewed in this plane—as a temptation for the mischievous savage—it becomes easy to suggest a moratorium on science, a period of abstinence during which humanity somehow will develop adequate spiritual or social resources for coping with the possibilities of inhuman uses of modern technical results. Here I need point out only the two main misunderstandings implied in this recurrent call for a moratorium.

First, science of course is not an occupation, such as working in a store or on an assembly line, that one may pursue or abandon at will. For a creative scientist, it is not a matter of free choice what he shall do. Indeed it is erroneous to think of him as advancing toward knowledge; it is, rather, knowledge which advances toward him, grasps him, and overwhelms him. Even the most superficial glance at the life and work of a Kepler, a Dalton, or a Pasteur would clarify this point. It would be well if in his education each person were shown by example that the driving power of creativity is as strong and as sacred for the scientist as for the artist.

The second point can be put equally briefly. In order to survive and to progress, mankind surely cannot ever know too much. Salvation can hardly be thought of as the reward for ignorance. Man has been given his mind in order that he may find out where he is, what he is, who he is, and how he may assume the responsibility for himself which is the only obligation incurred in gaining knowledge.

Indeed, it may well turn out that the technological advances in warfare have brought us to the point where society is at last compelled to curb the aggressions that in the past were condoned and even glorified. Organized warfare and genocide have been practiced throughout recorded history, but never until now have even the war lords openly expressed fear of war. In the search for the causes and prevention of aggression among nations, we shall, I am convinced, find scientific investigations to be a main source of understanding.

Ecological Disaster

A change in the average temperature of a pond or in the salinity of an ocean may shift the ecological balance and cause the death of a large number of plants and animals. The fifth prevalent image of science similarly holds that while neither science nor man may be inherently evil, the rise of science happened, as if by accident, to initiate a change in the balance of beliefs and ideas that now corrodes the only conceivable basis for a stable society. In the words of Jacques Maritain, the "deadly disease" science set off in society is "the denial of eternal truth and absolute values."

The main events leading to this state are usually presented as follows: the abandonment of geocentric astronomy implied the abandonment of the conception of the earth as the center of creation and of man as its ultimate purpose. Then purposive creation gave way to blind evolution. Space, time, and certainty were shown to have no absolute meaning. All a priori axioms were discovered to be merely arbitrary conveniences. Modern psychology and anthropology led to cultural relativism. Truth itself has been dissolved into probabilistic and indeterministic statements. Drawing upon analogy with the sciences, liberal philosophers have become increasingly relativistic, denying either the necessity or the possibility of postulating immutable verities, and so have undermined the old foundations of moral and social authority on which a stable society must be built.

It should be noted in passing that many applications of recent scientific concepts outside science merely reveal ignorance about science. For example, relativism in nonscientific fields is generally based on far-fetched analogies. Relativity theory, of course, does not find that truth depends on the point of view of the observer but, on the contrary, reformulates the laws of physics so that they hold good for every observer, no matter how he moves or where he stands. Its central meaning is that the most valued truths in science are wholly independent of the point of view. Ignorance of science is also the only excuse for adopting rapid changes within science as models for antitraditional attitudes outside science. In reality, no field of thought is more conservative than science. Each change necessarily encompasses previous knowledge. Science grows like a tree, ring by ring. Einstein did not prove the work of Newton wrong; he provided a larger setting within which some contradictions and asymmetries in the earlier physics disappeared.

But the image of science as an ecological disaster can be subjected to a more severe critique.[2] Regardless of science's part in the corrosion of absolute values, have those values really given us always a safe anchor? A priori absolutes abound all over the globe in completely contradictory varieties. Most of the horrors of history have been carried out under the banner of some absolutistic philosophy, from the Aztec mass sacrifices to the auto-da-fé of the Spanish Inquisition, from the massacre of the Huguenots to the Nazi gas chambers. It is far from clear that any society of the past did provide a meaningful and dignified life for more than a small fraction of its members. If, therefore, some of the new philosophies, inspired rightly or wrongly by science, point out that absolutes have a habit of changing in time and of contradicting one another, if they invite a re-examination of the bases of social authority and reject them when those bases prove false (as did the Colonists in this country), then one must not blame a relativistic philosophy for bringing out these faults. They were there all the time.

In the search for a new and sounder basis on which to build a stable world, science will be indispensable. We can hope to match the resources and structure of society to the needs and potentialities of people only if we know more about man. Already science has much to say that is valuable and important about human relationships and problems. From psychiatry to dietetics, from immunology to meteorology, from city planning to agricultural research, by far the largest part of our total scientific and technical effort today is concerned, indirectly or directly, with man—his needs, relationships, health, and comforts. Insofar as absolutes are to help guide mankind safely on the long and dangerous journey ahead, they surely should be at least strong enough to stand scrutiny against the background of developing factual knowledge.

Scientism

While the last four images implied a revulsion from science, scientism may be described as an addiction to science. Among the signs of scientism are the habit of dividing all thought into two categories, up-to-date scientific knowledge and nonsense; the view that the mathematical sciences and the large nuclear laboratory offer the only permissible models for successfully employing the mind or organizing effort; and the identification of science with technology, to which reference was made above.

One main source for this attitude is evidently the persuasive success of recent technical work. Another resides in the fact that we are passing through a period of revolutionary change in the nature of scientific activity—a change triggered by the perfecting and disseminating of the methods of basic research by teams of specialists with widely different training and interests. A generation ago the typical scientist worked alone or with a few students and colleagues. Today he often belongs to a sizable group working under a contract with a substantial annual budget. (In the research institute of one university more than 1500 scientists and technicians are grouped around a set of multimillion-dollar machines; the funds come from government agencies whose ultimate aim is national defense.) Science has thereby become a large-scale operation with a potential for immediate and world-wide effects. The results are a splendid increase in knowledge, and also side effects that are analogous to those of sudden and rapid urbanization—a strain on communication facilities, the rise of an administrative bureaucracy, the depersonalization of some human relationships.

To a large degree, all this is unavoidable. The new scientific revolution will justify itself by the flow of new knowledge and of material benefits that will no doubt follow. The danger—and this is the point where scientism enters—is that the fascination with the *mechanism* of this successful enterprise may change the scientist himself and society around him. For example, the unorthodox, often withdrawn individual, on whom most great scientific advances have depended in the past, does not fit well into the new system. And society will be increasingly faced with the seductive urging of scientism to adopt generally what is regarded—often erroneously—as the pattern of organization of the new science. The crash programs, the breakthrough pursuit, the megaton effect are becoming ruling ideas in complex fields such as education, where they may not be applicable.

Magic

Few nonscientists would suspect a hoax if it were suddenly announced that a stable chemical element lighter than hydrogen had been synthesized, or that a manned observation platform had been established at the surface of the sun. To most people it appears that science knows no inherent limitations. Thus, the seventh image depicts science as magic, and the scientist as wizard, *deus ex machina,* or oracle. The attitude toward the scientist on this plane ranges from terror to sentimental subservience, depending on what motives one ascribes to him.

Impotence of the Modern Intellectual

The prevalence of these false images is a main source of the alienation between the scientific and nonscientific elements in our culture, and therefore the failure of image is important business for all of us. Now to pin much of the blame on the insufficient instruction in science which the general student receives at all levels is quite justifiable. I have implied the need, and most people nowadays seem to come to this conclusion anyway. But this is not enough. We must consider the full implications of the discovery that not only the man in the street but almost all of our intellectual leaders today know at most very little about science. And here we come to the central point underlying the analysis made above: the chilling realization that our intellectuals, for the first time in history, are losing their hold of understanding upon the world.

The wrong images would be impossible were they not anchored in two kinds of ignorance. One kind is ignorance on the basic level, that of *facts*—what biology says about life, what chemistry and physics say about matter, what astronomy says about the development and structure of our galaxy, and so forth. The nonscientist realizes that the old commonsense foundations of thought about the world of nature have become obsolete during the last two generations. The ground is trembling under his feet; the simple interpretations of solidity, permanence, and reality have been washed away, and he is plunged into the nightmarish ocean of four-dimensional continua, probability amplitudes, indeterminacies, and so forth. He knows only two things about the basic conceptions of modern science: that he does not understand them, and that he is now so far separated from them that he will never find out what they mean.

On the second level of ignorance, the contemporary intellectual knows just as little of the way in which the main facts from the different sciences fit together in a picture of the world taken as a whole. He has had to leave behind him, one by one, those great syntheses which used to represent our intellectual and moral home—the world view of the book of Genesis, of Homer, of Dante, of Milton, of Goethe. In the mid-20th century he finds himself abandoned in a universe which is to him an unsolvable puzzle on either the factual or the philosophical level. Of all the bad effects of the separation of culture and scientific knowledge, this feeling of bewilderment and basic homelessness is the most terrifying. Here is the reason, it seems to me, for the ineffectiveness and self-denigration of our contemporary intellectuals. Nor are the scientists themselves protected from this fate, for it has always been, and must always be, the job of the humanist to construct and disseminate the meaningful total picture of the world.

To illustrate this point concretely we may examine a widely and properly respected work by E. A. Burtt, a scholar who warmly understands both the science and the philosophy of the sixteenth and seventeenth centuries. The reader is carried along by his authority and enthusiasm. And then, suddenly, one encounters a passage unlike any other in the book, an anguished cry from the heart: "It was of the greatest consequence for succeeding thought that now the great Newton's authority was squarely behind that view of the cosmos which saw in man a puny, irrelevant spectator (so far as a being, wholly imprisoned in a dark room, can be called such) of the vast mathematical system whose regular motions according to mechanical principles constituted the world of nature. The gloriously romantic universe of Dante and Milton, that set no bounds to the imagination of man as it played over space and time, had now been swept away. Space was identified with the realm of geometry, time with the continuity of number. The world that people had thought themselves living in—a world rich with color and sound, redolent with fragrance, filled with gladness, love and beauty, speaking everywhere of purposive harmony and creative ideals—was crowded now into minute corners in the brains of scattered organic beings. The really important world outside was a world hard, cold, colorless, silent, and dead; a world of quantity, a world of mathematically computable motions in mechanical regularity. The world of qualities as immediately perceived by man became just a curious and quite minor effect of that infinite machine beyond. In Newton, the Cartesian metaphysics, ambiguously interpreted and stripped of its distinctive claim for serious philosophical consideration, finally overthrew Aristotelianism and became the predominant world-view of modern times."[3]

For once, the curtain usually covering the dark fears modern science engenders is pulled away. This view of modern man as a puny, irrelevant spectator lost in a vast mathematical system—how far this is from the exaltation of man that Kepler found through scientific discovery: "Now man will at last measure the power of his mind on a true scale, and will realize that God, who founded everything in the world on the norm of quantity, also has endowed man with a mind which can comprehend these norms!" Was not the universe of Dante and Milton so powerful and "gloriously romantic" precisely because it incorporated, and thereby rendered meaningful, the contemporary scientific cosmology alongside the current moral and aesthetic conceptions? Leaving aside the question of whether Dante's and Milton's contemporaries by and large, were really living in a rich and fragrant world of gladness, love, and beauty, it is fair to speculate that if our new cosmos is felt to be cold, inglorious, and unromantic, it is not the new cosmology which is at fault, but the absence of new Dantes and Miltons.

And yet, Burtt correctly reflects the present dilemma. What his outburst tells us, in starkest and simplest form, is this: by having let the intellectual remain in terrified ignorance of modern science, we have forced him into a position of tragic impotence; he is blindfolded in a maze which he cannot traverse.

Once this is understood, the consequence also becomes plain. I find it remarkable that the intellectual today does not have even more distorted images and hostile responses with regard to science, that he has so far not turned much more fiercely against the source of apparent threats to his personal position and sanity[4]—in short, that the dissociation has not resulted in an even more severe cultural psychosis.

But this, I am convinced, is likely to be the result, for there is at present no countercyclical mechanism at work. Some other emergencies of a similar or related nature have been recognized and are being dealt with: we need more support for studies in humanities and social science, and the base of support is growing gratifyingly. We sorely need to give our young scientists more broad humanistic studies—and if I have not dwelled on this it is because, in principle, this can be done with existing programs and facilities, for the existing tools of study in the humanities, unlike the tools in science, are still in touch with our ordinary sensibilities. But hardly anything being done or planned now is adequate to deal with the far more serious problem, the cultural psychosis engendered by the separation of science and the rest of culture.

One may of course speculate as to how one could make science again a part of every intelligent man's educational equipment—not because science is more important than other fields, but because it is an important part of the whole jigsaw puzzle of knowledge. A plausible program would include sound and thorough work at every level of education—imaginative new programs and curricula; strengthened standards of achievement; extended college work in science, as used to be the rule in good colleges some 50 years ago; greater recognition of excellence; expansion of opportunity for adult education, including the presentation of factual and cultural aspects of science through the mass media. But while some efforts are being made here and there, few people have faced the real magnitude of the problem, or are aware of the large range and amount of scientific knowledge that is needed before one can "know science" in any sense at all. Moreover, while some time lag between new discoveries and their wider dissemination has always existed, the increase in degree of abstraction, and in tempo, of present-day science, coming precisely at a time of inadequate educational effort even by old standards, has begun to change the lag into a discontinuity.

This lapse, it must be repeated, is not the fault of the ordinary citizen; necessarily, he can only take his cue from the intellectuals—the scholars, writers, and teachers who deal professionally in ideas. It is among the latter that the crucial need lies. Every great age has been shaped by intellectuals of the stamp of Hobbes, Locke, Berkeley, Leibnitz, Voltaire, Montesquieu, Rousseau, Kant, Jefferson, and Franklin—all of whom would have been horrified by the proposition that cultivated men and women could dispense with a good grasp of the scientific aspect of the contemporary world picture. This tradition is broken; very few intellectuals are now able to act as informed mediators. Meanwhile, as science moves every day faster and further from the bases of ordinary understanding, the gulf grows, and any remedial action becomes more difficult and more unlikely.

To restore science to reciprocal contact with the concerns of most men—to bring science into an orbit about us instead of letting it escape from our intellectual tradition—that is the great challenge that intellectuals face today.

1. See, for example, the disturbing findings of Margaret Mead and Rhoda Metraux, *Image of the Scientist among High School Students*, SCIENCE, 126: 384, 1957.

2. See, for example, Charles Frankel, THE CASE FOR MODERN MAN, New York: Harper & Brothers, 1955 and 1956.

3. Edwin A. Burtt, THE METAPHYSICAL FOUNDATIONS OF MODERN PHYSICAL SCIENCE (New York: Harcourt, Brace & Co., 1927), pp. 236–237. [For an analysis of a recent book fundamentally hostile to science, see the review of Lewis Mumford's THE PENTAGON OF POWER (New York: Harcourt Brace Jovanovich, 1970) by the writer in the NEW YORK TIMES BOOK REVIEW, 13 December 1970, Section 7.]

4. For a striking recent example see the virulent attack on modern science in the final chapter of Arthur Koestler's THE SLEEPWALKERS, New York: Macmillan Co., 1959.

15 PHYSICS AND CULTURE: CRITERIA FOR CURRICULUM DESIGN

IN HIS speculative essay, *The Rule of Phase Applied to History,* dated January 1, 1909, the American historian Henry Adams came to a remarkable conclusion: ". . . the future of Thought, and therefore of History, lies in the hands of the physicist, and . . . the future historian must seek his education in the world of mathematical physics. A new generation must be brought up to think by new methods, and if our historical departments in the Universities cannot enter this next Phase, the physical departments will have to assume the task alone."

In arriving at this startling view, Henry Adams explains that he was guided by a desire to transfer to the study of history some of the conceptions developed in 1876–8 by Willard Gibbs, Professor of Mathematical Physics at Yale, in his famous paper *Equilibrium of Heterogeneous Substances,* and by other physicists and chemists who followed him. Just as the physicist Storey saw in the Phase Rule a means for putting into hierarchical order the sequence of "phases" consisting of solid, fluid, gas, electricity, ether, and space, so did Adams believe that

This essay is adapted from the opening address given at the Second International Conference on Physics Education held in Rio de Janeiro, Brazil, July 1963, and was published in the BULLETIN OF THE INSTITUTE OF PHYSICS AND THE PHYSICAL SOCIETY, 1963, pp. 321–329.

thought, too, in time had passed through different phases. Acknowledging himself to be a follower of Turgot, Littré, John Stuart Mill, Comte, and others who had also held that history obeys quasi-physical laws, Adams found confirmation for the correctness of his essentially prophetic and apocalyptic view of history in the apparently increasing rate of change of historic processes, and he thought they were analogous to the increasing motion of objects which are attracting one another by an inverse-square force.

I have cited this example not because physicists would believe that history obeys laws closely analogous to those of physics—indeed, a meeting of physicists would be the last place to find sympathy with the modern-day physiocrats. Rather I have spoken about Henry Adams, because as usual he had brilliant insights in this essay. He said that the ideas emerging from physics would continue to be, as they had again and again been since the seventeenth century, a central part of modern culture; and, second, he saw that science is both a major mechanism for change in culture as well as a way of understanding the change better.

Today, we would like to believe that these insights are generally shared by all men who have thought about the matter. But there are two main groups—not to mention the philistines who have already viewed any intellectual activity with suspicion—that would deny it. One represents, as it were, the reaction from the right, from the side of misguided traditionalism. They would say with Matthew Arnold that "culture is, or ought to be, the study and pursuit of perfection," and would then define the properties of perfection—for example, beauty and intelligence—in such a way that most scientific work stands exposed and condemned as soulless hackwork and the manipulation of trade-school mechanics. Or with T. S. Eliot, they would say that culture and religion are "different aspects of the same thing,"[1] and then, instead of noticing that science is also an "aspect of the same thing," they would define culture and religion in such a way that science, when it is mentioned at all, becomes identifiable with idolatry. [2]

A revealing quotation from an editorial in the English SUNDAY TELEGRAPH exhibits this attitude as clearly as we can wish:

> A free and prosperous society depends on the activities of three distinct classes—a political *élite*, trained by the study of the humanities to take broad and enlightened views about ends and means, a technical *élite*, willing to exer-

cise its skills in obedience to the community's will, and a proletariat with enough mechanical intelligence to respond to managerial direction.

The first of these conditions has not yet been entirely removed by the renunciation of the classics, the second will be secure so long as scientists are not encouraged to dabble in the humanities to develop a wish to govern society, and only the third . . . is strikingly absent . . . .[3]

The other, or left-wing opposition comes from the ranks of science itself. A relatively small but influential proportion of scientists would say that there simply is no meaning in the proposition that physics, for example, is more than what they and their best associates are actually doing at the blackboard or in the laboratory here and now. Unlike the first group of opponents who object that science has little place in culture, the second objects that culture has little validity compared to science.[4] Only politeness prevents them from dismissing with vocal impatience the suggestion that there are important links between what happens in the laboratory now and what happens, has happened, and will happen elsewhere—in the sculptor's studio, in the courtroom, in the study of a philosopher or an economist, on the stage, and even in the nursery where a child is asking his mother for help in making sense out of the world around him. And the major reason why some of these scientists can neglect the complex, tenuous, long-range links that attach to their science is that they are so successful doing what they are doing. The short-range forces, which they master, completely saturate their capability for forming and perceiving long-range connections.

One is reminded of the diagnosis C. P. Snow offered in his provocative book SCIENCE AND GOVERNMENT[5] for the reason why some scientists so single-mindedly stuck to a narrow decision or were satisfied with a narrow range of investigations. It was their success in one particular field or with the operation of one particular apparatus. Snow dubs these men "gadgeteers."

But on this occasion, I trust I shall be safe in assuming that one does not have to defend further the proposition of centrality of science in culture, either from the radical scientism of the left or the culture snobbism of the right. On this middle ground, we shall thus posit that the much-discussed cleavages of knowledge are all too often the unhappy results of erroneous definition. The controversy between T. S. Eliot and his critics, or between Snow and Leavis more recently, serve to remind us how necessary it is for each age to re-think what "culture" is in each of its multiple senses, what makes the culture of a people cohere, and what forces and mechanisms are at work to change it. In this light the important topic is not to what extent science is separated from other activities, but rather how we may define and transmit culture in such a way that the sciences are seen to be valid components of our culture. We therefore must here adopt, as one of our main tasks, the actual design of educational curricula stressing the coherence of physics and the other components of intellectual life.[6]

Threats to Coherence

Before any specific proposals can succeed, we should be able to deal with the major threats which every coherence-seeking program of science instruction will face. Two of these threats should now be discussed briefly.

The first is the rapidity with which the simplest terminology of the contemporary sciences is being removed from the natural language of the beginner. This makes obvious difficulties for the new learner—not to speak of the difficulty this student will have 10 or 20 years after he has passed through our classroom and, at the height of his career and ability, faces a science that is by then concerned with entirely changed problems, phrased in an entirely changed vocabulary. As we all know very well, for his teachers this is also a major and continuing problem. But on an even more fundamental level, there is the additional and obverse problem of the effect of the increasing vocabulary gap, not merely on the new learner, but on the language of science itself.

This is indicated by an obvious difference between even a most difficult and sophisticated natural language, such as Japanese, and even a relatively simple contemporary science, such as introductory quantum physics; the difference is clearly that the difficult natural language can be, after all, learned by any small child, whereas the simpler modern sciences turn out to be all but inaccessible to many of our most intelligent students. The point is of course that natural languages have developed in a very different way from scientific language. As Margaret Mead perceptively noted in an essay from which I wish to quote at some length, "it has been characteristic of all earlier forms of cultural transmission that new intellectual acquisitions—such as script, mathematical calculation, prosody—have been taught in face-to-face situations by adults who knew, to children and adolescents who did not yet know." Miss

Mead finds that while in the past even contemporary scientific language used to be so communicated, this situation has begun to change radically with the rise and acceleration of the sciences in recent times:

> As the several sciences have begun to grow at an unprecedented rate, the distance from one major advance to the next is often reduced from fifty to five years. And now, instead of teaching a widely selected, intelligent student audience, more and more the young scientists are communicating to each other, horizontally, in highly specialized languages, material so new that publication is not rapid enough to encompass it . . . . We are, in fact, in danger of developing—as other civilizations before us have developed—special esoteric groups who can communicate only with each other and who can accept as neophytes and apprentices only those individuals whose intellectual abilities, temperamental bents, and motivations are like their own. A schismogenic process is under way that is self-perpetuating and self-aggravating . . . .
>
> All of us who cherish the change in pace made possible by this new kind of horizontal, face-to-face, multimodal transmission, which works even across national boundaries, inevitably will guard jealously any attempt that would seem to slow up this intoxicating process. But now we must find new educational and communication devices that will not sacrifice this new high level of specialized communication and yet will protect our society and all the intellectual disciplines within it from the schismatic effects of too great a separation of thought patterns, language, and interest between the specialized practitioners of a scientific or humane discipline and those who are laymen in each particular field.[7]

This is, I believe, an important warning. It gives us an additional reason for insisting that our young students should receive early and full opportunity to learn the concepts and theories of modern science, and so to be brought to a state where the vocabulary and grammar of modern science, including some of the techniques of calculation, will no longer themselves be the main obstacle to an understanding of the proud achievements of our time. And Miss Mead concludes that such a program will also modify the threat that science will escape entirely from the area of natural discourse:

> The process of vertical communication of results arrived at by horizontal, face-to-face adult learning will alter the vocabulary and syntax of the *communicators*. Thus they will be the more able to transmit what they know, and they themselves will keep in closer touch with the other specialties of our

highly specialized societies. Any language taught only by adults to adults—or to children as if they were adults—becomes in certain respects "dead." It fails to enlist recruits, it may lose its productivity, and it serves in the end primarily to separate those who know it from those who do not. In contrast, any language that is taught to all children attains a multimodal comprehensiveness that makes it a suitable vehicle for the thought of not only the highly intelligent but also the moderately endowed person. By insisting that all children, not only those children who, by joining the ranks of a discipline, will accentuate its highly specialized style, should be taught recent advances in a particular discipline, we can set up an automatic corrective system for the dangerous intellectual divergencies of vocabulary and knowledge within our society. . . .[8]

We turn to the second threat to the achievement of coherence, and here we may well wish that Miss Mead's warning had been heard and heeded long ago. For the failure of "the others" to understand what the creative scientist now knows and does has had, I fear, some debilitating and even tragic consequences—particularly to the effectiveness and morale of our most valuable intellectuals outside science. For they are caught between their irrepressible desire really to understand this universe, and, on the other hand, their clearly recognized inability to make any sense out of the simplest vocabulary of modern science. I have stressed elsewhere[9] the chilling realization that our intellectuals, for the first time in history, may be losing their hold of understanding the world: "the nonscientist realizes that the old commonsense foundations of thought about the world of nature have become obsolete during the last two generations. The ground is trembling under his feet; the simple interpretations of solidity, permanence, and reality have been washed away, and he is plunged into the nightmarish ocean of four-dimensional continua, probability amplitudes, indeterminancies, and so forth. He knows only two things about the basic conceptions of modern science: that he does not understand them, and that he is now so far separated from them that he will never find out what they mean."

To take a concrete example, consider the recent, widely read book, THE SLEEPWALKERS, by Arthur Koestler. In it, Koestler tried to trace the rise of modern physics, and, with it, of modern philosophical thought, stemming from the work of Kepler, Galileo, Newton, and some of their contemporaries. This is indeed still a useful task to set oneself. Koestler has worked with devotion on his material. And, most important, he is of course the intelligent layman *par excellence* whom any scientist would be pleased and proud to have as his student in this evidently earnest search for an understanding of modern science.

And yet, something terrifying happened as Koestler came to the end of his book. He had still been able to see meaning and order in the physics of the seventeenth century. When he turned in the Epilogue to modern physics, all sense of understanding and coherence disappeared, and the incomprehensible modern conceptions seemed to rise around him on every side as threats to his sanity. As he summarizes his work, he finds that to a large degree "the story outlined in this book will be recognized as a story of the splitting-off, and subsequent isolated development, of various branches of knowledge and endeavour—sky-geometry, terrestrial physics, Platonic and scholastic theology—each leading to rigid orthodoxies, one-sided specializations, collective obsessions, whose mutual incompatibility was reflected in the symptoms of double-think and 'controlled schizophrenia.' "[10]

I believe it is important to consider this case as sympathetically as we can—to listen to the anguish of an intelligent man who has discovered that he cannot cope with the modern conceptions of physical reality. For what he is saying to us is what most people *would* say—if they were eloquent enough and interested enough in knowledge to be deeply disturbed by a state of unchangeable ignorance:

> Each of the "ultimate" and "irreducible" primary qualities of the world of physics proved in its turn to be an illusion. The hard atoms of matter went up in fireworks; the concepts of substance, force, of effects determined by causes, and ultimately the very framework of space and time turned out to be as illusory as the "tastes, odours and colours" which Galileo had treated so contemptuously. Each advance in physical theory, with its rich technological harvest, was bought by a loss in intelligibility . . . .
>
> Compared to the modern physicist's picture of the world, the Ptolemaic universe of epicycles and crystal spheres was a model of sanity. The chair on which I sit seems a hard fact, but I know that I sit on a nearly perfect vacuum . . . . A room with a few specks of dust floating in the air is overcrowded compared to the emptiness which I call a chair and on which my fundaments rest . . . .
>
> The list of these paradoxa could be continued indefinitely; in fact the new quantum-mechanics consist of nothing but paradoxa, for it has become an accepted truism among physicists that the sub-atomic structure of any object, including the chair I sit on, cannot be fitted into a framework of space and

time. Words like "substance" or "matter" have become void of meaning, or invested with simultaneous contradictory meanings . . . .

These waves, then, on which I sit, coming out of nothing, travelling through a non-medium in multi-dimensional non-space, are the ultimate answer modern physics has to offer to man's question after the nature of reality.[11]

And at the very end of the book, in its last, furiously splashing paragraph, I cannot but hear the cry of a drowning man, a cry for help that cannot leave one unconcerned if one believes that physics can and must be shown to play a valid, creative part within our culture:

The muddle of inspiration and delusion, of visionary insight and dogmatic blindness, of millennial obsessions and disciplined double-think, which this narrative has tried to retrace, may serve as a cautionary tale against the *hubris* of science—or rather of the philosophical outlook based on it. The dials on our laboratory panels are turning into another version of the shadows in the case. Our hypnotic enslavement to the numerical aspects of reality has dulled our perception of non-quantitative moral values; the resultant end-justifies-the-means ethics may be a major factor in our undoing. Conversely, the example of Plato's obsession with perfect spheres, of Aristotle's arrow propelled by the surrounding air, the forty-eight epicycles of Copernicus and his moral cowardice, Tycho's mania of grandeur, Kepler's sun-spokes, Galileo's confidence tricks, and Descartes' pituitary soul, may have some sobering effect on the worshippers of the new Baal, lording it over the moral vacuum with his electronic brain.[12]

This is the end of his road that had started with great hope and promise. What can *we* do about it?

Goals and Science Courses

There are several related areas of study that would alleviate the difficulties we have discussed and promote the sense of coherence which we seek to cultivate.

The first, obviously, is the study of modern science itself. Miss Mead's and Arthur Koestler's arguments both speak first of all for substantial attention to modern science in the school and college curriculum—and although I shall speak in what follows mostly about physics, the very same argument could be applied to other fundamental sciences.

I have little use for the attitude that one hears occasionally, to the effect that quantum physics and relativity should precede or even take the place of classical physics in introductory courses. I believe that this is pedagogically unsound for most students; rather, we must help their intuitions to bridge the distance from their natural Aristotelian bent to the strange new conceptions by means of classical (Galilean–Newtonian–Maxwellian) ideas. The growth of imagination recapitulates the growth of a field to a large extent.

But whether we all agree on the order of presentation or not, the main point should be beyond dispute here: the college student should, as part of his general education, arrive at an understanding of the main concepts and theories of modern physics. This includes certainly an introduction to quantum theory and relativity theory. Such a program necessitates careful consideration and use of the whole curriculum, from grammar school to college, in order to provide enough sophistication and tools, both in mathematics and in physics, to ensure that the minimum end-point achievement in physics be not superficial. This effort can, I feel, no longer be shirked. In our world, the essential notions of quantum physics and relativity are no longer optional for anyone who wishes to regard himself as generally educated, any more than are the essential notions of certain specifiable other fields.[13] It is also important to design impedance-matching devices in our curriculum by which a student from such a serious general education course *can* continue in more advanced physics courses if he wishes. Regardless of his field of final concentration or specialization, it should be possible for the talented and interested student to receive physics instruction at the highest level of which he is capable. The administrative problems for doing this are large at most colleges in the U.S.A., and worse at colleges following the European model. But this merely shows from another angle what all of us know already; that administrative difficulty is the *last* thing to consider and the *first* thing to stamp out ruthlessly—if any sound idea is to come to anything beyond mere talk.

Difficult though a program of serious physics instruction for all college students is to carry out, it represents in my view only half the necessary job. Even if we succeed in presenting fairly sophisticated physics to students who will not go on in the sciences, but have done nothing more, we shall have failed. For these students by and large do not come to us as we ourselves came to the study of physics. When we were trained as physicists, we demanded no more than good physics from our introductory courses. They, however, who do not intend to become *us* (sometimes a difficult point to remember!), want to see also what place physics has in the total reality, in the context of all intellectual endeavors; and unless we help them, nobody will, and they will know that they came to our shop erroneously.[14]

Here I must explain in more detail just why I believe it to be wrong to force a nonscience student to take, as is often done to "fulfill a science requirement," simply the regular introductory specialty course given by a science department. I start from a general belief[15] that the total orienting process of a young student in college, it seems to me, has at least five goals. If he or she is to emerge as an educated and sane person from our educational institutions, the student should be well on the road to recognizing which are his own talents, whatever they may be; second, he should know enough about his physical home, this universe, not to feel either overwhelmed by it or a total stranger in it; third, he should know how to be in fruitful relationship with his fellow men; fourth, he should know what the past means and what the probable future may be; and fifth, he should know the difference between, and the relative functions of, his mind and his soul.

Obviously, the study of science can't contribute equally, or even prominently, to every one of these goals. But my fundamental claim is that the student who will not go on in scientific studies can and should have science courses which attempt to contribute meaningfully to each of these general goals of education. This does not mean in the least that such courses have to be soft and simple; on the contrary, precisely because of their ambitious goals, they surely will have to make as taxing demands on the student (and, it should be added, on the instructor) as any he will encounter. Thus, to do justice to the first goal would mean seriously challenging, helping, testing, and watching the student, to enable him to discover his abilities in scientific work, including work in the laboratory, whether or not he himself has known of any such abilities. The second goal implies the all-but-impossible attempt to teach him, in the limited time available, enough basic and substantive material to show that the natural universe is fundamentally knowable.

The third goal implies that the student should hear and read at least occasionally about the social activity called science (which is not at all identical with the textbook content also called science); indeed, in such a course, we have a striking opportunity to show the plurality of roles played by the individual as a transformer of accumulated knowledge, as a member of a chain of teachers and students, as a collaborator in teams and social groups.

The fourth goal implies that such courses will not shrink from showing at the proper time that science has its historic tradition as well as its characteristic way of growing and, as it were, of anticipating the future. For science has grown, on the one hand, not by destroying old knowledge and by rebuilding on totally new lines, but rather like a tree, ring by ring, where the inner layers are invisible though still responsible for the strength of the whole structure of which only the outer part is seen. On the other hand, scientists differ from most other scholars in the degree of their optimism and future-directedness, a quality that at its best is the proper antidote to the existential despair so fashionable elsewhere.

The fifth goal would require us to convey to our students, at least on occasion, what has been thought to be the philosophical meaning of scientific knowledge. This is often said to be an unnecessary preoccupation, particularly by scientists—with the significant exception of the really great ones from Aristotle to Kepler, from Newton to Bohr and Heisenberg.

Orientation vs. Training Courses

It should be immediately obvious that these goals are not the ones animating the one type of science course that is so often the only one open to the nonscience student—the introductory departmental specialty course, directed primarily to an audience which is expected, correctly or incorrectly, to make science its profession. In such a course, the search for wider coherence and the orientation function are necessarily secondary. The classroom usually resembles a training ground at the foot of a large mountain that is to be conquered stage by stage by selected students in later years. Here, next to the person who has large, high-altitude lungs and who was born with climbing boots on his feet, there sit by administrative decree the eternal lowlander, the stolid farmer, the congenital subway rider, the dreaming sailor, and even the adventurous deep-sea diver. Silently, they listen and move through the mass of technical instructions guaranteed to pay off in the exhilarating climb to the top in which, alas, they will never take part.

Let us grant quickly one advantage the narrower training course has: we know *how to run it* for the benefit of students who really want the training—or at least we think we know, and we have learned to be satisfied with little. Almost any instructor is expected to do a tolerable job when given any one of the great number of similar, ready-made textbooks and a ton of laboratory equipment. The orientation course, on the other hand, is very much more ambitious, has not the backlog of folklore, of texts, of teaching tricks, of experienced staff, and is usually asked to live up to very high standards of performance. I suspect it is as hard at this time to give a truly successful orientation course as to give an altogether bad training course.

To escape this dilemma and yet to stress the wider context of science, some college courses have attempted to go away from science itself and have instead concentrated on study of *joint areas* involving science as one partner. I distinguish two kinds: the ". . . *of* science" courses, and the ". . . *and* science" courses.

The first of these types yields courses in history of science, sociology of science, philosophy of science, and so forth. They are characterized by using the activity and accomplishments of science as the raw materials, upon which to operate with the scholarly tools of history, or sociology, or philosophy. This is, of course, a worthy and useful procedure, but only if the raw material, namely science, has been fully understood. *Following* the study of science (e.g. in the syllabus of a science concentrator in his last year or two, or as a professional study in its own right, undergirded by thorough science courses) it can be an interesting field. But as an *alternative* to the study of science, such courses seem to me dubious if not misleading, and particularly so for the introductory student.[16]

The second of the joint-area studies, the ". . . *and* science" study, can be on history *and* science (as in the case of Henry Adams's essay), or psychology *and* science (as in the studies by Anne Roe and Bernice Eiduson), or science *and* public policy (A. H. Dupree, or Don K. Price, or J. Stefan Dupré and Sanford A. Lakoff), or science *and* art (Erwin Panofsky or James Ackerman). These do not use the achievement of science as raw material to as large a degree as the previous study area, but they are likely to develop the link between the two members of the couple without subjugating one to the other.

These studies, too, cannot be regarded as replacements for sound science study, but they can also be of great help in the coherent presentation of science in general education. Indeed, the specific suggestion I have to make is based on the idea of introducing into the conventional science course, which we all know well how to give, a multiplicity of references, illustrations, short discussions, longer reading and essay assignments, all based on various joint-area studies. Thereby we can build into our course the spirit as well as the substance of the coherence —making ties by which our students can clearly see from the beginning that physical science does not stand as an isolated and forbidding area, having no relationship or analogies with other fields. On the contrary, in this way they should be helped to see clearly what most scientists themselves are apt to take for granted without making the effort to verbalize it, and what is so easily lost in the haste of most "straight" science courses: that any part of present-day science has important connections with other achievements of human beings, with other sciences and with studies and activities other than science.

For the purpose of the discussion of such general education courses, it is useful to describe how a science specialty course is *traditionally* given at many colleges, in terms of a convenient model. The coverage to be attempted in a physics course is represented on a three-dimensional "map." (See Figure 1.) On the x axis lie the academic fields that make up the total academic educational experience of the student, arbitrarily arranged from the most quantitative (mathematics) to the least (humanities). The y axis may represent time from, say, 1600 to today, which, for introductory physics courses, corresponds very roughly to the development of the usual subdivision (mechanics to nuclear physics). The z axis represents depths of penetration attempted in the course on a given topic. On this map the pre-professional physics course may be represented roughly by a set of closely spaced or sometimes overlapping pyramids, based in a more or less narrow and well-defined strip in the xy plane.

When a physics student goes on to higher level science courses and finally to research, he eventually sees that the field he is studying does hang together. For as his study of physics and related sciences penetrates further along the axis of depth in Figure 1, he unavoidably finds that the separate "pyramids" representing mechanics, optics, etc., in Figure 1 are ultimately joined together along the depth dimension in a single interrelated corpus of knowledge that also stretches far to the right and left of the narrow strip under "physics." Although as an undergraduate physics student he will have taken mathematics, engineering, and chemistry courses in different departments and buildings, as soon as he tries to

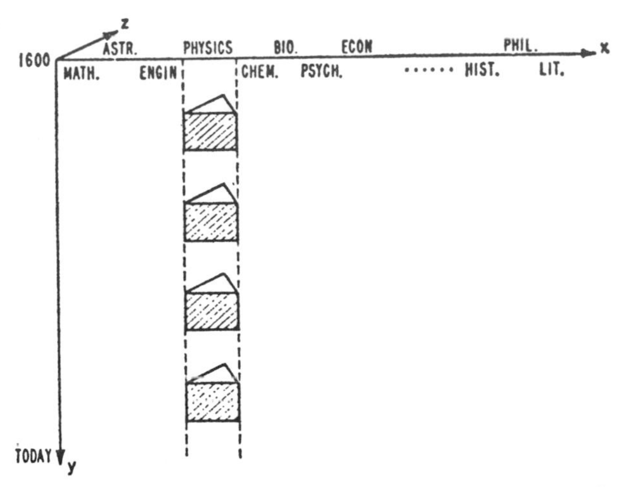

Figure 1. Traditional presentation of topics in introductory physics.

do any significant piece of research he discovers that the separation between neighboring sciences was a pedagogic, administrative convenience, which from the point of view of living science is just as artificial as the separation between the "pyramids" or beads in Figure 2.

Thus he will find that an experimental research project on, say, the dependence of molecular relaxation effects on pressure involves sooner or later material from every separate block in a physics course, but also brings in mathematical methods, metallurgy, electronic engineering, chemical thermodynamics (not to speak of the areas of politics and psychology which anyone must cope with if one wishes to obtain and correctly administer research grants in a busy department). No one who has engaged in actual scientific work can fail to have seen the intimate connection between physics and advances in other sciences and engineering, or between the advance made in pure physics and its social and other practical consequences. Indeed "pure" physics is an invention that exists only in the old-fashioned classroom. As soon as a real problem in physics, or any other field, is grasped, it appears that there hang from it connections to a number of expected and unexpected problems in fields that, by habit, we make our students think of as "belonging" to other professions.

The Connective Physics Course in General Education Programs

Now while our pre-professional students discover by later experience

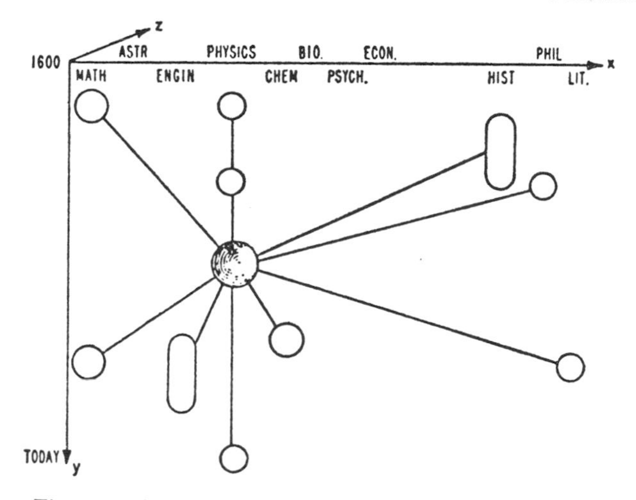

Figure 2. Connective presentation of a topic in physics.

the existence of these connections, the nonscience students do not have this opportunity. In order to do justice to our science, to their needs, and to the commitment which I have urged us to make to a coherent conception of culture, we therefore must stress with care these connections in physics courses for these students. I propose that in addition to the sound presentation of physics itself, we inject into such courses results from the study of certain joint areas, as discussed above. I urge that to start with, at least to some extent or at certain points in the physics course, the representation for a specific topic—for example, Newtonian mechanics—be changed from a pyramid to a "constellation" of related topics, roughly as shown for the *xy* plane in Figure 2. I should like to call it a *connective approach to the teaching of physics,* for it is a way of showing some of the interrelated complex of connections existing between any one specific field of study in physics (e.g., Newtonian mechanics, or thermodynamics, or electromagnetic radiation, or special relativity theory, or nuclear physics) on the one hand, and other fields of study (e.g. mathematics, history, philosophy) on the other.

Therefore a student in the class in which the special relativity theory has been or is being discussed should also be able to hear, read, or write an essay, about *some* of the following points (for which we must prepare more teaching aids, monographs, critical bibliographies, etc., to help the average, hard-pressed instructor):

(i) Galilean relativity (including classical arguments from kinematics and dynamics; meaning of transformation, invariance).

(ii) Classical concepts of space and time (including Newtonian absolutes, their critique by Ernst Mach).

(iii) Theories of luminiferous ether (including Descartes, Newton, Young, Fresnel, Fizeau experiment, Michelson experiment).

(iv) Electromagnetism (including theories of field in the work of Faraday, Maxwell, Lorentz).

(v) Einstein's early work (including background, analysis of first relativity papers, critical references to bibliographical material).

(vi) Other results (including outline of Minkowski representation and summary of, or references to, other sources dealing with cosmology, general relativity—perhaps along the lines of books by Einstein and Infeld or Eddington—and nuclear physics).

(vii) Epistemology (including Einstein's debt to Hume, Kant, Mach; positivism, operationalism, rationalism; use of the special relativity case to discuss such questions as these: What is the role of a conceptual scheme? Relation of science and common sense. Relation of observation or experiment and theoretical construction. Definition, axiom, and hypothesis in the formulation of laws of science. Criteria of scientific truth. Continuity and discontinuity in the growth of science).

(viii) Uses and abuses (including spread of "relativistic" views in philosophy and social studies. Effect of "relativistic" ideas on work of art, e.g. Durrell. Diffusion and misinterpretation of science in other fields. Ideological reasons for acceptance or rejection of theories, as in the case of the relativity theory in Hitler's Germany).

To repeat, these examples are meant to be only suggestive. An instructor would start to transform his course from a traditional physics course to a "connective" physics course by introducing for each of the five or six main topics of the course *a few items of this sort*, perhaps using no more than 10 per cent of the lecture time plus the assignment of certain reading and essays. While the block on relativity theory would be an excellent opportunity to show connections with philosophy, for other main blocks one could develop the relation of physics to engineering; science and policy making; the role of scientific societies; examples of the use of physics *in* chemistry, biology, etc.

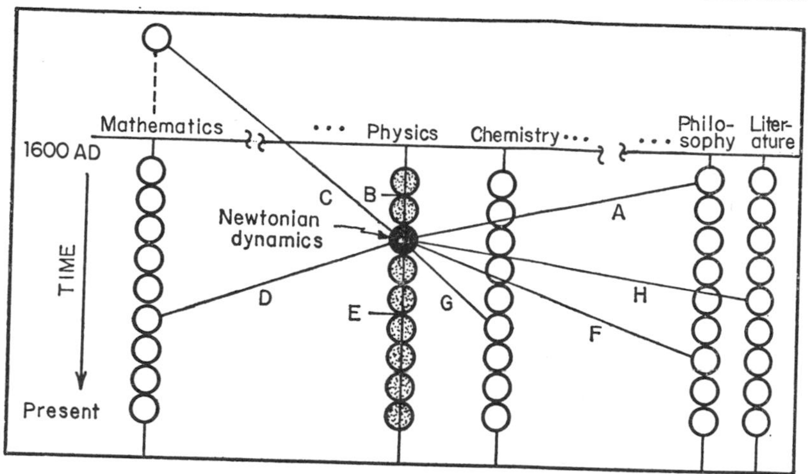

Figure 3. Newtonian dynamics is linked to earlier and later achievements, not only in physics but in other fields as well.

Think now of the opportunities that suggest themselves if in the constellation in Figure 2 the central, dark-shaded area is centered on (instead of relativity), in turn, Newtonian mechanics; thermodynamics; quantum theory; or nuclear physics. Figure 3 indicates the case for Newtonian mechanics, as applied to planetary motion, a subject that is usually one of the "beads" on the physics chain. Newton had studied theology and philosophy, and those ideas emerge in the *Principia* in his sections about the nature of time and space (see Figure 3, link A to philosophy). Within physics itself, Newton brought to a culmination the work of Kepler and Galileo (link B). Much of the established mathematics in Newton's work came from the Greeks (link C). New mathematics, particularly the basic ideas of calculus, were invented by Newton to aid his own progress, thereby advancing the progress of mathematics (link D).

Within physics, all who follow Newton will use his laws and approach (link E). His effects on the Deist theologians (link F), on John Dalton's atomic models in chemistry (link G), and on the artistic sensibilities of the eighteenth century in which Newton swayed the muses (link H) are quite easily documented.

The same kind of web extends around each of the chief topics of physics. Think of the link from philosophy to the work of Hans Christian

Oersted, André Ampère, and Faraday in electricity (through their interest in Naturphilosophie). Think of the link reaching from nuclear physics back along the chain to the classical physics of three centuries earlier (as in the determination of the neutron mass) and of the sideways links to biology, to engineering, and indeed to politics, owing to the various applications and byproducts of nuclear reactors, for example.

Such links exist between main achievements in many fields. Some of the topics and persons discussed in a physics course that draws attention to this state of affairs will also come up in other courses the student will be taking. If we drew all the links between fields on the intellectual map, we would see instead of separate strings of beads a tapestry, a fabric of ideas. This view of the relevance of science should deeply penetrate a course that intends to educate as well as to train. Science is then seen to be in dynamic interaction with the total intellectual activity of an age.

Beginning perhaps with only a few such illustrations, in time one can reorient one's course, until there emerges before the eyes of the student first the intimation and later the proof of physics as a member of a constellation of concerns, so different from the usual, artificial picture of physics as the isolated and stern subject that has nothing to contribute to anything but more physics.

I have, for some time, been working on such a course. To illustrate the variety of possibilities, I think you will be interested in topics students choose for their essays, written after they had studied the block on Newtonian dynamics, in the early part of the year. The overall theme assigned to the class for the first major paper has been "On the impact of the Newtonian synthesis on *X* in the eighteenth century," where *X* stands for philosophy or theology or political science or economics or law or psychology or fine arts— whichever is nearest to the interest or the prospective field of concentration of the student. To give better focus to this overall theme, the student has to define his attention as indicated by a subtitle of his choice, such as "The role of Newtonian mechanics in shaping the imagination of a certain eighteenth-century philosopher," or whatever. (The student is also free to apply for permission to work on another, similar general topic of his choice).

Through his essay the student is led to engage his own specialized interests, together with that which represents the major interest of the physical science course. He sees that the course tries to be something specific for him as an individual, rather than being an external object, to be accepted on its own unyielding terms. And, above all, he is trained to explore the cohesive links that in fact do exist between two fields of scholarship, namely physics and his own future field of professionalism. I must note here that we find again and again that students call this essay a high point of the course, and that the quality of papers is generally very good.

Here are some of the titles of papers, written by some of the students during the last academic year, on the impact of the achievements of physical science of the seventeenth century on other parts of Western culture.

On Biology:

"Aspects of Newtonian science incorporated by the leading eighteenth-century biologists"; e.g. La Mettrie's *L'Homme machine,* Hoffmann, Boerhaave, Trembley, Buffon, Lamarck, and some points on which we now see the adherence to the presumed Newtonian model of science misled them.

"Evolution from Newton to Darwin"; effect of success of mechanistic models, and of all-encompassing, nonhierarchical laws.

On Psychology:

"The English empirical psychologists"; the empirical rationalism in Locke's, Berkeley's, and Hume's theories of perception, and their debts to and differences with Newtonian method.

On Economics:

"Newtonian influence on Adam Smith"; the search for universal, natural force and law in economic systems.

"Theories of causality in eighteenth-century economics"; the extent to which these were socially and politically determined despite the attempt to be purely "scientific."

On Political, Social, and Legal Theory:

"The political theory of Thomas Jefferson"; on the acknowledged debt to seventeenth-century science.

"Henri Saint-Simon and Newtonism"; the development from ardent Newtonian to violent anti-Newtonian.

"The effect of science on legal theory," and "Science and the Commentaries on the Laws of England by Sir William Blackstone"; two fine studies.

On Philosophy and Theology:

A number of interesting essays, including "Bentham's *Principles of*

Morals and Legislation and its formal analogies with Newton's *Principia*"; the intentional and unintentional parallels.

"Science and Le Baron d'Holbach," on the most orthodox champion of unorthodoxy.

"Use of mechanics in arguments from design as a proof of God's existence in eighteenth-century theological discussions."

"The opposition to Newtonianism by Berkeley."

"Newton's epistemology and the ethics of Kant."

"Seventeenth-century mechanics and the eighteenth-century French materialists."

On Literature and the Arts:

"The conception of order in English poetry before and after Newton."

"The effect of science on the style and poetic themes of Alexander Pope."

"From *Paradise Lost* to *Essay on Man*."

"The reaction against the Newtonian synthesis: Blake and Wordsworth."

I should, finally, anticipate some questions that you will have concerning the proposal of developing Connective Physics Courses.

(*a*) I must re-emphasize that the addition of "external" material need not and must not detract from the soundness of the scientific content of the course; the latter is, of course, the *sine qua non* of any physics course given by an instructor whose fundamental loyalty properly belongs to the science in which he has been trained. In our introductory course for nonscientists[17] we develop and use the elements of the calculus for all students; those who already have some knowledge of the calculus can enter sections in which the calculus is used extensively from the beginning. This is decidedly not a course "about" physics rather than *in* physics. And I am convinced that this is possible regardless of where such a course is given, because it depends primarily on the attitude of the instructor and only secondarily on the quality of the student. Moreover, one must not neglect the effect of the continuing up-grading in student preparation for a serious physical science course; what we know to be possible with relatively unprepared students will be that much more possible with well-prepared ones.

(*b*) I have no illusion that such a course will ever be easy to teach, or necessarily will delight every student. The work I propose is not easy.

Of course not! Most easy things seem to have been tried already; and in physics instruction they clearly have not worked well enough. The instructor of this new course must build up his material patiently over a long period as he fashions his Connective Physics Course while teaching it. He must read widely, collect respected bibliographies—and learn new things always. His teaching assistants have to be especially trained—though our experience has been that this course attracts the most knowledgeable and loyal assistants, to whom this experience is as fresh and important as it is to the students themselves. Above all, the instructor must really be committed to his approach—otherwise he should not try it.

An obvious danger to guard against is dilettantism. For the preparation of illustrative material or for some essay bibliographies and advice on correction of some essays, the instructor should be prepared to seek assistance from faculty members who are not members of his own department, and who may be scholars in economics or literature, etc. But in what sense are we members of a "university," of a cultural "community," if this is not done simply? More positively: I would hope that the teaching of a Connective Physics Course helps to cement the bonds of colleagueship across departmental frontiers, so that, for example, there would arise at the college a "shop club" composed of scholars from different areas who share this interest.

(*c*) Even so, local talents should be supplemented by more generally available teaching aids—as I suggested, critical bibliographies, new brief monographs, etc. What is now needed is a collective, substantial effort to write, collect, and supply such teaching aids on a larger scale than a single college can hope to provide.

If this view of the function of physics instruction in general education can be made generally workable, in many colleges and in many lands, we shall have done justice both to the warning of Miss Mead to keep advanced vocabulary and beginning students in mutual contact, and also to the disguised plea of Koestler not to condemn bright men and women to permanent ignorance of the multifaceted reality seen by modern science. And if we are successful, we shall certainly have produced more than a mere educational improvement. For a curriculum which stresses the connective elements would permit a truly exciting new view of education, in which each student, in addition to penetration in depth of his own chosen field of specialization, will see the rest of the field of knowledge criss-crossed by connective links, with the same major elements appearing as members of different constellations. Thus, even as the Connective Physics Course, when centering on Newton's work, may also explore the links to Voltaire and Alexander Pope, so should the Philosophy or Literature course, when centering on Voltaire or Pope, be exploring the link to Newton from its own vantage point.

Some of this is of course already done; but by and large each department of learning presents its course now in an isolated, strip-like fashion analogous to Figure 1—the unfortunate result of too early and too singleminded specialization. It will be when many fields follow our lead and adopt a method sketched in the discussion of Figures 2 and 3 that our culture will be seen, by teachers as well as by students, to have the coherence which indeed already exists, but which so far has not been nurtured, conveyed, and championed enough in our time. Let us begin here.[18]

NOTES

1. T. S. Eliot, NOTES TOWARDS THE DEFINITION OF CULTURE, London: Faber & Faber, 1948.

2. That such feelings are not confined to intellectuals, but are more widely shared, has been shown in a number of studies, e.g., the extensive research by Donald D. O'Dowd and David C. Beardslee, COLLEGE STUDENT IMAGES OF A SELECTED GROUP OF PROFESSIONS AND OCCUPATIONS, Middletown, Conn.: Wesleyan University Press, 1960, some of the main conclusions are summarized by these authors in *The College-Student Image of the Scientist,* SCIENCE, 133:997–1001, 1961.

3. Cited in John H. Van Vleck, *The So-Called Age of Science*, CHERWELL-SIMON MEMORIAL LECTURES, 1961 and 1962 (Edinburgh: Oliver & Boyd, 1962), pp. 25–50. The editorial was dated March 11, 1962.

4. That this is not a new development is indicated by the attack John Woodhull of Columbia University levelled against the average college physics course of over fifty years ago: "These college students have a starvation course in measurement called physics. Their tutors, having just passed through the same course with excessive specialization, are suspicious of that expansive thing called culture. They affect to despise, not only the public, but all departments of learning other than their own. They surpass the theologians in narrowing down their lines of orthodoxy."

5. C. P. Snow, SCIENCE AND GOVERNMENT, Cambridge: Harvard University Press, 1961.

6. A discussion of detailed goals for such courses has been given in my article, *Science for Nonscientists: Criteria for College Programs*, TEACHERS COLLEGE RECORD, 64:497–509, 1963.

7. Margaret Mead, *Closing the Gap Between the Scientists and the Others*, DAEDALUS, 88: 142–143, 1959.

8. *Ibid.*, pp. 143–144.

9. Gerald Holton, *Modern Science and the Intellectual Tradition*, SCIENCE, 131: 1187–1193, 1960, and reprinted here as Essay 14.

10. Arthur Koestler, THE SLEEPWALKERS (London: Hutchinson, 1959), p. 518.

11. *Ibid.*, pp. 529–531.

12. *Ibid.*, pp. 541–542.

13. E.g., a biological science, an analytical social science, an historically oriented social study, a literary-humanistic field, and a creative-arts field—to indicate by only one phrase each of the other main subjects in which a sound general-education experience seems to be now essential for any college student.

14. We should remember a story C. N. Yang told not long ago to express the disenchantment of some physicists with the mathematician who is not interested in the realities of our concerns. "There is a story circulating among us," Professor Yang said, "describing the feelings of a physicist when he consults a mathematician. A man carried a large bundle of dirty clothes and searched for a laundry without success for a long time. He was greatly relieved when he finally found a shop displaying a sign 'Laundry done here' in the window. He went in and dumped the bundle on the counter. The man behind the counter said:

" 'What's this?'

" 'I want to have these laundered.'

" 'We don't do laundry here.'

" 'But you have a sign in the window advertising that you do laundry.'

" 'Oh! *That*! We only make signs.' "

We, too, are all too often only making signs, if we teach our nonphysics students as if they are going to be scientists.

15. The following six paragraphs are adapted from my article, *Science for Nonscientists: Criteria for College Programs*, mentioned in note 6.

16. This is not said to condemn another use of the history or philosophy or sociology of science, namely, to *aid* in the presentation of the material of science itself in a science course. On the contrary, I have long both preached and practised the doctrine that these aids are of interest and use to the student.

17. It has in the Harvard College Catalogue the designation "Natural Sciences 2: Foundation of Modern Physical Science," and is mostly elected by nonscience concentrators in the first or second year, and by premedical students for whom it counts in lieu of a traditional one-year physics course. The main topics covered are: Galilean and Newtonian Mechanics; Dynamics of the Solar System; Conservation of Mass and Energy; Origins of the Atomic Theory in Physics and Chemistry; Waves and Fields; Quantum Theory of Light and Matter; the Nature of Elementary Particles; Relativity Theory.

18. [The *Project Physics Course* materials for schools and colleges, based in large part on the approach outlined above, were subsequently developed under the supervision of F. James Rutherford, Fletcher Watson, and the writer; they were published by Holt, Rinehart, & Winston in 1970.]

INDEX